Soil Fertility Management
—for———
Sustainable Agriculture

Soil Fertility Management
—*for*—
Sustainable Agriculture

Rajendra Prasad
James F. Power

LEWIS PUBLISHERS

Boca Raton   New York

Library of Congress Cataloging-in-Publication Data

Prasad, Rajendra, 1936-
   Soil fertility management for sustainable agriculture / by Rajendra Prasad, James F. Power.
      p.   cm.
   Includes bibliographical references and index.
   ISBN 1-56670-254-2 (alk. paper)
   1. Soil fertility.   2. Fertilizers.   3. Sustainable agriculture.   I. Power, J. F.   II. Title.
S633.P83 1997
631.4′1—dc20   96-44795
                                                                                                      CIP

FOREWORD

The world population is expected to double within the next three to five decades, thus making the task of several national agricultural systems more difficult to provide needed food security. This is likely to be further complicated by environmental problems, which are cropping up due to intense use of chemicals. Therefore, sustainability of national agricultural systems is a major concern today. Management of soil fertility and soil health is the key to the development of sustainable agriculture.

Focusing attention on soil fertility in relation to sustainable agriculture is the need of the day. In view of this, it gives me great pleasure to write the foreword for this book entitled *Soil Fertility Management for Sustainable Agriculture* jointly authored by Dr. Rajendra Prasad, an eminent agronomist from India, and Dr. J. F. Power, a pioneer of soil fertility management from the U.S. Although the basic principles of soil fertility management for sustainable agriculture in temperate and tropical regions remain the same, the environmental factors, such as high temperature and heavy precipitation during short spells of time, make management practices more difficult under tropical conditions. It is thus gratifying to see a joint effort by two experienced researchers and teachers in producing a textbook on soil fertility using data from tropical as well as temperate regions and presenting the fundamentals of soil fertility for the benefit of students. This book is an excellent reflection of the Indo–U.S. Senior Scientists Panel Program, and I am sure it will prove useful to all concerned.

R. S. Paroda
Secretary to Government of India
Department of Agricultural Research and Education
Ministry of Agriculture
Indian Council of Agricultural Research
New Delhi

PREFACE

Sustainable agriculture is now on the agenda of agricultural institutions over the entire world. Most national governments are concerned with this issue. The problem is aggravated by environmental hazards associated with modern technology in the agriculture of advanced countries. There are also pressures to reduce the use of agricultural inputs, such as chemical fertilizers and pesticides, and to cut down on the use of farm machinery dependent on fossil fuels. With reduced chemical input, the United States can still produce enough food to meet its own needs, as well as the needs of several other parts of the world, but most developing countries with ever-increasing population pressure cannot afford to cut down the use of agricultural chemicals, especially chemical fertilizers. They must produce more and more food and yet maintain the fertility of soils. Soil fertility management for sustainable agriculture is therefore a primary concern.

There is a growing demand on young agricultural graduates coming out of U.S. universities to serve in developing countries in Asia, Africa, and South America. The soil, climate, and crop conditions in these continents are much different than those prevailing in the U.S. and Europe. Several countries have tropical and subtropical climates with large areas under oxisols, ultisols, and saline–alkali soils. Although there has been considerable research in the U.S. on such soils, not much has been placed in our textbooks on soil fertility. It is highly desirable that undergraduate students in agriculture in all universities have greater exposure to soil fertility problems in these other continental soils and in their management.

Also, most textbooks on soil fertility available to the millions of undergraduate and graduate students in developing countries of the world are based on scientific information obtained in advanced countries from temperate regions, such as the United States, United Kingdom and other European countries. While the basic principles of soil fertility apply to all soils, there is an urgent need for a textbook on soil fertility that incorporates available and useful scientific data obtained from developing countries.

The approach of this book is to present the principles for controlling soil fertility in the various climates of the world and to provide examples of soil fertility management for sustainable agriculture from both tropical and subtropical regions, as well as from the temperate regions.

ABOUT THE AUTHORS

 **Dr. Rajendra Prasad** is the former Chair of the Division of Agronomy, and now, ICAR National Professor at the Indian Agricultural Research Institute (IARI) in New Delhi, India. Dr. Prasad holds B.Sc. (Ag.) and M.Sc. (Ag) (1956) degrees from Agra University, Government Agricultural College, Kanpur Campus, and a Ph.D. degree (1961) from Mississippi State University. He is a member of Phi Kappa Phi. Dr. Prasad has published over 150 research papers and participated in a number of national and international seminars. He is a recipient of the Hooker Award of the Indian Agricultural Research Institute (New Delhi), the Rafi Ahmed Kidwai Prize of the Indian Council of Agricultural Research (New Delhi), and the Silver Jubilee Award of the Fertilizer Association of India (New Delhi) for his contributions in soil fertility and fertilizer nitrogen management. He is a Fellow of the Indian National Science Academy and National Academy of Agricultural Sciences (India). Dr. Prasad was the Secretary of the Indian Society of Agronomy for the period of 1975 to 1978 and was Vice President and Executive Chairman for 1979 and 1980. He is a member of editorial board of the *Journal of Agronomy and Crop Science*, Berlin. Dr. Prasad has been teaching a course on soil fertility and its management at IARI for the last 20 years and has guided over 30 graduate students in obtaining their Ph.D. degrees in agronomy. His students today occupy senior positions in state agricultural universities, institutes of the Indian Council of Agricultural Research and Council of Scientific and Industrial Research, and the fertilizer industry in India.

Dr. James F. Power is Research Leader for the Soil and Water Conservation Research Unit of the Agricultural Research Service (ARS) of the U.S. Department of Agriculture located at the University of Nebraska, Lincoln, Nebraska. He leads a group of nine ARS scientists, plus post-doctorates, visiting scientists, and graduate students, in a research program dedicated to improvement of sustainable agricultural production practices that maintain or enhance economic production and environmental quality. His B.S. (1951) and M.S. (1952) degrees in agronomy are from the University of Illinois, and his Ph.D. (1954) in soils is from Michigan State University. He has had over 40 years of employment as a Research Soil Scientist with ARS: 6 years at Sidney, Montana; 18 years at Mandan, North Dakota; and over 16 years at Lincoln, Nebraska. He is a member of a number of honorary societies and a Fellow in the Soil and Water Conservation Society, American Society of Agronomy, Soil Science Society of America, and American Institute of Chemists. In 1996, he was elected Fellow of the National Academy of Agricultural Sciences (India). He has held various offices and committee assignments in the above societies, as well as serving in several positions on the editorial staffs of their journals. In 1990 he was Scientist of the Year for the Northern Plains area of ARS. He has conducted extensive research on nitrogen nutrition and interactions with water availability for grassland soils, reclamation of lands disturbed by strip-mining, tillage methods (especially reduced and no-till) for grain production, and nitrogen cycling in cultivated ecosystems. He has served on the advisory committee for a number of graduate students. He and his graduate students have published over 200 peer-reviewed journal papers and book chapters.

ACKNOWLEDGMENTS

The author (Rajendra Prasad) is grateful to the Honorable Minister of Agriculture of the Government of India, Director General for the Indian Council of Agricultural Research, and the Director of the Indian Agricultural Research Institute, New Delhi for deputing him to the Agronomy Department at the University of Nebraska and to the Soil and Water Conservation Research Unit of the Agricultural Research Service, U.S. Department of Agriculture. The USDA-ARS group at Lincoln, Nebraska also provided financial assistance. Special thanks are due to my co-author, Dr. J.F. Power, Research Leader of the USDA-ARS group at the University of Nebraska, for his valuable guidance and assistance in writing this book. Thanks are due also to Dr. J.S. Schepers, Dr. J.W. Doran, Dr. G.E. Varval, Dr. W.W. Wilhelm, Dr. D.D. Francis, and Dr. B. Eghball of the USDA-ARS group for discussion and advice. Special thanks go to Ms. Pam Bushman for her dedication in entering the manuscript in the computer and for handling correspondence. In addition, thanks are due to my sons, Dheerendra and Neelendra, for their help and especially to my wife, Uma, for library assistance and help in preparing the subject index. Finally, my co-author and I are grateful to the large number of scientists and writers and their publishers for their kind permission to include their work in this text; we also thank Dr. R.S. Paroda, Director General, Indian Council of Agricultural Research, New Delhi for arrangements and for writing the foreword.

DEDICATION

The authors wish to dedicate this book to their wives

Uma Prasad
Marlene Power

CONTENTS

1   INTRODUCTION

1.1. SUSTAINABLE AGRICULTURE:
DEFINITIONS AND GOALS

The sustainability of agricultural production is a question of major concern for the human race because agriculture is the prime source of food for an increasing world population. Most projections indicate that world population will approximately double within another generation or two, making the task of approximately doubling food production with our finite resources even more critical. This task must be accomplished in a sustainable manner.

Maintenance and management of soil fertility is central to the development of sustainable food production systems. Sustainability is dependent to a large degree on recycling, to the extent possible, the inputs into a production system, thereby increasing efficiency of output per unit of resource input. The discipline of soil fertility defines and outlines the mechanisms by which nutrients contained in these inputs are transformed, made available to crops, and cycled through the production system. Thus the principles that regulate soil fertility are fundamental to the philosophy of sustainability.

Rodale (1988) has enumerated the following reasons for the current interest in sustainability in agricultural production:

1. Because nonrenewable resources are the basis of operation and productivity of modern (American) agriculture, it is feared that when these non-renewable resources are depleted, either food will become too expensive or productivity will decline.
2. The high level of production today contributes to environmental pollution in terms of soil erosion, degradation, and deforestation.
3. The escalation of pollution problems is often traceable to some agricultural practices.
4. There is concern about finding ways to rely more on internal farm resources and continuously enhancing them under rapid population growth and increasing pressures on the limited resources available.

1

5. It is likely that conventional technologies and agricultural production systems will be unsustainable in the future if agricultural production becomes the main source of energy and feed stocks.
6. There is the problem of whether the good life in rural areas can be maintained if family farms are replaced by large-scale industrialized farms that produce all the food.

The Board for International Food and Agricultural Development (BIFAD, 1988) task force gave the following definitions of sustainable agriculture, as developed by different sources:

1. The successful management of resources for agriculture to satisfy changing human needs, while maintaining or enhancing the natural resource base and avoiding environmental degradation.
2. The ability of an agricultural system to maintain production over time in the face of social and economic pressures.
3. One that should conserve and protect natural resources and allow for long-term economic growth by managing all exploited resources for sustainable yields.

Okigbo (1991) after analyzing the various definitions of sustainable agriculture put forth by various workers (Dover and Talbot, 1987; Knezek et al., 1988; Lynam and Hezdt, 1988) defined a sustainable agricultural production system as one that maintains an acceptable and increasing level of productivity that satisfies prevailing needs and is continuously adapted to meet the future needs for increasing the carrying capacity of the resource base and other worthwhile human needs. Sustainability can only by achieved when resources, inputs, and technologies are within the capabilities of the farmer to own, hire, and manage with increasing efficiency to achieve desirable levels of productivity in perpetuity with minimal or no adverse effects on resource base, human life, and environmental quality.

An agricultural production system is location specific, and it is uniquely determined on the basis of interacting physiochemical (soil, climate, radiation etc.), biological (crops, weeds, pests, beneficial organisms, etc.), managerial, and socioeconomic elements that satisfy specific objectives.

Croplands provide about 80% and range and fisheries 10% each of the total world food production (Swaminathan, 1986). Thus the soil has to bear most of the burden of production to meet world food needs. The estimate of the global food deficit by the International Food Policy Research Institute for the year 2000 A.D. is approximately 70 million Mg yr^{-1}. Thus even with all the advances made in agricultural sciences, millions of people may remain hungry by the end of the twentieth century. The question of sustainability therefore is of paramount importance. It should be recognized, however, that most often hunger results from political and social instability rather than from inability to produce food.

1.2. FACTORS DETERMINING SUSTAINABILITY

Of a large number of factors determining sustainability of agriculture in a region, population pressures and the availability of arable land are the most important. Population growth during 1963 to 1971 in North America, Europe, Latin America, Africa, and South Asia was 1.3, 0.8, 2.9, 2.9, and 2.8% yr^{-1}, respectively (United Nations, 1972). The population increase in developing countries in Asia, Africa, and South America between 1986 and 2000 A.D. will contribute to about 90% of the total global population increase (Population Reference Bureau, 1986). Also, the population percentage depending on agriculture as a livelihood in developing countries is much larger than in developed countries; in 1990 it was 68.3% in Africa, 63.6% in Southeast Asia, and 26.1% in Latin America as compared with 9.3% in Europe and 2.4% in North America. Despite such a large percentage of the population working in agriculture in developing countries, 12.5 million tons of cereals were provided to them in 1990 under World Food Program (1991). Regarding the availability of arable land, there is a possibility of cultivating an additional 200 million ha in North America, 290 million ha in South America, and 340 million ha in Africa, but little in Europe and Southeast Asia (Revelle, 1976; Buringh, 1981; Dudal, 1982; Lal and Stewart, 1992).

Considering population pressure and availability of arable land, the world can be classified under four classes as given below:

Class I. Regions having low population pressure and abundant per capita arable land such as North America and Oceania.

Class II. Regions having low population pressure but a shortage of per capita arable land such as Europe.

Class III. Regions having high population pressure and abundant per capita arable land such as Africa and South America.

Class IV. Regions having high population pressure and low per capita arable land such as Southeast Asia.

Sustainable agricultural practices would be different in different regions. In Class I areas, one can consider plans involving reduced use of fertilizers, pesticides, and other agricultural chemicals and greater dependence on natural resources such as organics — including legumes in rotation, use of farm waste as manure, and related practices. Sustainable agricultural plans for Class IV regions cannot avoid increased application of fertilizers, pesticides, and other manufactured inputs, despite using organic manures, green manuring, inclusion of legumes in crop rotation, and other resource-conserving practices.

1.3. SOIL FERTILITY

Whether the land is plentiful or in short supply, efficient soil fertility management is the key to sustainable agriculture.

Soil fertility research and management is primarily concerned with the essential plant nutrients—their amounts, availability to crop plants, chemical reactions that they undergo in soil, loss mechanisms, processes making them unavailable or less available to crop plants, and ways and means of replenishing them in these soils.

The objective of this text is to discuss various aspects of soil fertility management for a sustainable agriculture. We discuss the nutrient transformations that occur in the soil and how various management practices may be used to regulate and control these transformations. Special emphasis is placed on heavily populated, developing nations of the world, where the pressure on land is the greatest and the sustainability of prevailing agricultural systems is posing a serious problem.

REFERENCES

BIFAD (Board for International Food and Agricultural Development) Task Force. 1988. Environment and natural resources: strategies for sustainable agriculture. U.S. Agency for International Development, Washington, DC.

Buringh, P. 1981. *An Assessment of Losses and Degradation of Productive Agricultural Land in the World*. FAO, Rome, Italy.

Dover, M. and L.M. Talbot. 1987. To feed the earth: agro-ecology for sustainable development. World Resources Institute, Washington, DC.

Dudal, R. 1982. Land degradation in a world perspective. J. Soil Water Conserv. 37:245–247.

Knezek, B.D., O.B. Hesterman, and L. Wink. 1988. Exploring a new vision of agriculture. National Forum 68:10–13.

Lal, R. and B.A. Stewart. 1992. Need for land restoration. Adv. Soil Sci. 17:1–11.

Lynam, J.K. and R.W. Hezdt. 1988. Sense and sustainability: sustainability as an objective in international agricultural research. Paper prepared for CIP — Rockefeller Foundation Conference on Farmers and Food Systems, Lima, Peru, September 26–30, 1988 (Mimeo).

Okigbo, B.N. 1991. Development of sustainable agricultural production systems in Africa. International Institute of Tropical Agriculture, Ibadan, Nigeria, 66 pp.

Revelle, R. 1976. The resources available for agriculture. Scientific American 235:164–168.

Rodale, R. 1988. Agricultural systems: the importance of sustainability. National Forum 68:2–6.

Swaminathan, M.S. 1986. Building national and global nutrition security system, in *Global Aspects of Food Production*, Tycooly International, London, U.K., and International Rice Research Institute, Los Banos, Philippines, M.S. Swaminathan and S.K. Sinha, Eds., pp. 417–449.

United Nations. 1972. *Demographic Year Book*, Statistical Office of the United Nations, New York, p. 111.

World Food Program. 1991. *Food Aid Review*, Rome, Italy, 134 pp.

2   ESSENTIAL PLANT NUTRIENTS

2.1. CRITERIA FOR ESSENTIALITY

Plants differ from animals in that they use individual chemical elements and ions as their food. Arnon and Stout (1939) proposed the following criteria for the essentiality of a plant nutrient:

1. A deficiency of the element makes it impossible for a plant to complete its life cycle.
2. The deficiency is specific for the element in question.
3. The element is directly involved in the nutrition of the plant, for example, as a constituent of an essential metabolite or required for the action of an enzyme system.

Based on the above criteria 16 elements are considered as essential for the growth of higher plants. These are carbon (C), hydrogen (H), oxygen (O), nitrogen (N), phosphorus (P), potassium (K), calcium (Ca), magnesium (Mg), sulfur (S), iron (Fe), manganese (Mn), copper (Cu), zinc (Zn), molybdenum (Mo), boron (B), and chlorine (Cl).

There is a growing tendency among the plant nutritionists and soil scientists (Mengel and Kirkby, 1987; Tisdale, Nelson, and Beaton, 1985; Nicholas, 1961) to use a less restrictive definition of essentiality of a plant nutrient, for example, avoiding the necessity of establishing symptoms of deficiency and their correction by supplying a specific nutrient. By adapting such a definition, four more elements are added to the 16 elements listed above. These are sodium (Na), silicon (Si), cobalt (Co), and vanadium (V). The essentiality of Na for plants with C_4 photosynthetic pathway has been reported (Brownell, 1968). *Chenopodiaceae* and other plant species adapted to saline conditions take up Na in relatively high amounts (Flower et al., 1977). Crops that require Na for optimum growth include celery, spinach, sugar beet, and turnip (Lehr, 1953). Japanese, Chinese, and Korean researchers (Okuda and Takahashi, 1965; Tanaka and Park, 1966; Lian, 1976; Park, 1976) have established the essentiality of Si for rice. Si addition has also improved the growth of sugarcane (Ayres, 1966; Gascho, 1977). Cobalt is considered essential for microbial fixation of atmospheric nitrogen, namely, rhizobia, free-living nitrogen fixing bacteria, and blue-green algae (Ahmed and Evans, 1960; Johnson et al., 1966). Vanadium has been established as an essential element for some microorganisms (Nicholas, 1961). From the viewpoint of soil fertility management these can be listed as beneficial elements.

Important functions and deficiency symptoms of essential plant nutrients are given in Table 2.1.

**Table 2.1 Functions in Plants and Deficiency Symptoms
of Plant Nutrients**

Nutrient (forms absorbed)	Functions in plants	Deficiency symptoms
Primary nutrients		
Nitrogen (NH_4^+, NO_3^-)	Synthesis of amino acids, proteins, chlorophyll, nucleic acids, coenzymes	Light-green colored leaves, lower leaves turn yellow and die in severe deficiency
Phosphorus ($H_2PO_4^-$, HPO_4^-)	Metabolic transfer processes, ATP, ADP, photosynthesis, and respiration; component of phospholipids	Purplish leaves, especially on the margins
Potassium (K^+)	Involved in sugar and starch formation; lipid metabolism and nitrogen fixation; neutralizes organic acids	Marginal burning of leaves, curling of leaves
Secondary nutrients		
Calcium (Ca^{2+})	Component of cell walls; cell growth and cell division; cofactor for some enzymes	Failure in the development of terminal bud, dead spots in the mid rib of some plants. In corn, tip of the new leaves may be covered with a sticky, gelantinous material that causes them to adhere to one another

Table 2.1 Functions in Plants and Deficiency Symptoms of Plant Nutrients (Continued)

Nutrient (forms absorbed)	Functions in plants	Deficiency symptoms
Magnesium (Mg^{2+})	Component of chlorophyll, hence essential for food synthesis in plant	Light green leaves and yellowing of leaves similar to N. In rapeseed the leaves are cupped inward

Micro-nutrients

Nutrient (forms absorbed)	Functions in plants	Deficiency symptoms
Zinc (Zn^{2+})	Formation of auxins and chloroplasts; carbohydrate metabolism; stabilizing and structural orientation of membrane proteins	Stunted growth, pale to white coloration of young leaves — white bud and white streaks in leaves of corn; brownish red (rusty) discoloration of leaves in rice known popularly as "Khaira" disease in rice. Corn, beans, citrus, and rice are indicator plants for zinc deficiency
Manganese (Mn^{2+})	Photosynthesis-evolution of oxygen, oxidation-reduction processes, decarboxylation and hydrolysis reactions	Intervenous discoloration — green veins against a pale background; whitening and abscission of leaves, gray speck of oats, marsh spots of peas
Iron (Fe^{2+})	Structural component of cytochromes, perrichrome, and leghemoglobin and thus involved in oxidation — reduction reactions in respiration and photosynthesis	Yellowing or whitening of young leaves. In severe deficiency in rice nurseries, or direct-seeded rice or sorghum fields the entire plants may turn pale or white. Pale yellow interveinal chlorosis in stem
Copper (Cu^{2+})	Constituent of chlorophyll, catalyst for respiration, carbohydrate, and protein metabolism	Stunted growth, terminal leaf buds die, leaf tips become white, and leaves are narrowed and twisted

Table 2.1 Functions in Plants and Deficiency Symptoms
of Plant Nutrients (Continued)

Nutrient (forms absorbed)	Functions in plants	Deficiency symptoms
Boron (BO_3^-)	Involved in germination and pollen tube growth; fruiting, cell division, nitrogen metabolism	Terminal buds die, rosette formation, flower or fruit shedding
Molybdenum (MoO_4^-)	Essential component of nitrate reductase and nitrogenase, thus important in N fixation by legumes	Resembles N-deficiency symptom, whip tail disease of cauliflower
Chlorine (Cl^-)	Involved in the evolution of oxygen in photosystem II of photosynthesis, raising cell osmotic pressure	Chlorotic leaves, some leaf necrosis

From Grundon et al. (1987); Katyal and Randhawa (1983); Mengel and Kirkby (1987); Tisdale et al. (1985); Tanka and Yoshida (1970).

2.2. BASIS FOR CLASSIFICATION OF NUTRIENTS AS PRIMARY, SECONDARY, AND MICRONUTRIENTS

Of the 16 essential nutrients, C, H, and O are taken from the air and soil water, while the other 13 are supplied by the soil, and hence the study of their amounts, forms in which they are present, and reactions that they undergo individually or with each other form the core of soil fertility and its management.

2.3. PRIMARY NUTRIENTS, SECONDARY NUTRIENTS, AND MICRONUTRIENTS

On the basis of the amounts in which these 13 elements are taken up by the crop plants, they are classified as follows.

Primary nutrients (N, P, and K). These nutrients are taken up by the crop plants in the largest amount (Table 2.2) and are the nutrients most commonly applied almost each crop season unless organic farming is practiced. Production of materials containing one or more of these three nutrients is the goal of the World Fertilizer Industry, and the materials are known as fertilizers. Some fertilizers do contain some secondary and micronutrients also.

Secondary nutrients (Ca, Mg, and S). These nutrients are taken up in the next largest amounts (Table 2.2), next only to N and K, but uptake is

Table 2.2 Nutrient Uptake by Spring Wheat[a]

Primary Nutrients	kg ha^{-1} [b]	kg Mg^{-1} grain[c]
N	128	28.0
P	20	4.4
K	187	41.0
Secondary nutrients		
Ca	27	5.9
Mg	19	4.2
S	22	4.9
Micronutrients	**g ha^{-1}**	**g Mg^{-1} grain**
Fe	1800	400
Zn	500	100
Mn	500	100
Cu	150	30

[a]Wheat received 120 kg N ha^{-1} and yielded 4.6 t ha^{-1} grain and 6.9 t ha^{-1} straw; samples not analyzed for B, Mo, and Cl.

[b]Total for grain and straw.

[c]Obtained by dividing total (grain + straw) uptake by the grain yield.

From Joshep and Prasad. 1992. Fert. News 37(3): 33–35. With permission.

generally equal to or greater than P. Consequently, a point is being raised whether Ca, Mg, and S should be considered secondary nutrients or should be promoted as primary nutrients. Since most soils having a pH of 7.0 or above have abundant Ca and even Mg, these nutrients are generally not required to be added to crops each season. Some fertilizers marketed to supply the primary nutrients such as ammonium sulfate, ammonium sulfate nitrate, and ordinary superphosphate also contain large amounts of S. However, with the advent of high-analysis fertilizers such as anhydrous ammonia (83% N), urea (46% N), and diammonium phosphate (18% N and 46% P_2O_5), S deficiency is showing in several parts of the world and S is now considered the fourth most important fertilizer nutrient after N, P, and K.

Micronutrients (Fe, Mn, Zn, Cu, B, Mo, and Cl). These nutrients are taken up by the plants in very small amounts, generally reported in g ha^{-1} as compared with kg ha^{-1} in the case of primary and secondary nutrients. On deficient soils these nutrients have to be applied either to the soil or to the crop plants as a foliar spray. Micronutrients are also termed trace elements. Average concentrations of micronutrients in plants and their uptake relative to that for Mo, as estimated by Epstein (1972), are given in Table 2.3.

Table 2.3 Average Concentrations of Micronutrients in Plants

Micronutrient	Proportion according to dry weight		Relative number of atoms compared with Mo
	mol g⁻¹	mg kg⁻¹	
Mo	0.001	0.1	1
Cu	0.10	6.0	100
Zn	0.30	20	300
Mn	1.0	50	1000
B	2.0	20	2000
Fe	2.0	100	2000
Cl	3.0	100	3000

From Epstein. 1972. *Mineral Nutrition of Plants — Principles and Perspectives*, p. 63. With permission of John Wiley & Sons, New York.

2.4. FUNCTIONS OF ESSENTIAL NUTRIENTS IN PLANTS

Irrespective of the class to which they belong, all essential nutrients are equally important from the viewpoint of plant growth. For better crop production these elements need to be present in adequate amounts of available forms in soil(s). They must also be positionally available; that is, concentrations must be adequate within the soil volume in which the crop roots are active. Plant nutrient concentrations may be inadequate, insufficient, deficient, toxic, or excessive. By definition, when plant growth is satisfactory and optimum yields can be obtained, the amounts are adequate. However, if optimum yields cannot be reached because of nutrient deficiencies, the amounts are said to be insufficient. At nearly adequate levels of nutrient supply, deficiency symptoms are often difficult to detect visually. When a plant nutrient is present in amounts significantly lower than adequate levels, the deficiency symptoms in plants are then expressed and the nutrient is said to be deficient. When the amount of a plant nutrient is present in excess of the plant's needs, it may reach the level at which it causes the deficiency of other nutrients or results in environmental pollution. At such levels the nutrient is present in excessive amounts. When concentration of an essential nutrient or any other element is too high, it may become toxic to the plant. For example, under lowland conditions iron and manganese toxicities are often reported. Whether an element is abundant, insufficient, deficient, excessive, or toxic depends upon crop, soil, and soil-plant ecosystem. Critical deficiency and toxicity contents of various elements in the rice plant are given in Table 2.4. The margin between critical deficiency and toxicity limits is quite narrow for trace elements, especially Cu, B, and Fe. Soil fertility management for sustainable agriculture aims at maintaining essential plant nutrients in adequate amounts and includes plans for taking care of insufficiencies, deficiencies, excesses, or toxicities of essential plant nutrients.

**Table 2.4 Critical Deficiency and Toxicity Contents
of Various Elements in the Rice Plant**

Element	Deficiency (D) or toxicity (T)	Critical content	Plant part analyzed	Growth stage[a]
N	D	2.5%	leaf blade	T
P	D	0.1%	leaf blade	T
P	T	1.0%	straw	M
K	D	1.0%	straw	M
K	D	1.0%	leaf blade	T
Ca	D	0.15%	straw	M
Mg	D	0.10%	straw	M
S	D	0.10%	straw	M
Si	D	5.0%	straw	M
Fe	D	70 ppm	leaf blade	T
Fe	T	300 ppm	leaf blade	T
Zn	D	10 ppm	shoot	T
Zn	T	1500 ppm	straw	M
Mn	D	20 ppm	shoot	T
Mn	T	2500 ppm	shoot	T
B	D	3.4 ppm	straw	M
B	T	100 ppm	straw	M
Cu	D	<6 ppm	straw	M
Cu	T	30 ppm	straw	M
Al	T	300 ppm	shoot	T

[a]*M*, maturity; *T*, tillering.

From Tanaka and Yoshida. 1970. *Nutritional Disorders of Rice Plant*, p. 150. With permission of International Rice Research Institute, Manila, Philippines.

REFERENCES

Ahmed, S. and H.J. Evans. 1960. Cobalt: a micronutrient element for the growth of soybean plants and symbiotic conditions. Soil Sci. 90:205–210.

Arnon, D.I. and P.R. Stout. 1939. The essentiality of certain elements in minute quantity for plants with special reference to copper. Plant Physiol. 14:371–375.

Ayres, A.S. 1966. Calcium silicate slag as a growth stimulant for sugarcane on low silicon soils. Soil Sci. 101:216–227.

Brownell, P.F. 1968. Sodium as an essential micronutrient element for some higher plants. Plant Soil 28:161–164.

Epstein, E. 1972. *Mineral Nutrition of Plants — Principles and Perspectives*, John Wiley & Sons, New York.

Flower, T.J., P.F. Troke, and A.R. Yeo. 1977. The mechanism of salt tolerance in halophytes. Ann. Rev. Plant Physiol. 28:89–121.

Gascho, G.J. 1977. Response of sugarcane to calcium silicate slag. I. Mechanisms of response in Florida. Soil Crop Sci. Soc. Florida Proc. 37:55–58.

Grundon, N.J., D.G. Edwards, P.N. Taakar, C.J. Asher, and R.B. Clark. 1987. Nutritional disorders of grain sorghum, Australian Centre for International Agricultural Research (ACIAR) Canberra, A.C.t Mimeograph 2, 99 pp.

Johnson, G.V., P.A. Mayeux, and H.J. Evans. 1966. A cobalt requirement for symbiotic growth of *Azolla fibiculoides* in the absence of combined nitrogen. Plant Physiol. 41:852–855.

Joshep, P.A. and R. Prasad. 1992. Nutrient concentration and uptake by wheat. Fert. News 37(3):33–35.

Katyal, J.C. and N.S. Randhawa. 1983. *Micronutrients*. Food and Agricultural Organization of the United Nations, Rome, 82 pp.

Lehr, J.J. 1953. Sodium as a plant nutrient. J. Sci Fd. Agric. 4:460–471.

Lian, S. 1976. Silica fertilization of rice, in *The Fertility of Paddy Soils and Fertilizer Application for Rice*, Food & Fertilizer Technology Centre for the Asian and Pacific Region, Taipei, Taiwan, pp. 197–220.

Mengel, K. and E.A. Kirkby. 1987. *Principles of Plant Nutrition*, International Potash Research Institute, Berne, Switzerland, 687 pp.

Nicholas, D.J.D. 1961. Minor mineral nutrients. Ann. Rev. Plant Physiol. 11:314–324.

Okuda, A. and E. Takahashi. 1965. The role of silicon, in *The Mineral Nutrition of Rice Plant*, Johns Hopkins University Press, Baltimore, pp. 123–146.

Park, C.S. 1976. Silicate response to rice in Korea, in *The Fertility of Paddy Soils and Fertilizer Application for Rice*, Food and Fertilizer Technology Centre for the Asian and Pacific Region, Taipei, Taiwan, p. 221.

Tanaka, A. and S. Yoshida. 1970. *Nutritional Disorders of Rice Plant*, International Rice Research Institute, Manila, Philippines, p. 150.

Tanaka, A. and Y.D. Park. 1966. Significance of the absorption and distribution of silica in the growth of the rice plant. Soil Sci. Pl. Nutr. (Japan) 12:23–28.

Tisdale, S.L., W.L. Nelson, and J.D. Beaton. 1985. *Soil Fertility and Fertilizers*, 4th ed., Macmillan, New York, 756 pp.

3 SOIL, THE SUSTAINER

Soil, which is called pedosphere, can be envisioned as a chemical reservoir created at the earth's surface by the interaction of four great chemical reservoirs and powered by solar energy (Figure 3.1). Soil has sustained plants and animals since life began on the planet Earth. Soil is made up of all three physical forms of a matter, namely, solid, liquid, and gas. On a volume basis nearly half the soil is solid, while the other half is made up of soil, water and air. The amount of air in a soil depends upon its water content; at optimum water content for the growth of most upland plants, water and air may each make up about 30 and 20% of the soil volume, respectively (Figure 3.2A). Tillage practices can influence the proportion of water and gases in surface soil. In rice paddies, however, where water floods the soil, the only oxygen present is that dissolved in soil water (Figure 3.2B).

As regards the solid phase, 95% or more of it is mineral (inorganic) in nature, while the remaining 5% or less is organic in nature. However, in temperate and cooler regions of the world soil organic matter may be 5 to 10% or even more of the solid phase, while in warm tropical and subtropical soils organic matter content could be much less than 5%. Thus the proportion of mineral and organic matter differs considerably from soil to soil, depending particularly on the climate of the region.

3.1. SOIL ORGANIC MATTER

Soil organic matter originates from plant and animal residues, which are generally present in various stages of decomposition, namely, from fresh additions to well-decayed soil humus. Although a detailed discussion on soil organic matter is provided later (Chapter 5), it needs to be mentioned that soil organic matter controls several soil physical and chemical properties. Soil organic matter increases the water-holding capacity of soils and is a source of several essential plant nutrients, especially N, S, and P. It is also a source of energy for soil microorganisms. Some general properties of soil organic matter and associated effects on soil properties are given in Table 3.1. Were it not for soil organic matter, there would hardly be life in soil. Management of crop

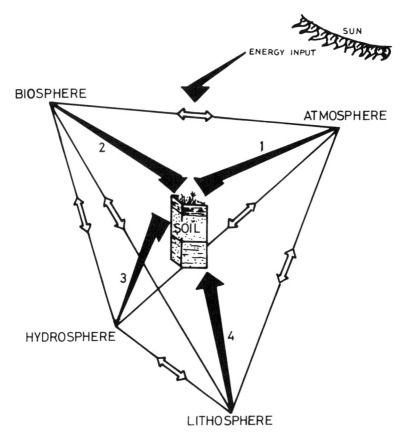

Figure 3.1. The interactions of the four great chemical reservoirs at the earth's surface, which, powered by solar energy, produce soil. (From Chesworth, 1991. Micronutrients in Agriculture, J.J. Mortvedt, F.R. Cox, L.M. Shumah, and R.M. Welch, Eds. With permission of Soil Science Society of America, Madison, WI.)

residues and returning farm wastes to cultivated fields form essential components of soil fertility management for sustainable agriculture.

3.2. SOIL WATER

Sustainability of the agriculture of a country or region depends much on how soil water is managed. Associated major problems require the management of both surface and underground water. Water infiltrates the pores between soil particles and is held with varying degrees of tenacity. Soil water can be measured directly by weight loss on drying, or by using a soil water tensiometer, gypsum or nylon blocks, or a neutron or time-domain reflectance (TDR) probe. The tenacity or soil tension with which water is held by soil particles increases as the soil water content decreases. Water tension in soil at

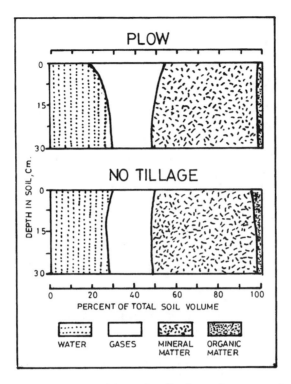

Figure 3.2A. Volume composition of soil when plant growth is in good condition. Tillage increased oxygen concentration in surface (7–8 cm) soil layer. (From Doran, 1982. Crops and Soils Magazine 34:10–12. With permission of ASA.)

any moment controls movement of soil water in soil and its use by plants. When tension is between 0.01 and 0.03 M Pa*, water moves to lower layers due to gravitational pull. Also when soil water tension is 1.5 M Pa or above, the adhesive force is so strong that plant roots can hardly extract water from soil. At approximately this water tension, most plants permanently wilt and stop growing. Soil water between about 0.01 and 1.5 M Pa is considered available to plants.

In addition to its essentiality for life *per se* soil water also serves as a carrier of plant nutrients. All plant nutrients are absorbable by plant roots after these come in solution. Thus the management of soil water forms an integral part of soil fertility management.

3.3. SOIL AIR

The content and composition of soil air is determined by soil-water relationships. Soil air differs from that in the atmosphere in several aspects. Soil

---

* The earlier unit used for soil moisture tension was atmospheres: 1 atmosphere = 0.1 M Pa.

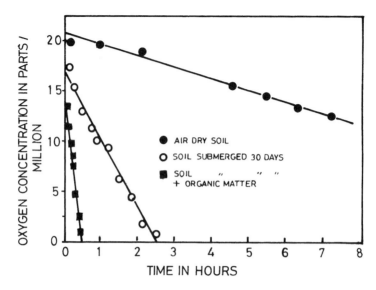

**Figure 3.2B. The oxygen concentration of Corwley silt loam after satura-
tion with oxygen as affected by submerging the soil and by
the addition of organic matter. (From Patrick and Sturgis,
1955. Soil Sci. Soc. Am. Proc. 19:59–62. With permission of
Soil Science Society of America, Madison, WI.)**

air is much more moist and contains several hundredfold greater carbon dioxide concentration than the atmosphere. As a result oxygen content in soil air
decreases considerably and may be 10 to 12% or less as compared with 21%
for the normal atmosphere. Air moves through the soil pores primarily by
diffusion, so diffusion rates are many times greater in air-filled than water-
filled pores.

Because of the presence of pores between soil particles, the soils have
two kinds of densities, namely, bulk density and true or particle density. Bulk
density is defined as the mass (weight) per unit volume of soil; this volume
includes both solids and pores. True or particle density of soil, on the other
hand, is the mass (weight) per unit volume of soil particles. The relationship
between bulk density (BD), particle density (PD), and pore space in soil is
expressed below:

$$\text{Pore space } (\%) = 100 \, (1 - \text{BD/PD}) \tag{1}$$

In most mineral soils particle density is normally about 2.65 Mg m^{-3}, nearly
twice that of bulk density. Bulk density in soils generally varies from 1.0 to
1.8 Mg m^{-3}. Bulk density in very compact soils may be as high as 2.0 Mg m^{-3}.
Soils with lower bulk density are easy to cultivate and manage. Addition of
farmyard manure and crop residues over years can lower the bulk density.

Table 3.1 General Properties of Soil Organic Matter and Associated Effects on Soil Properties

Property	Remarks	Effect on Soil
Color	The typical dark color of many soils is caused by organic matter	May facilitate warming
Water retention	Organic matter can hold up to 20 times its weight in water	Helps prevent drying and shrinking. May significantly improve the water-retaining properties of sandy soils
Combination with clay minerals	Cements soil particles into structural units called aggregates	Permits exchange of gases, stabilizes structure, increases permeability
Chelation	Forms stable complexes with Cu^{2+}, Mn^{2+}, Zn^{2+}, and other polyvalent cations	May enhance the availability of micronutrients to higher plants
Solubility in water	Insolubility of organic matter results from its association with clay. Also, salts of divalent and trivalent cations with organic matter are insoluble. Isolated organic matter is partly soluble in water	Little organic matter is lost by leaching
Buffer action	Organic matter exhibits buffering in slightly acid, neutral, and alkaline ranges	Helps to maintain a uniform reaction in the soil
Cation exchange	Total acidities of isolated fractions of humus range from 300 to 1400 cmol kg^{-1}	May increase the cation exchange capacity (CEC) of soil. From 20 to 70% of the CEC of many soils (e.g., Mollisols) is caused by organic matter
Mineralization	Decomposition of organic matter yields CO_2, NH_4^+, NO_3^-, PO_4^{3-}, and SO_4^{2-}	A source of nutrient elements available for plant growth
Combines with organic molecules	Affects bioactivity, persistence and biodegradability of pesticides	Modifies application rate of pesticides for effective control

From Stevenson. 1982. *Humus Chemistry*, p. 18. With permission of John Wiley & Sons, New York.

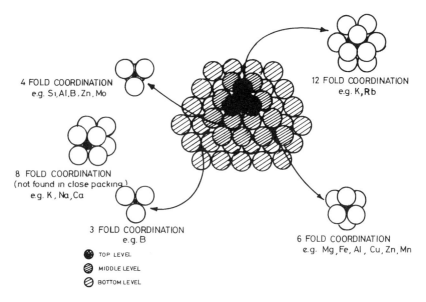

Figure 3.3. Close packing of oxygen anions as a model for the crust of the earth, showing typical coordination structures for the plant nutrients and other elements. (From Chesworth. 1991. Micronutrients in Agriculture, J.J. Mortvedt, F.R. Cox, L.M. Shumah, and R.M. Welch, Eds. With permission of Soil Science Society of America, Madison, WI.)

3.4. SOIL MINERAL MATTER

Oxygen and Si make up most of the soil mineral matter. Ninety percent of the volume of soil particles is O; the O atoms are hexagonally packed (each oxygen atom touches six of its neighbors). Other elements such as Si, Al, and Fe fill the voids left in oxygen packing. The position of these ions depends upon their ionic radii and radius ratio of the concerned ion and oxygen, which also determines their coordination number (number of oxygen ions shared) (Figure 3.3) (Table 3.2).

3.4.1. Soil Texture

Soil mineral particles can vary in size from coarse (over 2 mm) to very fine (less than 2 μm). Depending upon their size, soil particles are divided into gravel, sand, silt, and clay, which are known as soil separates. Regarding particle size for different soil separates, there is considerable variation in the limits set for sand and silt particles by different classifications. Limits set by the International Society of Soil Science and the U. S. Department of Agriculture are given in Table 3.3.

Table 3.2 Most Common Elements and Their Volume in Particulate Matter of Earth's Crust, Their Ionic Radii, Coordination Number, and Frequently Encountered Configuration

Element	Volume %	Ionic radii (nm)[a]	Coordination number	Configuration
O^{2-}	89.84	0.140	—	
Si^{4+}	2.37	0.039	4	Tetrahedral
Al^{3+}	1.24	0.051	4, 6	Tetrahedral Octahedral
Fe^{3+}	0.79	0.074	6	Octahedral
Mg^{2+}	0.60	0.066	6	Octahedral
Ca^{2+}	1.39	0.099	8	
Na^+	1.84	0.097	8	Cubic
K^+	1.84	0.133	8–12	
Mn^{4+}	0.01	0.060	6	
Ti^{4+}	0.08	0.068	6	

[a] 1 nm $= 10^{-9}$ m $= 10^{-6}$ mm $= 10^{-3}$ µm.

From Hurlbut and Klein (1977); Schulze (1989).

A particle size analysis is done after removing soil organic matter, usually by oxidation with hydrogen peroxide or hypobromate solutions. Standard sieves are used to mechanically separate out the very fine sand and larger particles from finer particles. The silt and clay fractions are then determined by measuring the rate of settling of these particles from their suspension in water. After the amounts of different soil particles are determined, a definite textural class name (such as sandy loam or clay loam) can be given to a soil using the diagram given in Figure 3.4. Because the size of mineral particles in a soil is not readily subject to change* by soil management practices, the soil texture (texture class) is an important and permanent characteristic of a soil and gives a general picture of the soil's physical properties such as density, porosity, consistency, water holding capacity, and tilth.

3.4.2. Soil Structure

In nature the soil mineral particles do not exist separately. They are bound to each other by oxides and hydroxides of iron, organic substances excreted by plant roots, root pressures, decomposition products of plant residues, microbial cells and fungal hyphae, and excretory products of microorganisms

---

* The only way to change soil texture of a field is by adding large amounts of soils of different textures brought in from an external source. Rice growers in Asian countries in the upper regions of an undulating topography bring heavy clayey soil from the lower regions or soil at the bottom of the dried ponds and add and incorporate it in the soil of their field to improve texture. Also, flooding or wind erosion sometimes deposits materials of a different texture on the soil surface.

Table 3.3 Classification of Soil Mineral Particles According to Size

Soil separate	International Society of Soil Science (mm)	U.S. Department of Agriculture (mm)
Gravel	2.0 or more	2.0 or more
Sand		
Very coarse		2.0–1.0
Coarse	2.0–0.2	1.0–0.5
Medium		0.5–0.25
Fine	0.2–0.02	0.25–0.1
Very fine		0.1–0.05
Silt	0.02–0.002	0.05–0.002
Clay	0.002 or less	0.002 or less

From USDA (1951).

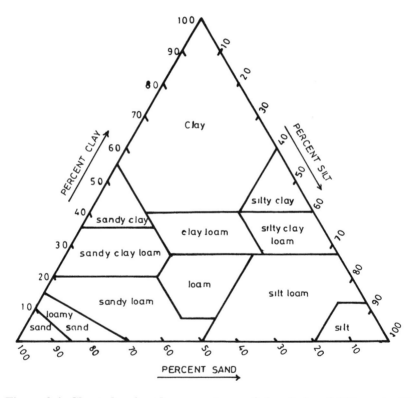

Figure 3.4. Chart showing the percentages of clay (below 0.002 mm), silt (0.002 to 0.05 mm), and sand (0.05 to 2.0 mm) in the basic soil textural classes. (From USDA, 1951.)

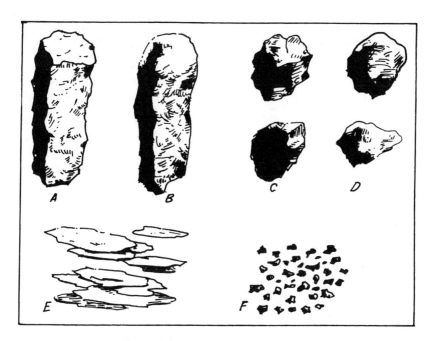

Figure 3.5. Drawings illustrating some of the types of soil structure: A, prismatic; B, columnar; C, angular blocky; D, subangular blocky; E, platy; and F, granular. (From USDA, 1951.)

and gelatinous substances secreted by earthworms. These are aggregated into larger units (or peds) of different sizes and shapes that determine the soil structure. In general, the following four kinds of structures (Figure 3.5) are found in the soils.

Platy.   Aggregates are arranged in thin horizontal plates (E in Fig. 3.5). Such structure is found in soils rich in 2:1 layer silicates such as black, cotton soils in India.

Columnar or prismatic.   Aggregates are vertically oriented in pillar-like structures, which vary in diameter from a few to as much as 15 cm or more. When the tops of the prisms are rounded, the structure is said to be columnar (B in Fig. 3.5); however, when the tops are plane and level it is said to be prismatic (A in Fig. 3.5). Such structures are found in the subsurface soils of arid and semiarid regions, and in soil derived from wind-blown loess.

Blocky.   The aggregates are blocky in shape, more or less equal in three dimensions; one side may be 1 to 10 cm in thickness (C and D in Figure 3.5).

Table 3.4 Relationship Between Land Use Pattern, Organic Carbon,
and Water-Stable Aggregates (0.25 mm or More Diameter)

Land use	Soil depth (cm)	Organic carbon (%)	Water stable aggregates (% by wt of soil)
Bare soil	0–15	0.21	11.7
	15–30	0.17	6.6
Corn (after 5-yr cultivation)	0–15	0.30	19.3
	15–30	0.22	1.6
Corn and black gram	0–15	0.34	28.2
(after 5-yr cultivation)	15–30	0.29	15.7
Afforestation (*Ecualyptus*	0–15	0.58	47.0
tereticornis:			
after 2-yr plantation)	15–30	0.56	57.4
Grassland (*Cenchrus ciliaris*:	0–15	0.61	50.7
after 12 yr of growing)	15–30	0.59	42.2

From Bhatia and Srivastava. 1984. J. Indian Soc. Soil Sci. 32:201–204. With permission.

Granular or crumb.   Aggregates are round in shape (F in Figure 3.5). When aggregates are nonporous, the structure is said to be granular, while the porous aggregates (found in organic matter–rich surface soil) give crumb structure. For most crops crumb structure is highly desirable.

While the texture of a soil cannot be easily changed by soil management practices, structure can certainly be altered. An example of this is the destruction of good soil structure by flooding and puddling the soil as practiced in rice cultivation. Thus soil structure (as judged by stable aggregates) can be modified by the land use pattern (Table 3.4). Compaction, such as occurs when heavy machinery is used on wet soils, also destroys soil aggregates.

3.5. SOIL COLLOIDS

A description of soil composition will not be complete without a mention of soil colloids. Finer mineral and organic matter particles form soil colloids.

The word colloid was coined by Graham (1861) from the Greek "κωλλα," which means glue or gluelike materials. A suspension of clay particles or fine organic matter particles would look very much like glue. As a matter fact any substance can acquire colloidal properties if it is broken down to a very fine size; the generally agreed value is 1 μ (micron) (10^{-4} cm or 10^{-3} mm). This size of particle cannot be seen under light microscope, but can be seen under an electron microscope. Still finer size will lead to atomic or molecular stage.

The most important and unique property acquired by a substance when it is broken down to very fine particle size is a many-fold increase in surface area, which is responsible for the great adsorption capacity of colloids.

Table 3.5 Examples of Colloidal Systems

Dispersed phase	Dispersion medium	Colloidal systems
Liquid	Gas	Fog, liquid sprays
Solid	Gas	Smoke, dust
Liquid	Liquid	Milk
Solid	Liquid	Gold sol
Solid	Solid	Colored glass

From Shaw (1968).

The enormous increase in surface area can be visualized when a cube of 1 cm per side (surface area 6 cm^2) is divided into cubes of 1 μ (10^{-4} cm) per side. The surface area of these cubes will be 6×10^4 cm^2. Thus just by reducing the particle size the surface area has been increased 10,000 fold.

Another point that needs to be known about colloids is that colloids are not a substance but a system. Each colloidal system has two components, namely, a dispersed phase (the substance making up the particles) and a dispersion medium. In the case of soil colloids, the clay mineral particles, hydroxides of Al and Fe, and the fine organic matter particles are the dispersed phase and soil water is the dispersion medium. This is an example of solid-liquid colloidal systems. Some other examples of colloidal systems are given in Table 3.5. Colloidal systems where the dispersed phase has an affinity for the dispersion medium are known as lyophyllic (hydrophyllic when water is the dispersion medium), while those where the dispersed phase does not have an affinity for the dispersion medium are known as lyophobic (hydrophobic when water is the dispersion medium). The soil colloidal system is an example of a hydrophillic colloid, while gold sol is an example of hydrophobic colloid. Several properties of the clay minerals, soil organic matter, and hydroxides of iron and aluminum discussed in later chapters are due to their colloidal nature.

3.6. SOIL LIVING ORGANISMS

An examination of a sample of fresh garden soil first with the naked eye and then under a magnifying glass will show it is teeming with different kinds of organisms, which would include earthworms, ants, spiders, mites, and others (Tables 3.6 and 3.7). Microscopic examinations would reveal the presence of nematodes, protozoa, fungi, algae, actinomycetes, and bacteria of thousands of species. A teaspoon of soil may contain as many as a billion organisms. These macro- and microorganisms are responsible for the decomposition of freshly added organic matter and several biological processes of immense importance in soil fertility management. Agricultural management practices greatly influence the number and species of various soil macro- and micro-organisms present. Soil biology is an interesting area of soil research (Burges and Raw, 1967) and has yielded considerable information that is used in soil

Table 3.6 Frequently Occurring Groups of Soil Organisms

Animals (fauna)

I. Macro
    Herbivores and Detritivores
        Mice, squirrels, gophers, shrews, woodchucks
        Ants, beetles, grubs, etc.
        Millipeds, sowbugs, slugs, and snails
        Earthworms
    Largely predators
        Moles
        Insects (ants, beetles, etc.)
        Centipedes
        Spiders
II. Micro
    Protozoa
    Nematodes

Plants (flora)

I.   Algae: green, blue-green, diatoms
II. Fungi: mushroom fungi, yeasts, molds, VAM mycorrhizae
III. Actinomycetes
IV. Bacteria: aerobic, anaerobic, autotrophic, heterotrophic

Adapted from Brady (1990).

fertility management. Essentially all soil nitrogen that becomes available for crop utilization must first undergo various microbiological transformations. This also holds true for organic P and S sources.

Cultivation practices can considerably influence the population of soil organisms. For example, House and Parmele (1985) from Athens, Georgia, reported 2202 and 637 earthworms m^{-2} in no-till and conventionally tilled plots under a sorghum-rye cropping system. Ground beetles, spiders, and other macroarthropods, as well as microarthropods, were also frequently more prevalent under no-till than under conventional tillage. The no-till method enhances growth of soil organisms because of reduced water loss (more optimum soil water content), amelioration of temperature extremes and fluctuations, and the presence of a relatively continuous substrate for decomposers.

Table 3.7 Average Standing Crop and Energetic Parameters for Microorganisms, Mesofauna, and Earthworms in a Lucerne Ley and Georgia No-Tillage Agroecosystem

	Naked amoebas	Flagellates	Ciliates	Bacteria	Fungi	Microbivorous nematodes	Collembola	Mites	Enchytraeids	Earthworms
Typical size in soil	30 μm	10 μm	80 μm	0.5–1 × 1–2 μm	Ø 2.5 μm 1.0–5.5 μm	Ø ~40 μm	Ø 5000 μm	Ø 1000 μm	Ø 1000 μm	Ø 5000 μm
Mode of living	In water films on surfaces	Free-swimming in water films	Free-swimming in water films	On surfaces	Free and on surfaces	In water films free and on surfaces	Free	Free	Free	Free in soil
Biomass (kg DW/ha)				500–750[b]	700–2700[c]	1.5–4[d]	0.2–0.5[d]	2–8[d]	1–8[d]	25–50[d]
% active	95%	5%	<1%	15–30	2–10	0–100	80–100	80–100	?	0–100
Estimated turnover times/season	0–100	10–50[a]		2–3	0.75	2–4	2–3	2–3	?	3
No. of bacteria/division × 10⁻³	3–8	0.6–1	20–2000							
Minimum generation time in soil (hr)	2–4	2–4		0.5	4–8	120	720	720	170	720

Note: DW, dry weight; *Ø*, diameter

[a] MPN technique.

[b] Direct counts plus size-class estimations.

[c] Direct estimation of total hyphal length and diameter.

[d] Extractions and sorting.

From Coleman et al. (1993).

REFERENCES

Bhatia, K.S. and A.K. Srivastava. 1984. Studies on soil characteristics related to erodibility under different types of land use. J. Indian Soc. Soil Sci. 32:201–204.

Brady, N.C. 1990. *The Nature and Properties of Soils*, 10th ed., Macmillan, New York.

Burges, A. and F. Raw. 1967. *Soil Biology*, Academic Press, London, 532 pp.

Chesworth, W. 1991. Geochemistry of micronutrients, in *Micronutrients in Agriculture*, J.J. Mortvedt, F.R. Cox, L.M. Shuman, and R.M. Welch, Eds., Soil Science Society of America, Madison, WI, pp. 1–30.

Coleman, D.C., P.F. Hendrix, M.H. Bearer, W.X. Cheng, and D.A. Crossley, Jr. 1993. Microbial and faunal interactions as they affect soil organic matter dynamics in subtropical agroecosystems, in *Soil Biodata, Nutrient Cycling and Farming Systems*, M.G. Paoletti, W. Foissner, and D. Coleman, Eds., CRC/Lewis, Boca Raton, FL, pp. 1–14.

Doran, J.W. 1982. Tilling changes soil — differences not just physical; chemical and biological too. Crops and Soils Mag., 34:10–12.

Graham, T. 1861. Liquid diffusion applied to analysis. Phil. Trans. Royal Soc. (London) 151:183–224.

House, G.J. and R. W. Parmele. 1985. Comparison of soil arthropods and earthworms from conventional and non-tillage agro-ecosystems. Soil Tillage Res. 5:352–360.

Hurlbut, C.S. Jr. and C. Klein. 1977. *Manual of Mineralogy* (after James D. Dana), 19th ed., John Wiley & Sons, New York.

Patrick, W.H. Jr. and M.B. Sturgis. 1955. Concentration and movement of oxygen as related to absorption of ammonium and nitrate nitrogen by rice. Soil Sci. Soc. Am. Proc. 19:59–62.

Schulze, D.G. 1989. An introduction to soil mineralogy, in *Minerals in Soil Environment*, 2nd ed., J.B. Dixon and S.B. Weed, Eds., Soil Sci. Soc. Am., Madison, WI, Book Ser. No. 1, pp. 2–34.

Shaw, D.J. 1968. *Introduction to Colloid and Surface Chemistry*, Butterworth, London, 190 pp.

Stevenson, F.J. 1982. *Humus Chemistry*, John Wiley & Sons, New York, 443 pp.

USDA. 1951. *Soil Survey Manual*, U.S. Department of Agriculture Handbook, 18, 503 pp.

4   SOIL COLLOIDS

Soil mineral colloids, which include crystalline and noncrystalline silicate minerals and hydrated oxides of aluminum and iron (also manganese in some soils), are responsible for many of the physical, physicochemical, and chemical properties exhibited by the soils. Soil colloids are particularly important in determining the capability of a soil to absorb and hold plant nutrients (both cations and anions), thereby restricting their rate of loss by leaching. An understanding of the structure and properties of colloids is therefore important in soil fertility management.

4.1. CLAY MINERALS

As the name indicates clay minerals are found in the clay fraction of the soil. All clay minerals are made up of layers of Si-tetrahedra and Al-octahedra (Figure 4.1) and are therefore known as layer silicates. Broadly, there are two major kinds, namely, minerals having one layer each of Si-tetrahedra and Al-octahedra (1:1 layer silicates) and those having one layer of Al-octahedra and two layers of Si-tetrahedra (2:1 layer silicates).

As shown in Figure 4.1, the dimensions of the void created by packing of four O^{2-} ions in tetrahedral configuration can accommodate one Si^{4+} ion. However, in the formation of a clay mineral or in its weathering, sometimes an Al^{3+} ion may substitute for an Si^{4+} ion. Since the charge on the Si-tetrahedral layer is neutral with Si^{4+}, substitution by Al^{3+} lowers the positive charge by one unit or it creates the negative charge of one unit. Such ionic substitution with cations of similar size is known as isomorphous substitution. Since the charge created by isomorphous substitution is due to the change in structure of the clay mineral, this is known as permanent change. Ionic substitution of Al^{3+} in the octahedral layers by Mg^{2+} or Fe^{2+} may also occur creating different kinds of clay minerals that also exhibit different physical and chemical properties. The ideal 2:1 layer silicates have a total cation charge of 22 (100 + 2 OH for pyrophillite) (see Table 4.2); +6 with octahedral sheet and +16 with the tetrahedral sheets. The total cation charge of +6 in the octahedral layer can be satisfied with two trivalent cations (Al^{3+}) or three

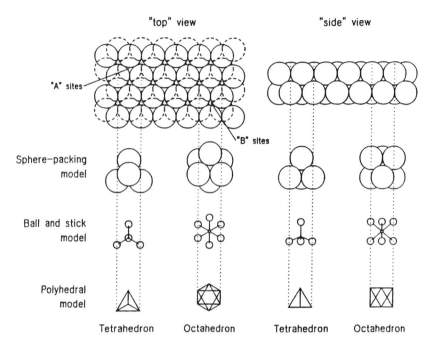

Figure 4.1. Octahedra and tetrahedra as a consequence of two planes of close-packaged spheres and three ways of representing octahedra and tetrahedra. (Adapted from Schulze, 1989.)

divalent cations (Mg^{2+}, Fe^{2+}). When a trivalent ion such as Al^{3+} is involved, only two out of three sites are filled in the octahedral layer and the arrangement or mineral is called dioctahedral. However, when a divalent ion such as Mg^{2+} or Fe^{2+} is involved, all three sites are filled and such an arrangement or mineral is known as trioctahedral.

Schematic diagrams of 1:1 and 2:1 layer silicates are given in Figure 4.2. In 1:1 layer silicates there are three planes of anions: (1) basal plane of O's, formed by the bases of a Si-tetrahedral sheet; (2) a central plane of OH's of Al-octahedral sheet and shared O's of tetrahedral sheet; and (3) a top plane of OH's of Al^{3+}-octahedral sheet. The tetrahedral and octahedral cations occupy their appropriate places. In 2:1 layer silicates there are four planes of anions (O, OH). Names and a brief description of important clay minerals follow.

4.1.1. The 1:1 Layer Silicates

4.1.1.1. Kaolinite

Kaolinite is the most common member of this subgroup. It is a dioctahedral 1:1 layer silicate and contains Al^{3+} in the octahedral and Si^{4+} in the

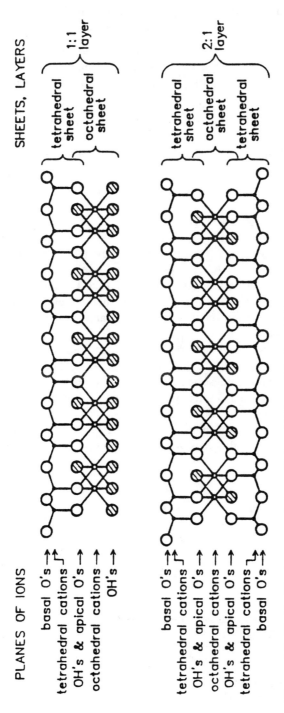

Figure 4.2. Phyllosilicate nomenclature. (Adapted from Schulze, 1989.)

**Table 4.1 Particle Size, C-Axis Spacing and Surface Area
of Some Clay Minerals**

Minerals	Particle size (μm)	External surface (m² g⁻¹)	Internal surface[a] (m² g⁻¹)	C-axis spacing (nm)
Kaolinite	0.5–5.0	10–30	—	0.7
Smectite	0.01–1.0	70–120	550–650	1.0–2.0
Illite	0.2–2.0	70–100	—	1.0
Vermiculite	0.1–0.5	50–100	500–600	1.0–1.5
Chlorite	0.1–2.0	70–100	—	1.4

[a]This is a property of expanding-layer silicates.

tetrahedral sites (Figure 4.3). The Unit cell* formula is $Al_2^{3+}Si_2^{4+}O_5^{2-}(OH)_4^-$, which is charge neutral (14 positive and 14 negative charges). Consequently, the cation exchange capacity (CEC) is very low (2 to 5 cmol kg⁻¹)** and generally due to OH^- on broken edges. Soils having kaolinites as the dominant mineral are generally less fertile. Thus a 1:1 layer is electrically neutral, and adjacent layers are held together by electrostatic binding between the basal O's of the tetrahedral sheet and the OH's of the adjacent octahedral sheet.

Kaolinite crystals are hexagonal in shape and range from 0.10 to 5 μm across; mostly in the 0.2 to 2 μm range (Table 4.1). The C-axis (interlayer) spacing is 0.7 μm or 7 angstroms.

Kaolinite is formed in soils by weathering of 2:1 layer silicates and other minerals such as feldspars. Soil parent materials such as clayey sediments may also contain kaolinite. Kaolinite is found in more weathered soils such as ultisols and oxisols and is thus abundant in tropical and subtropical regions of the world.

4.1.1.2. Halloysite

In halloysite the 1:1 layers are separated by a layer of H_2O molecules (Fig. 4.3). Halloysite often occurs as tubular or spheroidal particles. It is usually found in soils formed from volcanic deposits, soils in the Andept suborder. Halloysite easily weathers to kaolinite.

---

* A unit cell is the smallest particle of a mineral containing all its physical and chemical properties. The 3-dimensional arrangement of the unit cells defines the crystal lattice of a mineral and determines its shape and size. In clay minerals the units repeat themselves in a horizontal plane forming the layers and thus the name layer silicates.

** The earlier unit for expressing CEC was meq 100 g⁻¹. Since one cmol equals 10 milliequivalents, the numerical value remains the same. Thus 1 meq 100 g⁻¹ is the same as 1 cmol kg⁻¹.

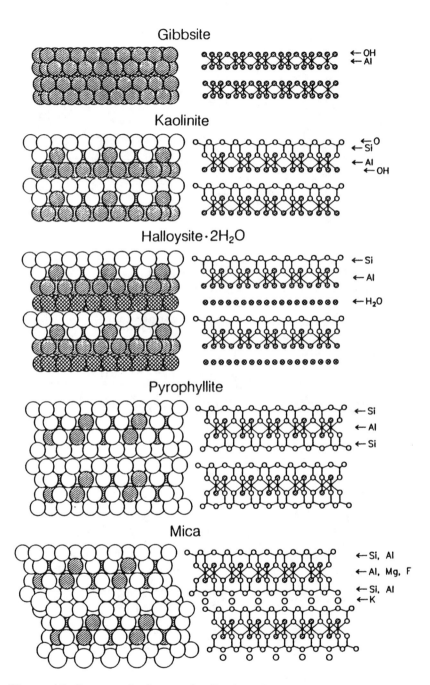

Figure 4.3. Structural scheme of soil minerals based on octahedral and tetrahedral sheets. (Adapted from Schulze, 1989.)

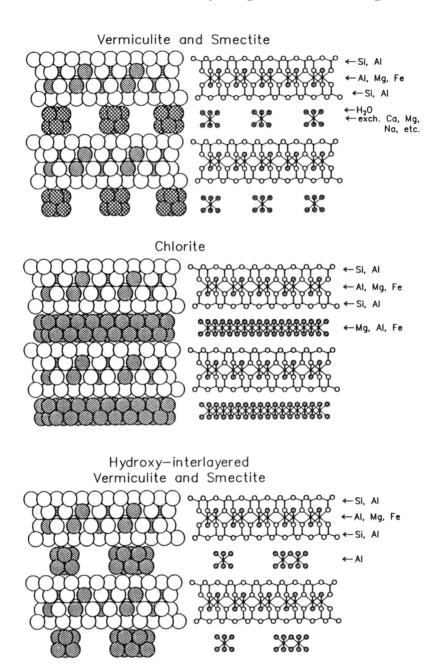

Figure 4.3. (continued).

4.1.2. The 2:1 Layer Silicates

As already explained, 2:1 layer silicates have one octahedral layer sandwiched between two tetrahedral layers. Different minerals in this group differ in the amount of isomorphous substitution and consequent negative charge. The most widespread group of minerals is smectites, which include montmorillonite, beidellite, nontronite, etc. Smectites are responsible for the high cation exchange capacity of Vertisols. Also these minerals have a great capacity to adsorb water and swell in the vertical (C-axis) direction and are therefore known as expanding clay minerals. The swelling on wetting and shrinking on drying property of Vertisols is due to smectites.

4.1.2.1. Pyrophyllite and Talc

Pyrophyllite is an ideal dioctahedral 2:1 layer silicate with Al in the octahedral sheet, while talc is trioctahedral with Mg in the octahedral sheet (Table 4.2). Since there is no isomorphous substitution, there is little negative charge on the mineral particles and these minerals are nonexpanding. Talc and pyrophyllite occur only rarely in soils.

4.1.2.2. Micas

Micas have structure similar to pyrophyllite or talc with the following distinct additional qualities:

1. There is considerable isomorphic substitution of Si^{4+} by Al^{3+}, Mg^{2+} or Fe^{2+} in the tetrahedral layer; nearly 20 to 25% sites have Al^{3+}, Mg^{2+}, and Fe^{2+}. This leads to a unit negative charge per unit formula (Table 4.2).
2. This excess negative charge is balanced by K^+ ions, which snugly fit in the hexagonal void created between two 2:1 layers. Since most of the negative charge is balanced by K^+, micas are nonexpanding in nature.
3. The micas have oxygen planes on both sides of their units. The six oxygens forming a hexagonal arrangement provide space just about the same size as oxygen (2.8 Å in diameter). Since the K^+ has a radius of 1.33 Å, it is the ion that snugly fits in the space. The next best choice is NH_4^+ ion with a radius of 1.48 Å. Thus soils having mica-derived minerals have the capacity to fix (immobilize) K^+ or NH_4^+ ions between the layers. Fixation and release of K^+ and NH_4^+ ions is important from the viewpoint of the availability of these ions to growing crop plants.

When the ion substituting for Si^{4+} in the tetrahedra is Al^{3+}, the mica is known as muscovite, while the mica containing Mg^{2+} in tetrahedral layer is

Table 4.2 Ideal Formula and Unit Negative Charge of Common Clay Minerals

Mineral	Ideal formula	Charge per unit formula	Fixed interlayer component	Net negative charge (cmol kg^{-1})
1:1 Layer silicates				
Kaolinite	$Al_2Si_2O_5(OH)_4$	0	None	2–5
Antigorite	$Mg_3Si_2O_5(OH)_4$	0	None	
2:1 Layer silicates				
Pyrophyllite	$Al_2Si_4O_{10}(OH)_2$	0	None	
Talc	$Mg_3Si_4O_{10}(OH)_2$	0	None	
Muscovite (Mica)	$Al_2(Si_3Al)^{-1}O_{10}(OH)_2$	−1.0	K^+ 1.0	
Illite	$Al_2(Si_{3.2}Al_{0.8})^{-0.8}O_{10}(OH)_2$	−0.8	K^+ 0.7	15–40
Vermiculite	$(Al_{1.7}Mg_{0.3})^{-0.3}(Si_{3.6}Al_{0.4})^{-0.4}O_{10}(OH)_2$	−0.7	XH_2O	100–180
or	$(Mg_{2.7}Fe^{3+}_{0.3})^{+0.3}(Si_3Al)^{-1.0}O_{10}(OH)_2$	−0.7	XH_2O	
Smectites				
Montmorillonite	$(Al_{1.7}Mg_{0.3})^{-0.3}(Si_{3.9}Al_{0.1})^{-0.1}O_{10}(OH)_2$	−0.4	None	80–120
Beidellite	$(Al_2)(Si_{3.6}Al_{0.4})^{-0.4}O_{10}(OH)_2$	−0.4	None	80–120
Nontronite	$(Fe^{3+}_2)(Si_{3.6}Al_{0.4})^{-0.4}O_{10}(OH)_2$	−0.4	None	80–120
2:1:1 Layer silicates				
Chlorite	$(Mg_{2.6}Fe^{3+}_{0.4})^{+0.4}(Si_{2.5}Al/Fe_{1.5})^{-1.5}O_{10}(OH)_2$[a]		$Mg_2Al(OH)_6^{+1}$	15–40

[a]Unit negative charge is neutralized by unit positive charge on interlayer component.

known as phlogopite. Biotite mica on the other hand has Fe^{2+} and some Mg^{2+} as the substituting ions.

Micas in the clay fraction of soils and sediments often have poorer crystallinity, lower K and higher H_2O content than other 2:1 clays. Micas with Al^{3+} as the substituting ion are referred to as illite. Glauconite is similar to illite, but contains more Fe^{3+} and less Al^{3+}. The illites are dioctahedral and nonexpandible, and a part of their octahedral charge is balanced in the tetrahedral layer by additional Si beyond the ideal Si:Al ratio of 3:1 for muscovite. A representative formula is $K_{0.75}$ $(Al_{1.75} RS_{0.25}^{2+})$ $(Si_{3.5} Al_{0.5})$ O_{10} $(OH)_2$, and the interlayer charge varies from 0.6 to 0.8 where R could be Mg^{2+} or Fe^{2+} (Fanning et al., 1989).

Micas weather to other minerals, particularly to vermiculites and smectites, and K^+ released during weathering is an important source of K for plants. Therefore soils derived from rocks rich in micas such as those in the Indo-Gangetic plains of India and Pakistan are rich in plant-available K.

4.1.2.3. Vermiculite

It is a 2:1 layer expanding silicate mineral with less ionic substitution than in micas. Thus the negative charge per unit mineral is less than 1; it is near 0.7* (range 0.6 to 0.9) (Table 4.2). In dioctahedral vermiculite, which is most abundant, the negative charge is nearly equally distributed in the octahedral and tetrahedral sheets. In the trioctahedral vermiculites, the octahedral sheet has a negative charge of –1.0, while the tetrahedral sheet has a positive charge (due to Fe^{3+} substituting for Mg^{2+}) of +0.3. Vermiculite has thus a net negative charge of 0.7. This charge attracts and holds cations such as Ca^{2+} and Mg^{2+} in the interlayer (Figure 4.3). These cations can be changed for other cations in soil solution, and the phenomenon is known as cation exchange. Cation exchange capacity (CEC) is an important property of a clay mineral and is equivalent to the net negative charge. Thus vermiculite has a CEC** of 100 to 180 cmol kg^{-1}. This high CEC gives vermiculites a high affinity for weakly hydrated ions such as K^+, NH_4^+ and Ca^{2+}. Consequently, soils rich in vermiculite have high K-fixation capacity. Vermiculites occur widely in soils.

---

* These figures are near average. In nature there is likely to be considerable variation. This applies to all the minerals discussed in this chapter.

** Because cation exchange capacity (CEC) of the clay minerals is an important property, it is imperative to understand how it is computed. Let us take the example of pyrophyllite. Its ideal formula is $Al_2Si_4O_{10}$ $(OH)_2$ (Table 4.2), having a formula weight of 360. If an Si^{4+} is replaced by an Al^{3+}, the formula weight changes to 359 and there will be a unit negative charge, which is equal to 1 mol (or 100 centimoles abbreviated as cmol). For 1000 g pyrophyllite the charge will be

$\dfrac{1000}{359} \times 100 = 278$ cmol kg^{-1}. This is the basis on which one estimates the CEC of a clay mineral.

Thus montmorillonite, having a net negative charge of 0.4, is likely to have CEC of 111 cmol kg^{-1}. Since different samples of various clay minerals differ in their net negative charge, it is conventional to give a range rather than a single value to CEC of a clay mineral.

Because of hydrated cations occupying the interlayer spaces, vermiculites have considerable expansion capacity and a large internal surface (500 to 600 m^2 g^{-1}). The C-axis spacing is 1.0 to 1.5 nm depending upon the degree of hydration.

4.1.2.4. Smectites

The smectites are also 2:1 layer expanding silicates with a net negative charge of about 0.4, less than that for vermiculite (range 0.25 to 0.6). The dioctahedral (Al containing) smectite montmorillonite is the most common member in soils. In this mineral the negative charge is mostly created by the substitution of Mg^{2+} for Al^{3+} in the octahedral sheet, though there may be some charge on the tetrahedral sheet. In beidellite, which is also dioctahedral, the negative charge is due to ionic substitution of Al^{3+} for Si^{4+} in the tetrahedral sheet. This is also true for nontronite, which has Fe^{3+} in place of Al^{3+} in the octahedral sheet.

Smectites, because of their enormous expanding nature, have a very large internal surface (550 to 650 m^2 g^{-1} or more) and CEC (80 to 120 cmol kg^{-1}). They are the major clay mineral in vertisols such as black cotton soils of India and the blacklands of Texas. The shrink-swell behavior (swelling when wet and shrinking when dry) of vertisols, due to smectites, is a major problem in the management of these soils.

4.1.2.5. Chlorite

Chlorite is a 2:1:1 layer silicate. The additional layer occupies the inter-layer space between 2:1 layers. This layer consists of Mg-Al hydroxides with the general formula $Mg_2Al\ (OH)_6{}^+$ and has a unit positive charge that nearly balances the negative charge on the 2:1 layer. The net negative charge on 2:1:1 layer is thus hardly 0.1, giving it a low CEC (15 to 40 cmol kg^{-1}), similar to that of kaolinite.

Chlorites are infrequent in soils, and when present, they make up a small fraction of clay minerals. Chlorites are primary minerals and weather to form vermiculites and smectites.

4.1.2.6. Polygorskite and Sepiolite

These minerals are found in soils of arid and semiarid environments and have a fibrous morphology created by adjacent tetrahedral bands in a tetrahedral sheet pointing in the opposite direction.

4.1.3. Interstratified Silicate Minerals

Silicate minerals in soils do not always occur as discrete particles of mica, vermiculite, smectite, or kaolinite, but generally one, two, or three silicate minerals occur in mixtures or mixed layers and are known as interstratified minerals. The interstratified minerals may be two-component or three-component systems, for example, chlorite-vermiculite, chlorite-smectite, kaolinite-smectite, etc. The sequence of layers can be either regular or random, the latter being more common in soils.

4.1.4. Noncrystalline Silicates

Noncrystalline silicates include allophanes and imgolite. They are hydrated aluminum silicates (Demumbrum and Chesters, 1964) and are amorphous in nature and occur as a major constituent of young soils formed from volcanic ash. They are commonly found in Japan, New Zealand, New Guinea, Western Victoria, and South Australia (Norrish and Pickering, 1983) and in the B in the horizons of podzols and pozolized soils (Farmer, 1982). Allophanes have a SiO_2/Al_2O_3 ratio between 1.3 and 2.0 (this ratio for kaolinite is 2.0) and consists of spherules of 35 to 40 Å in outer diameter (Norrish and Pickering, 1983). Imgolite, on the other hand, has a tubular structure. Allophanes in B horizons of podzols are proto-imgolite in nature and consist of fragments of the imgolite tube wall (Figure 4.4) (Farmer, 1987). The CEC is 20 to 60 cmol kg^{-1}.

Allophanes have a very large surface area, values for which vary greatly depending on how it is determined. Aomine and Otsuka (1968) obtained values ranging from about 50 m^{2}g^{-1} using N_2 to 900 m^{2}g^{-1} using ethylene glycol. Soils containing much allophane may have very high water content (>100%), but they do not swell on wetting or shrink on drying. Loss of water is irreversible.

4.2. OXIDE MINERALS

Much of the research on the ionic charge of soil particles was carried out in the past on crystalline silicate minerals, which dominate the temperate region soils where most information in soil science has developed. However, in the second half of the twentieth century considerable interest has arisen in soils of the humid tropics (Van Raij and Peech, 1972; Morais et al., 1976; Keng and Uehara, 1974; El Swaify and Sayegh, 1975; Gallez et al., 1976), which contain significant quantities of oxide minerals (oxides and hydroxides of Fe, Al, Mn, Ti, and Si). These soils come under the order ultisols and oxisols and occupy large tracts of level land in Latin America, Africa, and Southeast Asia. The high oxide content of these soils gives them the colors and names by which they have been known for time immemorial such as yellow soils, red soils, terra rosa, krasnozenes, etc. These soils have good physical conditions and are well drained, but have impoverished fertility due to excessive leaching of bases.

Oxide minerals differ from silicate minerals in the kind of ionic charge they possess. While silicate minerals carry "permanent charge" created by

a

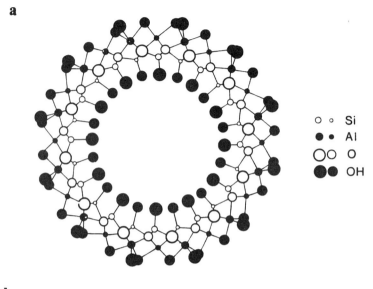

b

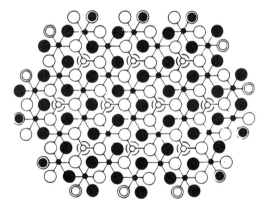

Successive atomic layers :

(1)⬤ OH   ◉H₂O   (2)• Al

(3) ○ O,OH ◎H₂O   (4)○ Si

(5) ⊃ OH

Figure 4.4. a. Cross section of an imogolite tube, showing atoms at two levels (large and small symbols). b. Proto-imogolite allophane is thought to consist of fragments of the imogolite wall, such as that shown here in plan. (From Farmer, 1987. Trans. 13th Int. Cong. Soil. Sci. 5:63–68. With permission of the International Society of Soil Science, Vienna, Austria.)

isomorphous substitution of metallic ions, most of the charge in oxide minerals is pH dependent and is therefore known as "variable charge." Variable charge is also present in small amounts in silicate minerals and humus. At some pH value an oxide mineral can have equal quantities of H^+ and OH^- ions, so that the net surface charge is zero in the absence of any specifically adsorbed ions; this pH is known as zero point of charge (ZPC) or isoelectric point. The ZPC for the common soil oxides is usually between pH 6.5 to 10.4 (Parks, 1965). Under acid conditions, an oxide is positively charged and can have an appreciable anion exchange capacity. This imparts to oxide minerals an ability to adsorb OH^- ions and therefore high buffering capacity. Consequently, soils having oxide minerals require large amounts of liming material. On the other hand, under alkaline conditions oxide minerals acquire a negative charge and exhibit cation exchange capacity as shown below:

(From Parks and deBruyn, 1962)

The surface charge density for goethites and hematites at pH 4 in 1 M NaCl was between 1.8 and 4.9 μeq m^{-2} (Atkinson et al., 1967; Atkinson, 1969). The specific surface area for goethites is 14 to 77 m^2 g^{-1}, while that for hematite ranges between 34 and 45 m^2 g^{-1}. Under warm climatic conditions with high rainfall organic matter decomposes rapidly, leaving oxide surfaces with a positive charge and incapable of retaining nutrient cations against leaching losses; such soils thus have impoverished soil fertility. Under such conditions application of silicate amendments might be useful, since silicate ions can be specifically adsorbed onto oxides, with a resultant increase in negative charge (Taylor et al., 1983).

The most common oxide minerals in soils are of Fe, Mn, Al, and Si. In some soils (not under cultivation) the oxides of Ti may also be present.

4.2.1. Iron Oxides

Iron oxides may be evenly distributed throughout the profile or may be concentrated in a particular horizon. They can exist as coatings on sand grains and crack faces, as constituents in cemented aggregates, or as discrete fine particles.

Iron is fairly mobile in the soil profile because of its soluble reduced Fe^{2+} form. Also it forms complexes with several anions. Iron therefore plays a vital role in the structure and fabric of a soil by cementing aggregates or by formation of concretionary layers, or iron-stone cappings. The most common soil iron oxides are goethite, lepidocrocite, hematite, maghemite, ferrihydrite, and magnetite. Magnetite (Fe_3O_4) exists as a residual primary mineral, while the other five minerals are products of pedogenesis.

Goethite (α - Fe OOH) is the most common mineral and is present in most soil types under diverse climatic conditions. It is the "yellow ochre" pigment responsible for a yellow-brown color in many soils, for example, yellow earths; it produces the yellow mottles in laterites (Davey et al., 1975).

Lepidocrocite (γ - Fe OOH) is less common than goethite. It is a common mineral in hydromorphic soils of the humid temperate regions, where it is often associated with goethite (Schwertmann and Fitzpatrick, 1977).

Hematite (α - Fe_2O_3) is a common mineral in the soils of arid and tropical regions, where the decomposition of organic matter is so rapid that it does not promote goethite formation. It is the "red ochre" responsible for red coloration of many soils such as terra rosas and red earths. It occurs often in association with goethite. In x-ray diffraction the presence of goethite or hematite can be recognized by a common line at 2.69 to 2.70 Å (Taylor et al., 1983).

Maghemite (γ - Fe_2O_3) is common in highly weathered soils of the tropics and subtropics, especially in soils formed on basic igneous rocks. This mineral generally contains a small percentage of ferrous iron but less than its isomorph magnetite, which generally contains close to the theoretical one-third of its total iron in the ferrous form (Fasiska, 1967).

Ferrihydrite (HFe_5O_8, $4H_2O$) is often found in drainage ditches and is often considered as having been formed by microbial decomposition of soluble Fe-organic compounds and subsequent hydrolysis and precipitation (Taylor et al., 1983).

The iron in soils originates from the weathering of iron-bearing silicates in which iron is generally present in the divalent state. Once separated from the silicate mineral, ferrous iron is subsequently precipitated and oxidized or oxidized and precipitated; the sequence of processes is important. In general, where oxidation precedes hydrolysis, the α -iron oxides, goethite and hematite, form. However, when hydrolysis and precipitation occur before oxidation, the R-oxides, lepidocrocite and maghemite, often occur. Possible pathways of iron oxide formation under near pedogenic conditions as proposed by Schwerrt-mann and Taylor (1977) are shown in Figure 4.5. These processes may occur *in situ* where iron is released from silicates or after transportation in subsurface waters to other parts of the profile or landscape.

The strong affinity that iron oxides have for organic anions is of great importance from the viewpoint of soil fertility and crop production. Important points to be brought out are as follows:

1. The beneficial effect of organic manures, in part, may be due to the production of organic anions on decomposition, providing sites for phosphate ions to sorb on iron oxides. Regarding adsorption of

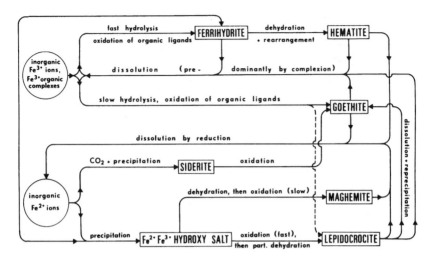

Figure 4.5. Possible pathways of iron oxide formation under near pedogenic conditions. (From Schwertmann and Taylor, 1977. *Minerals in Soil Environments,* **J.B. Dixon and S.B. Weed, Eds., p. 160. With permission of Soil Science Society of America, Madison, WI.)**

phosphate on iron oxides, both mono (a) and bidentate (chelated) (b) complexes may exist, explaining different bonding strengths (b>a) and the partial irreversibility of phosphate sorption (Hingston et al., 1974).

2. Also, as explained earlier, since the charge on iron oxides is pH dependent, both anions and cations are retained. Furthermore, the adsorption also depends upon the concentration of different ions in soil solution. Results from a study by Bowden et al. (1973) are shown in Figure 4.6. Phosphate adsorption was greater in the acidic pH range and declined as pH increased; adsorption of silicate followed the reverse pattern, being greater at alkaline pH. Zn adsorption was greatest between pH 7 and 8.

3. Complex formation of iron oxides with organic anions makes iron more mobile.

4. Iron oxide-organic anion complexes improve soil structure as judged by the formation of aggregates (Table 4.3).

5. The cation exchange capacity of soil organic matter is reduced by iron oxide-organic anion complex formation, and therefore the ability of the surface soils to retain plant nutrients is reduced.

6. Figure 4.6 also shows the effects of pH on adsorption of 2,4-D, which was maximum at pH 3. Thus many herbicides and pesticides lose their efficiency by reaction with iron oxides, and they may persist undecomposed for long periods in such soils.

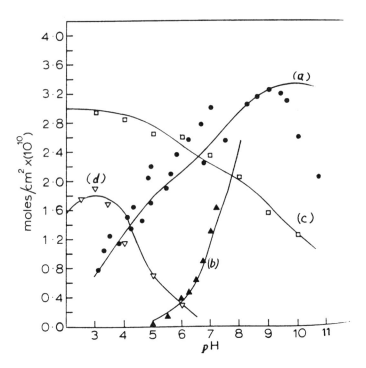

Figure 4.6. Adsorption on goethite as a function of pH; (a) silicate, $8 \times 10^{-4} M$ (0.1 M NaCl); (b) zinc, $1 \times 10^{-4} M$ (0.1 M NaCl); (c) phosphate, $3.2 \times 10^{-4} M$ (0.01 M NaCl); (d) 2,4-D, $3 \times 10^{-4} M$ (0.01 M NaCl); —, present model. Points are from experimental data: ●, silicate; ▲, zinc; □, phosphate; ▽, 2,4-D. (From Bowden et al. 1973. Nature (London) 245:81–82. With permission.)

4.2.2. Manganese Oxides

Manganese occurs in a number of oxidation states, but only the cationic oxidation states (positive) 2, 3, and 4 occur in nature. A large number of oxides and hydroxides occur in nature, where there is an extensive substitution of Mn (II) and Mn (III) for Mn (IV) and of OH^- for O^{2-}. The structures of manganese oxides can change with a change in the oxidation state of Mn. This and their intimate occurrence with iron oxides in soils make the determination of chemical composition of manganese oxides difficult.

As is the case with the iron oxides, manganese oxides also occur in soils as deposits in cracks and veins and as coatings and stains or as a cementing material in soil aggregates. Manganese oxides may also occur roughly as spherical nodules up to about 2 cm in diameter. These nodules contain considerable amounts of iron oxides, as well as some other minerals, and many have a concentric, layered structure suggestive of seasonal growth. Burns and

Table 4.3 The Influence of Synthetic Iron Oxide Addition to a Loess Soil (86% Silt) on the Formation of 1–2 mm Water-Stable Aggregates after Air Drying

Fe-oxide	Amount added (% Fe)	Oxalate solubility[a], (% of total Fe)	Aggregates (1–2 mm, %)
No addition	0.0	—	18
Ferrihydrite	11.0	100.0	48
Ferrihydrite	2.3	100.0	83
Lepidocrocite	1.0	68.0	53
Goethite	1.0	0.1	17
Goethite	0.6	93.0	77
Hematite	1.0	0.1	20

[a]Taken as a measure of degree of crystallinity.

From Schahabi and Schwertmann, 1970. Z. Pflanzen. Dueng Bodenk. 125:193–204.

Table 4.4 Some Manganese Oxide Minerals

Mineral name	Other names	Composition
Pyrolusite	β-MnO_2, polianite	MnO_2
Ramsdellite	—	MnO_2
Nsutite	γ-MnO_2, ρ-MnO_2	$Mn(O, OH)_2$
Birnessite	δ–MnO_2, manganous manganite, 7 Å manganite	Variable
Todorokite	—	Variable
Cryptomelane	α-MnO_2	$K_2Mn_8O_{16}$
Hollandite	α-MnO_2	$Ba_2Mn_8O_{16}$
Coronadite	α-MnO_2	$Pb_2Mn_8O_{16}$
Psilomelane	romanechite	$BaMn_9O_{16}(OH)_4$
Lithiophorite	—	$(Al, Li)MnO_2(OH)_2$
Groutite	α-MnOOH	MnOOH
Manganite	γ-MnOOH	MnOOH
Hausmannite	—	$Mn^{2+}Mn_2^{3+} + O_4$

From Taylor et al. 1983. *Soils — Australian Viewpoint,* Academic Press, San Diego, CA, pp. 309–334. With permission of Academic Press, Inc.

Burns (1975) suggested that nucleation and growth of many manganese nodules is initiated by a deposit of ferrihydrite on a nucleating agent, followed by epitaxial growth of δ MnO_2 and ferrihydrite on each other.

Although a large number of Mn oxide minerals are known (Table 4.4), only a few occur in soils. It seems likely for soils that either δ MnO_2 or birnessite are formed first and that these minerals may recrystallize subsequently, under suitable conditions, to lithiophortite or todorokite or the cryptomelane group (Taylor et al., 1983).

A large amount of Mn in soil solution, as under submerged rice paddies, can lead to Mn-toxicity problems; manganese toxicity can be a constraint to plant growth on acid oxisols (Fageria et al., 1990).

4.2.3. Aluminum Hydroxides and Oxyhydroxides

A number of crystalline aluminum oxides or rather hydroxides and oxy-hydroxides are found in nature. Hsu (1977) suggested the term hydroxides for polymorphs of $Al(OH)_3$ and oxyhydroxides for polymorphs of $AlOOH$, and these terms are used in this chapter.

Crystalline Al-hydroxides exist in three polymorphs, namely, gibbsite, bayerite, and norstrandite. Only gibbsite is common in soils and bauxite deposits, although the presence of bayerite (Benter et al., 1963) and norstrandite (Wall et al., 1962) has been reported. Gibbsite is a major mineral in andosols in Japan (Wada and Aomine, 1966) and ultisols and oxisols of the tropics and subtropics and temperate regions (Rich et al., 1959). Bauxite deposits, dominated by gibbsite and to a lesser degree by boehmite or diaspore, are products of extremely advanced weathering (Hsu, 1977). All three poly-morphs of Al-hydroxides are composed of the same fundamental units, which is two planes of close-packed OH- octahedra with Al^{3+} sandwiched between them. The Al^{3+} ions occupy two-thirds of the octahedral positions. The three polymorphs differ in the stacking of fundamental units. Gibbsite crystallites are usually well developed in the x and y (horizontal) directions, but limited in the z (vertical) direction and therefore appear as hexagonal plates. While bayerite crystallites usually take the form of a pyramid, nordstandite crystal-lites are often long, rectangular prisms.

Crystalline Al-oxyhydroxides are present in two polymorphs, namely, diaspore and boehmite. Both the polymorphs have double chains of HO-Al-O; the structural difference is in the arrangement of these double chains.

Large amounts of noncrystalline hydroxides and oxyhydroxides are also present in nature. These materials do not have a definite composition or structure, but exhibit properties similar to those of the crystalline Al-hydrox-ides and oxyhydroxides.

Anion adsorption and fixation properties of Al-hydroxides and oxyhy-droxides are similar to those of iron oxides. The affinity of both iron oxides and Al hydroxides and oxyhydroxides for phosphate is well established (Wild, 1950). In addition to phosphate, sulfate is also adsorbed by gibbsite (Chao et al., 1964), but can be replaced by phosphate, which is held more tightly.

The role of Al-hydroxides and oxyhydroxides as cementing agents for soil aggregates is well documented (Saini et al., 1966; El Swaify and Emerson, 1975). Al-hydroxides decrease their positive charge, as well as their cementing effectiveness, with increasing pH. Iron oxides have a stronger tendency to crystallize than Al-hydroxides; the former is thus likely to be less effective than the latter in cementing effectiveness. For example, Saini et al.

(1966) found that in Canadian soils (B horizons of podzols) the partial regression coefficient relating Al-oxides with aggregation was 1.84 times larger than that for iron oxides.

4.2.4. Silicon Oxides

Silica is the most abundant oxide in the lithosphere. The most common forms of silica found in soils are quartz, cristobalite, tridymite and the poorly crystalline opals. Thermodynamically, quartz is the only stable variety.

Because of its large particle size, and its relatively unreactive surfaces, quartz plays a negligible role in plant nutrition. It, however, has a definite role in determining soil texture and structure. Quartz can be identified by an X-ray diffraction peak at 3.34 Å.

4.2.5. Titanium Oxides

Three polymorphic forms of TiO_2 are found in soils, namely, rutile, anatase and brookite (Taylor et al., 1983). In addition, three combined oxides or titanites are: Ilmetite ($FeTiO_3$), Sphene ($CaTiSiO_5$) and Perovskite ($CaTiO_3$). Titanium is not an essential plant nutrient and these oxides have thus no effect on the nutritional status of a soil.

REFERENCES

Aomine, S. and H. Otsuka. 1968. Surface of soil allophanic clays. Trans. 9th Int. Cong. Soil Sci. (Adelaide) 1:731–737.

Atkinson, R.J. 1969. Crystal morphology and surface reactivity of goethite. Ph.D. thesis, University of Western Australia, Nedlands, W.A. (Schwertmann and Taylor, 1977).

Atkinson, R.J., A.M. Posner, and J.P. Quirk. 1967. Adsorption of potential determining ions at the ferric oxide aqueous electrolyte interface. J. Phys. Chem. 71:550–557.

Bowden, J.W., M.D.A. Bolland, A.M. Posner, and J.P. Quirk. 1973. Generalized model for anion and cation adsorption at oxide surfaces. Nature (London) 245:81–82.

Bentor, Y.K., S. Gross, and L. Miller. 1963. Some unusual minerals from the "mottled zone" complex, Israel. Am. Minerol. 48:924–930.

Burns, R.G. and V.M. Burns. 1975. Mechanism for nucleation and growth of manganese nodules. Nature (London) 255:130–131.

Chao, T.T., M.E. Howard, and S.C. Fang. 1964. Iron and aluminum coatings in relation to sulphate adsorption characteristics of a soil. Soil Sci. Soc. Am. Proc. 28:632–635.

Davey, B.G., J.D. Russell, and M.J. Wilson. 1975. Iron oxide and clay minerals and their relation to colors of red and yellow podzolic soils near Sydney, Australia. Geoderma 14:125–138.

DeMumbrum, L.E. and G. Chesters. 1964. Isolation and characterization of some soil allophanes. Soil Sci Soc. Am. Proc. 28:355–359.

El Swaify, S.A. and A.H. Sayegh. 1975. Charge characteristics of an oxisol and an inceptisol from Hawaii. Soil Sci. 120:49–56.

El Swaify, S.A. and W.W. Emerson. 1975. Changes in physical properties of soil clays due to precipitated aluminum and iron hydroxides. I. Swelling and aggregate stability after drying. Soil Sci. Soc. Am. Proc. 39:1056–63.

Fageria, N.K., V.C. Baligar, and D.G. Edwards. 1990. Soil—plant nutrient relationships at low pH stress, in *Crops as Enhancers of Nutrient Use*, V.C. Baligar and R.r. Duncan, Eds., Academic Press, Harcourt Brace Jovanovich, San Diego, pp. 475–507.

Fanning, D.S., V.Z. Keramidas, and M.A. El-Desoky. 1989. Micas, in *Minerals in Soil Environments*, J.B. Dixon and S.B. Weeds, Eds., Soil Sci. Soc. Am., Madison, WI, Book Series No. 1, pp. 551–634.

Farmer, V.C. 1982. Significance of the presence of allophane and imogolite in podzol Bs horizons for podzolization mechanism: a review. Soil Sci. Pl. Nutr. 28:571-578.

Farmer, V.C. 1984. Distribution of allophane and organic matter in podzol B horizons. J. Soil Sci. 35:452–458.

Farmer, V.C. 1987. Natural and synthetic allophane and imogolite: a synergistic relationship. Trans 13th Int. Cong. Soil Sci. 5:63–68.

Fasiska, E.J. 1967. Structural aspects of the oxides and oxyhydrates of iron. Corros. Sci. 7:833–839.

Gallez, A., A.S.R. Ino, and A.J. Herbillon. 1976. Surface and charge characteristics of selected soils in the tropics. Soil Sci. Soc. Am. J. 40:601–608.

Hingston, F.J., A.M. Posner, and J.P. Quirk. 1974. Anion adsorption by goethite and gibbsite. II. J. Soil Sci. 25:16–26.

Hsu, P.H. 1977. Aluminum hydroxides and oxyhydra-oxides, in *Minerals in Soil Environment*, J.B. Dixon and S.B. Weed, Eds., Soil Science Society of America, Madison, WI, pp. 99–143.

Keng, J.C.W. and G. Uehara. 1974. Chemistry, mineralogy and taxonomy of oxisols and ultisols. Proc. Soil Crop Sci. Soc. Fla. 33:119–126.

Morais, F.I., A.L. Page, and L.J. Lund. 1976. The effect of pH, salt concentration, and nature of electrolytes on the charge characteristics of Brazilian tropical soils. Soil Sci. Soc. Am. J. 40:521–527.

Norrish, K. and J.G. Pickering. 1983. Clay minerals, in *Soil—An Australian Viewpoint*, CSIR/Academic Press, Adelaide, Australia, pp. 281–308.

Parfitt, R.L., A.R. Frazer, and V.C. Farmer. 1977. Adsorption on hydrous oxides. III. Fulvic acid and humic acid on goethite, gibbsite and imogolite. J. Soil Sci. 28:289–296.

Parks, G.A. 1965. The isoelectric points of solid oxides, solid hydroxides and aqueous hydroxy complex systems. Chem. Rev. 65:177–198.

Parks, G.A. and P.L. deBruyn. 1962. The zero point of charge of oxides. J. Phys. Chem. 66:967–973.

Rich, C.I., L.F. Seatz, and G.W. Kunz. 1959. Certain properties of selected southeastern United States soils and mineralogical procedures for their study. Southern Regional Bull. 61 for Cooperative Regional Research Project S-14/. Virginia Agriculture Experimental Station, Blacksburg.

Saini, G.R., A.A. McLean, and J.J. Doyle. 1966. The influence of some physical and chemical properties on soil aggregation and response to VAMA. Can. J. Soil Sci. 46:155–160.

Schahabi, S. and U. Schwertmann. 1970. Der einflub von synthetischen eisenoxiden auf die aggregation zweier bo bodenhorizonte. Z. Pflanzen. Dueng. Bodenk. 125:193–204.

Schultze, D.G. 1989. An introduction to mineralogy, in *Minerals in Soil Environments*, J.B. Dixon and S.B. Weed, Eds., Soil Sci. Soc. Am. Madison, WI, Book Series No. 1, pp. 1–33.

Schwertmann, U. and R.M. Taylor. 1977. Iron oxides, in *Minerals in Soil Environments*, J.B. Dixon and S.B. Weed, Eds., Soil Sci. Soc. Am., Madison, WI, pp. 145–180.

Schwertmann, U. and R.W. Fitzpatrick. 1977. Occurrence of lepidocrocite and its association with goethite in Natal soils. Soil Sci. Soc. Am. J. 41:1013–1018.

Taylor, R.M., R.M. McKenzie, A.W. Fordham, and G.P. Gillman. 1983. Oxide minerals, in *Soils—Australian Viewpoint*, CSIRO/Academic Press, Melbourne, pp. 309–334.

Van Raij, B. and M. Peech. 1972. Electrochemical properties of some oxisols and ultisols of the tropics. Soil Sci. Soc. Am. Proc. 36:587–593.

Wada, K. and S. Aomine. 1966. Occurrence of gibbsite in weathering of volcanic materials at Kuroishibaru, Kumamoto. Soil Sci. Plant Nutr. 12:151–157.

Wall., J.R.D., E.B. Wolfenden, E.H. Beard, and T. Deans. 1962. Nordstrandite in soil from West Sarawak, Borneo. Nature (London), 196:264–265.

Wild, A. 1950. The retention of phosphate by soil: a review. J. Soil Sci. 1:221–238.

5   ORGANIC MATTER

On the basis of organic matter content, soils are characterized as mineral or organic. Mineral soils form most of our cultivated land and may contain from a mere trace to 20 to 30% organic matter (Table 5.1). On the contrary, organic soils contain 80% or more organic matter. Marshes, bogs, and swamps provide conditions suitable for the accumulation of large amounts of organic matter. Organic soils or histosols are found in cold glacial regions such as parts of Minnesota, Wisconsin, and Michigan in the United States, large areas in Russia and Canada, and in several countries of Europe, namely, Germany, Holland, Norway, Sweden, Poland, and the United Kingdom. Histosols also occur in warmer regions of the world such as Kari soils in Kerala, India. The discussion on soil organic matter in this chapter deals mainly with the organic matter of mineral soils.

Most of the soil organic matter originates from plant tissue; animal tissue is also derived from plant products. Plant residues contain 60 to 90% moisture, and the rest is dry matter. On a weight basis, the dry matter is mostly carbon and oxygen (about 40% each), with less than 10% each of hydrogen and inorganic elements (ash). Therefore C, H, and O dominate the organic matter in soil. Other elements such as S, N, P, K, Ca, Mg, and micronutrients, although present in small amounts, are important from the viewpoint of soil fertility management. Soil organic matter is the primary source of native nutrients from a soil.

When plant residues are added to a soil, various organic compounds undergo decomposition. Sugars, starches, and proteins decompose rapidly, while cellulose, fats, waxes, and lignins decompose slowly; lignin decomposes very slowly. There are five major kinds of products from the decomposition of plant or animal residues.

1. CO_2—Evolution of CO_2 can be easily observed. The rate of decomposition of various plant and animal residues is determined by measuring the rate of evolution of CO_2.

**Table 5.1 Carbon and Nitrogen Contents of Soils
under Various Moisture-Temperature Interactions[a]**

Temperature–moisture interaction zone	Carbon ($Mg\ ha^{-1}\ m^{-1}$)	Nitrogen ($Mg\ ha^{-1}\ m^{-1}$)	Carbon/ nitrogen
Boreal dry bush	102	6.3	16
Boreal wet forest	150	9.8	15
Boreal rain forest	320	15.0	22
Cool temperature desert bush	99	7.8	13
Cool temperature grassland	133	10.3	13
Cool temperature wet forest	120	6.3	19
Cool temperature rain forest	200	8.0	25
Warm temperature desert bush	60	3.0	20
Warm temperature moist forest	93	6.5	14
Warm temperature wet forest	270	18.0	14
Warm temperature rain forest	270	7.0	38
Tropical desert bush	10	0.5	20
Tropical dry forest	100	8.9	11
Tropical wet forest	145	6.6	22
Tropical rain forest	180	6.0	30

[a]Adapted from Zinke et al. (1984) and Paul and Clark (1989).

2. Heat or energy—This can be easily measured. Energy released is utilized by the soil microorganisms. Plant tissue has a heat value approximating 4 to 5 K calories per gram of air-dry substance.
3. Water—This is released as a product of several enzymatic oxidations of organic carbon compounds.
4. Plant nutrients—N, S, P, Ca, Mg, K, etc., are released, which is the primary goal of applying farmyard manure or other organic residues to a field. Some of the released nutrients may again be immobilized or fixed by accompanying microbiological/chemical reactions. The much talked about "organic farming" aims at meeting all or most plant nutrient needs by the use of farm residues and biological nitrogen fixation.
5. Soil humus—This is a combination of the residues of the added organic matter, as well as resynthesized microbial tissue that is resistant to microbial action. Humus is an important component of soil and plays a major role in determining the physical and physico-chemical properties of a soil.

5.1. HUMUS, ITS STRUCTURE AND PROPERTIES

Humus is a complex mixture of dark-brown, amorphous and colloidal substances modified from the original plant tissue or synthesized by the various soil organisms and is resistant to microbial decomposition. The contribution

**Table 5.2 Contribution of Microbial Biomass
to Soil Organic Matter in 0–10 cm Soil Layer[a]**

| Region | Treatment | Microbial biomass | |
		mg kg^{-2}	% Soil organic
U.S.	Wheat-fallow	266	2.2
	Annual crop[b]	421	2.8
	Grass pastures	1158	5.2

[a]Plots established on Walla Walla soil at the Columbia Basin
Agricultural Research Center, Pendelton, Oregon, in 1931;
samples collected in 1987.

[b]Continuous wheat and wheat-pea rotation.

From Collins et al., 1992. Soil Sci. Soc. Am. J. 56:783–788. With
permission of the Soil Science Society of America, Madison, WI.

of microbial biomass to soil organic matter can vary from 2 to 5% (Table 5.2).
Microbial biomass is greatest in soils receiving manure or green manuring.
Also grass pastures have much more microbial biomass, which contributes to
increased soil organic C.

The general scheme of analyzing humus involves extraction with an alkaline solution followed by reaction with an acid. The portion of humus that
cannot be extracted by an alkaline solution is known as humin. The portion
of the alkaline extract that precipitates on acidification is known as humic
acid, while that which remains in acid solution is known as fulvic acid
(Figure 5.1). The molecular weight ranges from several hundred for fulvic
acids to over 300,000 for humic acids and humin (Stevenson, 1982).

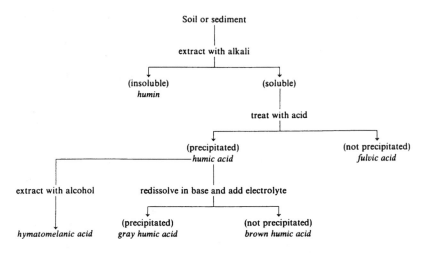

Figure 5.1. Scheme for separation of soil humus. (From Stevenson, 1982.
***Humus Chemistry: Genesis, Composition, Reactions*, p. 43.**
With permission of John Wiley & Sons.)

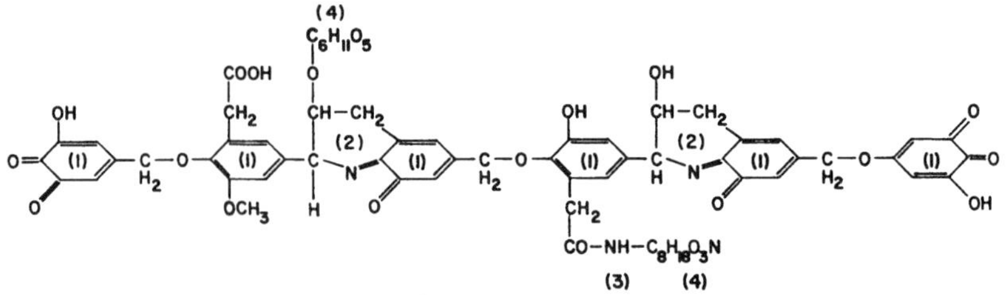

Figure 5.2. The structure of humic acid molecules after Dragunov. (From
Kononova et al., 1961. *Soil Organic Matter — Its Nature Its
Role in Soil Formation and in Soil Fertility*, p. 65.

Very complicated structural formulas for the humic acid molecule have
been proposed; one formula, given by Dragunov in 1948 (Kononova et al.,
1961), is shown in Fig. 5.2. Orlov (1985) has suggested a probable scheme
of a structural cell of humic acid from a chernozem (Figure 5.3). As is evident
from Figures 5.2 and 5.3, the chemical formulas for different fractions of soil
humus are difficult to write due to the extremely complex nature of compounds.

Humus has properties distinctly different from the original plant tissue
and has its own identity as a natural body, though exceedingly variable and
heterogenous. Some of the properties exhibited by soil humus, which are well
known and important from the viewpoint of soil fertility management, include
the following:

1. Humus particles become bonded to clay or other silicate surfaces,
   leading to the formation of clay-humus complexes.
2. Humus stores and releases soil N.
3. Humus possesses buffering capacity
4. Humus possesses cation exchange capacity
5. Humus possesses anion exchange capacity
6. Humus adsorbs pesticides and other agricultural chemicals.

The core material in soil humus is lignin, which is made up of basic units
of a 6-carbon benzene ring with a 3-carbon side chain attached to it, giving a
basic formula (C6-C3) n, where n is the number of monomers linked together
to form the highly polymerized molecule. The breakdown of the lignin molecule
leads to the release of C6-C3 units (Stevenson, 1982). A common C6-C3
monomer in lignin is coniferyl alcohol, which can be transformed to phenols
(i.e, phenolic acids) and quinones as shown in Figure 5.4.

Soil humus is therefore characterized by having benzene type monomers.
Stevenson and Olsen (1989) have proposed a dimer structural unit as a building
block for soil humus. This dimer unit is shown in Figure 5.5 and is made up
of two benzene based monomers, one of them having an amino group ($-NH_2$).

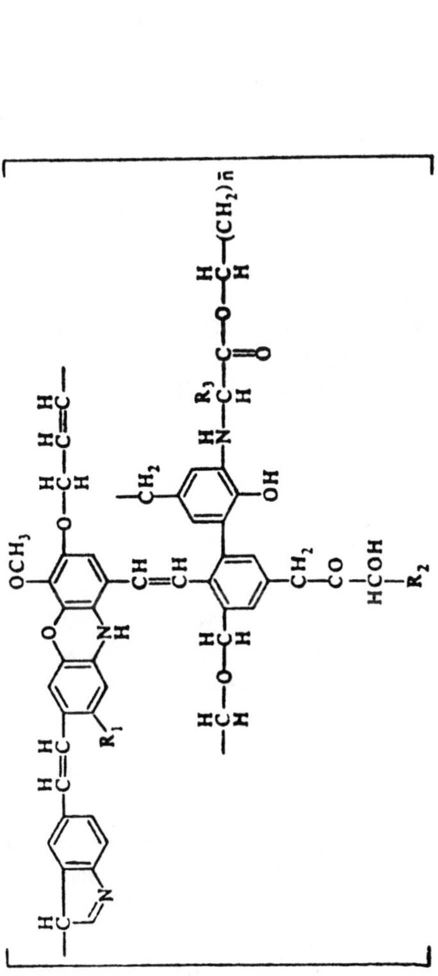

Figure 5.3. Probable scheme of a structural cell of humic acid. (Adapted from Orlov, 1985. *Humus Acids of Soils*, p. 378.

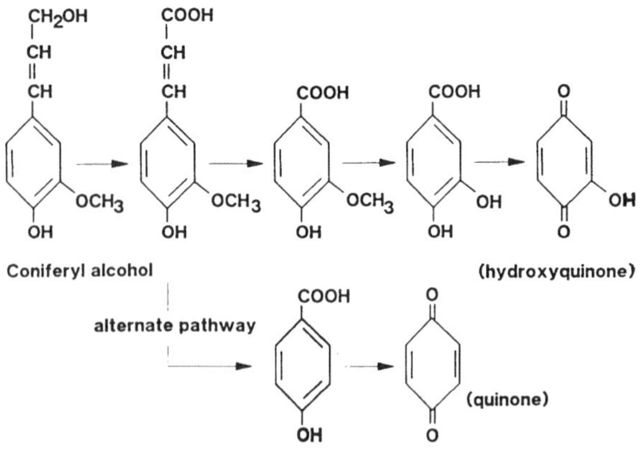

Figure 5.4. Biochemical conversion of coniferyl alcohol, a lignin binding unit to phenols and quinones. (From Stevenson and Olsen, 1989. J. Agron. Educ. 18:84–88. With permission of the American Society of Agronomy, Madison, WI.)

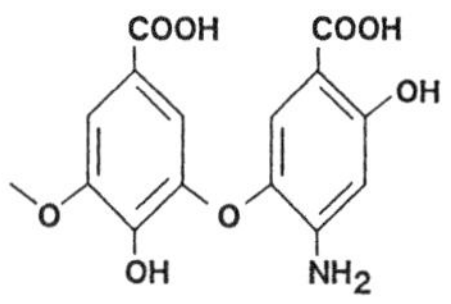

Figure 5.5. A dimer structural unit as the building block of soil humus. (From Stevenson and Olsen, 1989. J. Agron. Educ. 18:84–88. With permission of the American Society of Agronomy, Madison, WI.)

The dimer in Figure 5.5 has a molecular weight of approximately 250 and includes the more important functional groups found in humic substances. The C:N ratio of the dimer is 14:1, which is of the order of that found for soil humus. The chemical composition, C:N ratio, and some other properties of humic acids and fulvic acids are given in Table 5.3.

Because the types of monomers could be numerous, and the ways in which they combine is great, structures of the resulting macromolecules are exceedingly variable and complex.

The dimer structure would explain why nitrogen is released on decomposition of soil humus. It also conveniently explains why the cation exchange capacity (CEC) of soil organic matter increases with an increase in soil pH (Figure 5.6). In addition to the exchange of cations, the closely associated COOH and OH groups also permit chelation of heavy metal ions. Similarly, organic molecules of herbicides and other pesticides become attached to OH or COOH radicals, explaining why soil organic matter or humus can adsorb pesticides, and necessitate the application of higher doses on soils having

Table 5.3 Chemical Composition of Humic Acids (HA) and Fulvic Acids (FA) From Some Soils

Soil	HA or FA	C	H	O	N	C:N	Total acidity	Carboxyl (-COOH)	Phenolic/enolic (-OH)
		(%)						(mmol g^{-1})	
Alfisol	HA	56.8	5.3	33.3	4.6	12.3	6.8	3.9	2.8
Inceptisol	HA	51.4	5.8	38.7	6.0	8.6	6.0	2.4	3.6
Spodosol	HA	49.0	4.6	45.7	0.7	70.0	12.0	9.2	2.8
Spodosol	FA	50.9	3.3	44.7	0.7	72.7	12.4	9.1	3.3
Ultisol	HA	48.7	4.8	42.7	3.8	12.8	8.7	2.7	6.0
Ultisol	FA	40.6	4.1	53.9	1.4	29.0	10.2	8.8	1.4
Histosol	HA	57.0	4.5	34.8	3.3	17.3	7.2	3.1	4.2
Histosol	FA	54.5	5.3	37.6	1.9	28.7	8.6	4.0	4.6

Adapted from Hayes (1991).

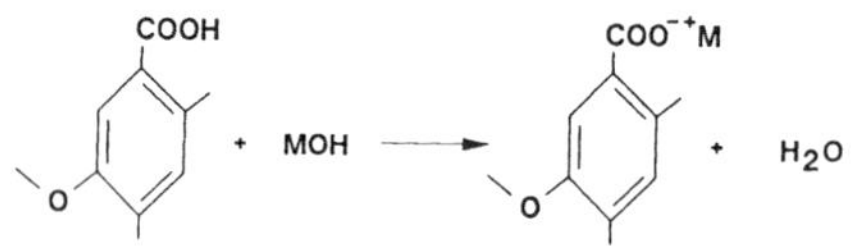

Figure 5.6. Effect of pH on CEC of dimer. (From Stevenson and Olsen, 1989. J. Agron. Educ. 18:84–88. With permission of the American Society of Agronomy, Madison, WI.)

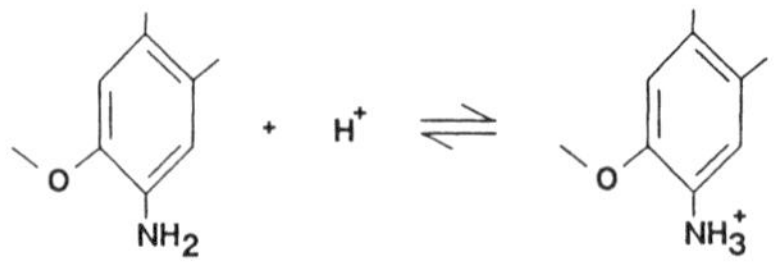

Figure 5.7. Anion exchange properties of dimer under acid conditions. (From Stevenson and Olsen, 1989. J. Agron. Educ. 18:84–88. With permission of the American Society of Agronomy, Madison, WI.)

higher organic matter content. The dimer structure also easily explains why the humus has anion exchange properties (Figure 5.7).

The tendency of humus to be bonded to clay and other silicate surfaces can easily be illustrated through linkages with polyvalent cations (M) and H-bonding (Figure 5.8).

Thus several properties of humus or soil organic matter can be explained using the dimer unit proposed by Stevenson (1982). However, it may be pointed out that this is an oversimplified view; the nature and properties of soil organic matter is complex and there is still much more to be learned.

5.2. C:N RATIO

C:N ratio of soil organic matter is an important property. Since C makes up a large and rather definite proportion of soil organic matter, it is not surprising that the C:N ratio of soils tends to be fairly constant, generally between 8:1 to 15:1, the median being between 10:1 to 12:1. Temperature and precipitation influence the C:N ratio in soil organic matter. When rainfall is constant, the C:N ratio is lower in warmer than in cooler regions. Similarly, when annual temperatures are about the same, the C:N ratio tends to be lower in drier regions.

Since the C:N ratio of a soil tends to be constant, the C:N ratio of added organic residues to soil assumes considerable importance. When a residue having a high C:N ratio such as wheat straw is added to soil, there is a sudden increase in the evolution of CO_2 due to increased microbial activity. Concomitant to this is the depression in soil nitrates (Figure 5.9). This is well demonstrated by yellow crop plants with stunted growth in those parts of the

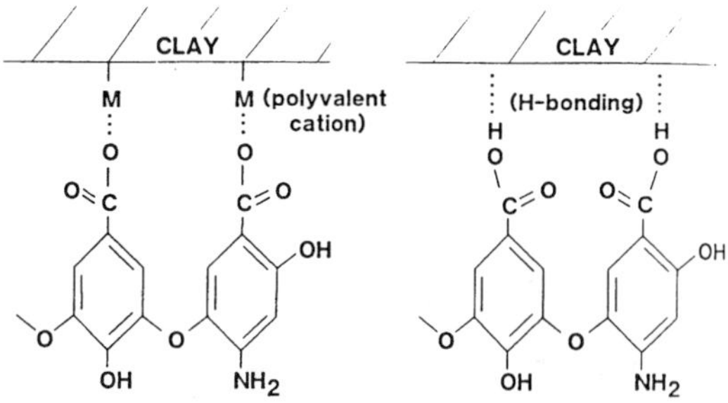

Figure 5.8. Binding of dimer unit to clay particles. (From Stevenson and Olsen, 1989. J. Agron. Educ. 18:84–88. With permission of the American Society of Agronomy, Madison, WI.)

field where straw has accumulated, such as areas used for threshing the previous crop in developing countries, where combine harvesting is not yet practiced. After much of the C of the added residues has been oxidized, CO_2 evolution slows down and nitrification restarts. This does not happen when leguminous crop residues having narrow C:N ratios (20 to 30:1) are added to the soil. In such instances, nitrates tend to accumulate after a short period of immobilization because C availability limits microbial activity more than N availability.

5.3. FACTORS AFFECTING THE ORGANIC-MATTER CONTENT OF SOIL

5.3.1. Climate

Climate is of major importance in determining the organic matter content of soil. Of the different components of climate, temperature and precipitation are the most important. Jenny (1930) observed that for each decrease of 10°C in annual temperature, the average organic-matter content of soil increased two to three times. The C:N ratio of the organic matter also increased as temperature decreased. Jenny further observed that at a constant temperature the organic-matter content of soils of the central United States increased logarithmically with the increase in available water supply. Similar conclusions were drawn by Jenny and Raychaudheri (1960) for their studies on Indian soil. The ideal trend of organic soil N for large areas of loamy grassland soils of the United States is shown in Figure 5.10. According to Jenny, the order of importance of different soil-forming factors in determining the organic-matter content of soil is as follows: climate > vegetation > topography > parent material > age. Theng et al. (1989), however, opined that the overall importance of the environmental factors determining soil C content is in the following order: rainfall > pH> clay content > temperature for tropical regions and rainfall > pH > temperature > clay content for

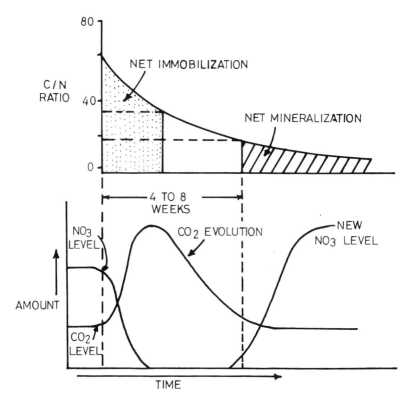

Figure 5.9. Changes in nitrate levels of soil during the decomposition of low-nitrogen crop residues. (From Stevenson, 1986. *Cycles of Soil — Carbon, Nitrogen, Phosphorus, Sulfur and Micronutrients***, pp. 13 and 166. With permission of John Wiley & Sons, New York.)**

temperate regions. The relationships between rainfall and soil organic C for tropical and temperate regions of Australia are shown in Figure 5.11 (Thang et al., 1989).

5.3.2. Texture

Under similar climatic conditions the organic-matter content in fine-textured soils is two to four times that of coarse-textured soil. While some of this could be attributed to increased oxidation in coarse-textured soils, two other important factors are (1) the formation of clay-humus complexes and (2) the formation of metal organic-matter complexes. Again, soils having 2:1 expanding-layer silicates have more organic matter than the soils having 1:1 nonexpanding-layer silicates.

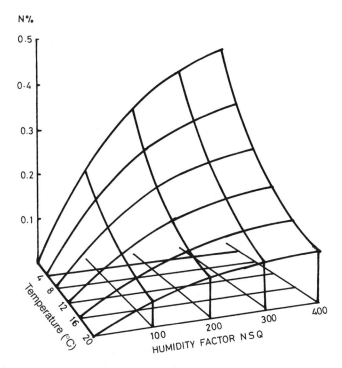

Figure 5.10. The nitrogen content of loamy grassland soils in the United States as a function of annual temperature and annual humidity factor (N.S.Q.). The curves express the approximate idealized trend of soil nitrogen of large areas. (From Jenny, 1930. Mo. Agric. Exp. Sta. Res. Bull. 324:66. With permission of University of Missouri, Columbia, MO.)

5.3.3. Agricultural Practices

Cultivation.   When a forest land is brought under cultivation, the organic-matter content starts declining, and the decline depends upon climatic factors and the intensity of cultivation. In a classical study at Missouri (Jenny, 1933) about 25% of the soil organic N was lost in 20 years and 35% in 60 years. The data of Hobbs and Brown (1957) for a few locations in Kansas (Figure 5.12) show the loss of 50% N in some soils after 40 years of cultivation. In another study more than 50% of soil organic matter was lost in 50 years of cultivation (Bauer and Black, 1981). In long-term studies covering 60 to 80 years, carbon and nitrogen losses from Canadian prairie surface soils due to cultivation ranged from 50 to 60% and 40 to 60%, respectively (Campbell et al., 1976). Similarly, in an Australian study (Dalal and Meyer, 1986) carbon N nitrogen losses averaged 36% in soils cultivated for 20 to 70 years.

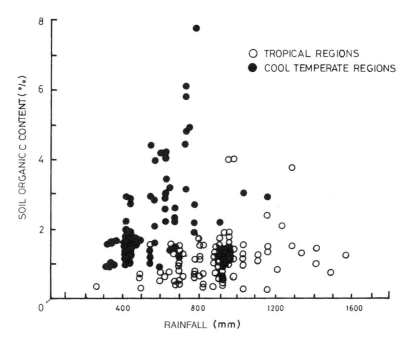

Figure 5.11. Relationship between organic carbon content and rainfall for ultisols (Krasnozems) from tropical, and cool, temperate regions of Australia. (From Theng et al., 1989. *Dynamics of Soil Organic Matter in Tropical Ecosystems*, p. 6. With permission of the University of Hawaii.)

Fallowing. Increasing the frequency of fallowing generally increases the loss of organic matter from the soil (Rasmussen and Collins, 1991). Haas et al. (1957) reported that cropland in grain-fallow rotation lost more organic N than did continuously cropped land at 13 or 14 locations throughout the midwestern United States; the average loss after 30 to 43 years of cultivation was 24% in the continuous small grain versus 29% with the grain-fallow cropping system. Similarly, Monreal and Janzen (1993) reported that after 80 years of cropping at Lethbridge in Alberta, Canada, the organic C loss was 23% under fallow-wheat, 21% under fallow-wheat-wheat, and only 17% under continuous wheat. In general, those cropping systems that return the most C to the soil usually exhibit the least loss of soil organic matter.

Tillage. Available information on tillage effects on soil organic matter and N availability leads to the following conclusions: (1) tillage of all kinds leads to a decrease in soil organic C and N as compared with native sod; (2) incorporation of crop residues as compared with their removal usually increases organic-matter content of soil; (3) leaving a crop residue cover on

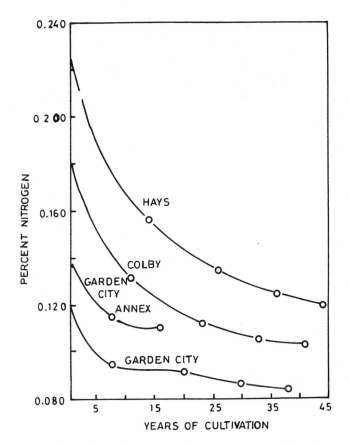

Figure 5.12. Effect of cultivation on nitrogen content of surface soils of Hays, Colby, and Garden City, Kansas. (From Hobbs and Brown, 1957. Agron. J. 49:259. With permission of the ASA, Madison, WI.)

the soil surface leads to an accumulation of organic matter and total N in the surface soil; and (4) burning of crop residues often produces a variable effect on soil organic-matter content, depending on soil depth, tillage practices, degree of burning, time, and other factors (Prasad and Power, 1991). Effects of conservation tillage on the soil organic-matter content are summarized in Table 5.4. Most of these data show a higher organic-matter content in soils under conservation tillage. Tillage generally increases oxygen content of soil air and exposes protected organic matter to microbial action; thus the more a soil is tilled, the greater the loss of organic matter.

In many Asian and African countries rice paddies are prepared by puddling, which is cultivation by a plough or harrow in a flooded field. Lal (1986) showed that puddling reduced the organic-matter content from 2.2 to 2.5% in

Table 5.4 Effect of Conservation Tillage
on Organic C and N in Soil

Location and soil	Annual precipitation (mm)	Soil depth (cm)	Length of study (yr)	Tillage system[a]	Increase (%/yr)	
					C	N
South Africa						
Haploxeralf	412	10	10	TT	5.6	3.4
Haploxeralf	412	10	10	NT	7.3	5.1
Germany						
"Podsol"		30	5	NT	3.2	1.4
"Podsol"		30	5	NT	2.4	1.6
"Podsol"		30	6	NT	1.3	1.3
Australia						
Western						
Psamment	345	15	9	NT	1.6	—
Alfisol	307	15	9	NT	0.7	—
Alfisol	389	15	9	NT	1.4	—
Queensland						
Pellustert	698	10	6	NT	1.2	1.3
Canada						
Saskatchewan						
Chernozem		15	6	NT	6.7	2.8
United States						
North Dakota						
Haploboroll	375	45	25	SM	1.8	1.3
Haploboroll	375	45	25	SM	−0.1	0.1
Argiboroll	375	45	25	SM	0.5	0.4
Kansas						
Haplustoll		15	11	NT	0.7	0.6
Nebraska						
Haplustoll	446	9	15	NT	2.8	2.4
	446	10	15	NT	1.2	1.0
Oregon						
Haploxeroll	416	15	44	SM	0.3	0.4
Washington						
Haploxeroll	560	5	10	NT	1.9	2.0

**Table 5.4 Effect of Conservation Tillage
on Organic C and N in Soil (Continued)**

Location and soil	Annual precipitation (mm)	Soil depth (cm)	Length of study (yr)	Tillage system[a]	Increase (%/yr) C	N
Mean					2.2	1.7
Minimum					–0.1	0.1
Maximum					7.3	5.1

[a]TT, tine-till; NT, no-till; SM, stubble-mulch

Adapted from Rasmussen and Collins (1991).

no-till plots to 1.5 to 1.9% under puddling in a 0- to 5-cm layer of surface soil over 6 years.

5.3.4. Fertilizers and Manures

Continued application of organic manures is well known to increase soil organic matter (Rasmussen and Collins, 1991). The application of chemical fertilizer leads to better development of roots, and even if grain and straw is removed, additional root mass can increase soil organic-matter content. Data from long-term manurial experiments in India (Nambiar et al., 1992) are given in Table 5.5. At only three out of eight centers was there a decrease in soil organic-matter with fertilizer application; at the other five centers there was an increase in soil organic matter due to fertilizer application. Thus the myth that fertilization leads to impoverishing of soil fertility is not true, at least for soil organic matter. Nevertheless, the increase in soil organic matter was most when FYM (farmyard manure) was applied along with chemical fertilizer.

**Table 5.5 Changes in Soil Organic C (%) as Affected by
Fertilizer and Manure (Over the Years 1971–1989)**

Location	Soil class	Cropping system	Initial (1971)	NPK (1989)	NPK + FYM (1989)
Barrackpore	Eutrochrept sl	R-W-J[a]	0.71	0.46	0.53
Ludhiana	Ustochrepts ls	M-W-C	0.21	0.28	0.40
Jabalpur	Chromustert c	S-W-M	0.57	0.59	1.11
Hyderabad	Tropaquept scl	R-R	0.51	0.70	1.28
Ranchi	Haplustalf sic	S-P/T-W	0.45	0.33	0.38
Bhubaneswar	Tropaquept s	R-R	0.27	0.56	0.80
Palampur	Hapludalf sil	M-W	0.79	0.96	1.57
Pantnagar	Hapludoll sicl	R-W-C	1.48	0.86	1.45

[a]R, rice; W, wheat; J, jute; C, cowpea (fodder); M, maize; S, soybean; P, potato; T, toria.

From Nambiar et al., 1992. Annual Report 1988–89 of (ICAR) Indian Agricultural Research Institute, New Delhi.

REFERENCES

Bauer, A. and A.L. Black. 1981. Soil carbon, nitrogen, and bulk density comparisons in two cropland tillage systems after 25 years and in virgin grassland. Soil Sci. Soc. Am. J. 45:1166–1170.

Campbell, C.A., E.A. Paul, and W.B. McGill. 1976. Effect of cultivation and cropping on the amounts and forms of soil N, in Western Canadian Nitrogen Symposium Proceedings, Calgary, Alberta, pp. 7–101.

Collins, H.P., P.E. Rasmussen, and C.L. Douglas. 1992. Crop rotation and residue management effects on soil C and microbial dynamics. Soil Sci. Soc. Am. J. 56:783–788.

Dalal, R.C. and R.J. Mayer. 1986. Long-term trends in fertility of soils under continuous cultivation and cereal cropping in southern Queensland. II. Total organic carbon and its rate of loss from the soil profile. Aust. J. Soil Res. 24:281–282.

Haas, H.J., C.E. Evans, and E.F. Miles. 1957. Nitrogen and carbon changes in Great Plains soil as influenced by cropping and soil treatments. U.S. Dept. Agric. Tech. Bull. 1164:1–111.

Hayes, M.H.B. 1991. Influence of the acid/base status on the formation and interactions of acids and bases in soils, in *Soil Acidity*, B. Ulrich and M.E. Summer, Eds., Springer-Verlag, Berlin, pp. 80–96.

Hobbs, J.A. and P.L. Brown. 1957. Nitrogen changes in cultivated dryland soils. Agron. J. 49:257–260.

Jenny, H. 1930. A study on the influence of climate upon the nitrogen and organic matter content of the soil. Mo. Agric. Exp. Sta. Res. Bull. 152:66.

Jenny, H. 1933. Soil fertility losses under Missouri condition. Mo. Agric. Exp. Sta. Res. Bull. 324:1–10.

Jenny, H. and S.P. Raychandhari. 1960. Effect of climate and cultivation on nitrogen and organic matter reserves in Indian soils. Indian Council of Agricultural Research, New Delhi, 126 pp.

Kononova, M.M., T.Z. Nawakowski, and G.A. Greenwood. 1961. *Soil Organic Matter — Its Nature, Its Role in Soil Formation and in Soil Fertility*, Pergamon Press, New York, 450 pp.

Lal, R. 1986. Effects of 6 years of continuous no-till or puddling systems on soil properties and rice (Oryza sativa) yield of a loamy soil. Soil and Tillage Res. 8:181–200.

Monreal, C.M. and H.H. Janzen. 1993. Soil organic carbon dynamics after 80 years of cropping a dark brown chernozem. Can. J. Soil Sci. 73:133–136.

Nambiar, K.K.M, P.N. Soni, M.R. Vats, D.K. Sehgal, and D.K. Mehta. 1992. Annual Report 1987–88 and 1988–89. All India Coordinated Research Projects on Long-Term Fertilizer Experiments (ICAR), Indian Agricultural Research Institute, New Delhi.

Orlov, D.S. 1985. Humus Acids of Soils, Amerind Publishing, New Delhi, p. 378.

Paul, E.A. and F.E. Clark. 1989. *Soil Microbiology and Biochemistry*, Academic Press, San Diego, 273 pp.

Prasad, R. and J.F. Power. 1991. Crop residue management. Adv. Soil Sci. 15:205–251.

Rasmussen, P.E. and H.P. Collins. 1991. Long-term impacts of tillage, fertilization and crop residues on soil organic matter in temperate semi-arid regions. Adv. Agron. 45:93–134.

Stevenson, F.J. 1982. *Humus Chemistry: Genesis, Composition, Reactions*, John Wiley & Sons, New York.

Stevenson, F.J. 1986. *Cycles of Soil — Carbon, Nitrogen, Phosphorus, Sulfur and Micronutrients*, John Wiley & Sons, New York.

Stevenson, F.J. and R.A. Olsen. 1989. A simplified representation of the chemical nature and reactions of soil humus. J. Agron. Educ. 18:84–88.

Theng, B.K.G., K.R. Tate, P. Sollins, N. Moris, N. Nadkarni, and R.L. Tate III. 1989. Constituents of organic matter in temperate and tropical soils, in *Dynamics of Soil Organic Matter in Tropical Ecosystems*, D.C. Coleman, J.M. Oades, and G. Uehara, Eds., College of Tropical Agriculture and Human Resources, University of Hawaii, Honolulu, pp. 5–32.

Zinke, P.J., A.G. Stanenberger, W.M. Post, W.R. Emanual, and J.S. Olson. 1984. Worldwide organic soil carbon data, Publ. 2217, Environ. Sci. Div., Oak Ridge National Lab., Oak Ridge, TN.

6   SOIL ACIDITY

Soil acidity controls the solubility and precipitation of chemical compounds of all essential plant nutrients and is therefore a deciding factor on their availability. Soil acidity has a far-reaching influence on soil fertility and plant growth. For example, in strongly acidic soils Ca, Mg, P, B, and Mo become deficient, while Mn and Fe may reach toxic limits (Figure 6.1). Similarly, the availability of Cu, Fe, Zn, and Mn is reduced in alkaline calcareous soils. In mineral soils the pH (discussed later in this chapter) for the greatest availability of the most nutrients is 6.5 (Figure 6.1A), while in organic soils (peats and mucks) the optimum pH is about 5.5 (Figure 6.1B). In addition, in strongly acidic soils (pH < 5) Al-toxicity poses a serious problem (Foy, 1992). An understanding of the nature of soil acidity and its management therefore forms an integral part of soil fertility.

6.1. ACIDS

Although more precise definitions of acids and bases are available, the one that defines acids as substances yielding hydrogen ions (protons) when dissolved in water is the most practical and useful. Acids do differ in their rate of release of H^+ ions (dissociation constant). Those that release H^+ ions readily are known as strong acids. Examples are hydrochloric, sulfuric, and nitric acids. Acids that release their H^+ ions rather slowly are known as weak acids, and examples are acetic, carbonic, and boric acids. The simplest way to determine if an acid is strong or weak is to titrate it with a base and record the pH change (Figure 6.2). If near neutrality pH abruptly rises from a very low to a very high value, the acid is strong. On the other hand, if the pH change is gradual, the acid is weak. The rate of release of H^+ ions is measured by the dissociation constant of an acid. Dissociation constants of some acids are given in Table 6.1.

6.2. THE pH CONCEPT

Weak acids release few H^+ ions; the H^+ ion activity (measured as moles per liter) may typically be about 0.0001 M or even less. Writing such values

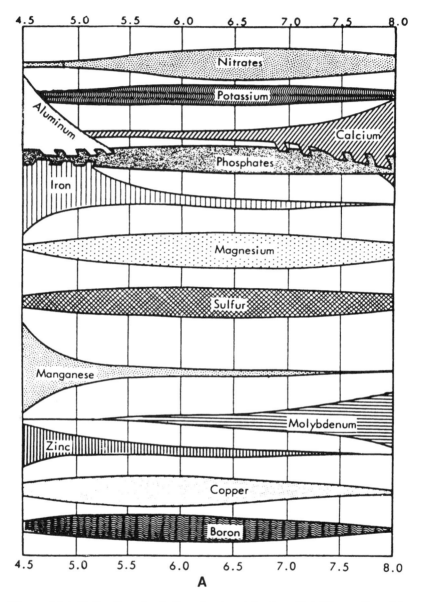

Figure 6.1A. Soil pH and relative plant nutrient availability — the wider
the bar, the greater the plant availability. (From Rasnake,
M. 1973. University of Kentucky, College of Agriculture,
Cooperative Service. AGR-19, p. 2.

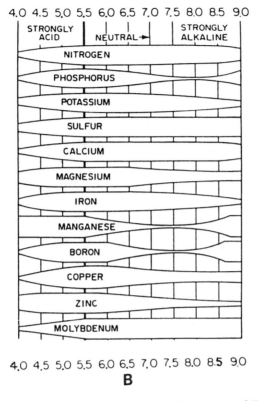

Figure 6.1B. (From Brady. *The Nature and Properties of Soils***, 11th ed., © 1990. With permission of Prentice-Hall, Inc., Upper Saddle River, N.J.**

of H⁺ ion activity could be very cumbersome. Sorenson, a Swedish chemist, therefore developed the concept of pH, which is the logarithm of the reciprocal of the H⁺ ion activity.* Thus

$$pH = \log \frac{1}{aH^+} = -\log aH^+$$

where aH⁺ is the hydrogen ion activity in moles per liter.

6.3. DETERMINATION OF SOIL pH

Soil pH is generally determined in soil slurries with soil to water ratios of 1:1 to 1:2.5. For example, 10 g of soil is added to 10 ml of distilled water in a beaker and stirred. The pH is then recorded with a pH meter (with a glass

---

* The best way to understand this concept is to prepare 0.0001 *M* and 0.00001 *M* solutions of HCl and determine their pH on a pH meter. pH values recorded will be very near theoretical values of 4 and 5.

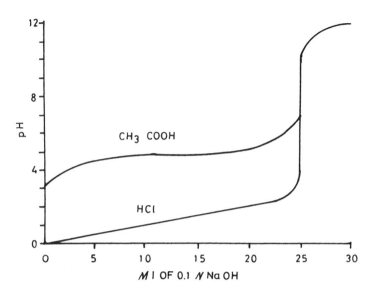

Figure 6.2. Titration of 0.10 N CH₃COOH and 0.10 N HCl with 0.10 N NaOH. (From Tisdale et al., 1993. *Soil Fertility and Fertilizers*, 5th ed., p. 366. With permission of Prentice-Hall, Inc., Upper Saddle River, NJ.)

Table 6.1 Ionization Constants of Some Acids

Acid		Ka	pKa[a]
Formic		1.772×10^{-4}	3.75
Acetic		1.754×10^{-5}	4.76
Propionic		1.336×10^{-5}	4.87
n-Butyric		1.515×10^{-5}	4.82
Lactic		1.374×10^{-4}	3.86
Carbonic[b]	K_2	4.69×10^{-11}	10.33
Boric		5.79×10^{-10}	9.24
Phosphoric	K_1	7.516×10^{-3}	2.12
	K_2	6.226×10^{-8}	7.21
Sulfuric	K_2	1.01×10^{-2}	2.00

[a] $pKa = \log Ka; \; Ka = \dfrac{aH^+ \, aA^-}{aHA}$

[b] Carbonic and sulfuric acids have two H⁺ to release; K_2 refers to the dissociation constant of the second H⁺.

From Daniels and Alberty, 1967. *Physical Chemistry*, 3rd ed., p. 767. With permission of John Wiley & Sons, Inc.)

and a calomel electrode or a combined electrode) before the suspension settles. If the soil to solution ratio is increased to 1:5 or 1:10, the pH value will normally increase if the pH recorded in a 1:1 suspension is less than 6.5. However, if the pH in 1:1 soil to solution ratio was 8.5 or above (as in alkali soils), the value recorded in 1:5 or 1:10 soil to solution ratio would decrease. The reason that the pH in acid soils goes up on increasing the soil to solution ratio is because the released H^+ ions are distributed in a larger volume of water (dilution effect). On the contrary, the concentration of Na^+ (and associated OH^- ions) present in alkali soils goes down on increasing soil to solution ratio and the pH value is lowered.

To overcome this variation in determining soil pH due to change in soil to solution ratio, pH may be determined in 1 M KCl or 0.01 M $CaCl_2$ solution. Such strong electrolytes (KCl or $CaCl_2$) bring more exchangeable H in solution, and the pH value determined is generally 0.5 to 1.0 less than in aqueous suspensions. Data on the lowering of soil pH as affected by its determination in some electrolyte solutions, vis-a-vis determination in 1:1 soil to water ratio for some acid soils from Andaman Island, India, are given in Table 6.2. pH determined in electrolyte solutions does not change due to an increase in soil to solution ratio. It should therefore be clearly understood that soil pH values depend very much upon how they are determined.

Although the use of KCl or $CaCl_2$ solutions reduces variation in soil pH measurement, the fact remains that in nature crop plants growing on soils encounter pH values determined in soil-water suspensions. This variability in soil pH has practical applications in crop production. This is at least one reason why rice grows well on acid sulfate soils, where the pH as determined in the laboratory is 5.0 or less. Due to submerged conditions under which rice is grown there is considerable dilution of H^+ ions in the soil solution and therefore the soil pH encountered by rice roots under field conditions is increased. Similarly, rice grows well on submerged alkali soils. The effect of submergence on the pH of some soils is shown in Figure 6.3. This would explain why soil acidity did not receive much attention in several regions of the world where acid soils exist but lowland rice is the main crop.

6.4. ACTIVE AND POTENTIAL ACIDITY

When an acid soil is suspended in water and pH is determined, one can calculate the concentration of H^+ ions present in the soil at that point and at that water content. For example, if the pH is 4.0 and the soil water content is 20%, one can determine H^+ ions present in the soil as shown below:

$$H^+ \text{ ions} = 2.2 \times 10^6 \times 20 \times 10^{-2} \times 10^{-4} \text{ moles } H^+ \text{ L}^{-1} = 44 \text{ g ha}^{-1}$$

where 2.2×10^6 kg is taken as the weight of a surface (0 to 15 cm) furrow slice of soil on one hectare of land.

Table 6.2 Changes in pH Values in Different Electrolyte Solutions

Location/soil group	pH in H_2O	Change in pH from pH H_2O in different electrolyte solutions			
		KCl	KNO_3	K_2SO_4	KH_2PO_4
School Line (Tropfluvents)	5.5	-0.8	0.9	-0.5	-0.8
Garacharma (Troporthents)	4.4	-0.4	0.5	-0.5	-1.9
Pahargaon (Troporthents)	4.6	-0.8	0.9	-0.6	-1.4
Rangachang (Fluventic Quartzipsamments)	6.3	-0.2	0.2	-0.8	-0.4
Tushnabad (Umbric Fluventic Haplaquepts)	6.2	-1.5	1.9	-1.0	-0.3
Wandoor (Fluventic Dystrochrepts)	5.5	-1.7	-1.9	-0.3	-0.7
Shoal Bay - 9 (Fluventic Dystrochrepts)	5.0	-1.2	1.4	+0.3	-1.0
Shoal Bay -12 (Fluventic Dystrochrepts)	5.2	-1.3	1.3	+0.2	-1.0
Shoal Bay -19 (Fluventic Dystrochrepts)	4.9	-0.9	1.0	-1.2	-1.0
Manarghat (Troporthents)	4.9	-1.3	1.0	-0.7	-1.4
Paithankhari (Fluventic Sulfaquents)	4.9	0.8	0.9	0.8	-1.3
Chouldari (Tropofluvents)	5.8	-1.3	1.2	-0.8	-0.4
Sipighat (Troporthents)	4.6	-0.7	0.9	-0.5	-1.7
Guptapara (Tropofluents)	5.8	-1.5	1.2	-0.6	-0.5
Wandoor (New) (Fluventic Dystrochrepts)	5.2	-0.4	0.4	-0.5	-0.6

From Mongia and Bandypadhyay, 1991. J. Indian Soc. Soil Sci. 39:351–354. With permission.

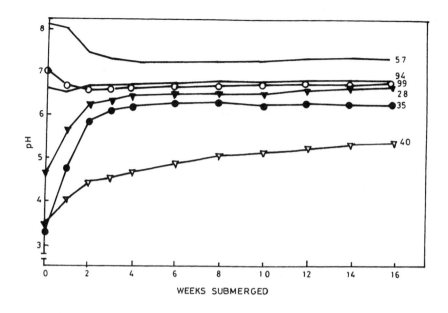

Soil No.	Texture	pH	O.M. %	Fe %	Min %
28	clay	4.9	2.9	4.70	0.08
35	clay	3.4	6.6	2.60	0.01
40	clay	3.8	7.2	0.08	0.00
57	clay loam	8.7	2.2	0.63	0.07
94	clay	6.7	2.6	0.96	0.09
99	clay loam	7.7	4.8	1.55	0.08

Figure 6.3. Kinetics of the solution pH of six submerged soils. (After Ponnamperuma, 1976. *The Fertility of Paddy Soils and Fertilizer Applications for Rice*, Food and Fertilizer Technology Centre for the Asian and Pacific Region, Tapei, Taiwan. pp. 1–27.)

A very small amount of base would be needed to neutralize this acidity. However, when neutralized, the soil will still record essentially the same pH value. This occurs because more H^+ ions will be released by soil to maintain equilibrium. Soil has much more stored H^+ ions than is determined by measuring soil pH. Such stored acidity is known as "potential acidity," while the acidity determined as soil pH is termed as "active acidity." This behavior of acid soils is similar to that of weak acids, which also have much more potential acidity than the active acidity. This is why there is no sharp end point in the titration of acid soils with bases (Figure 6.4). On the other hand, potential and active acidity in strong acids is nearly the same.

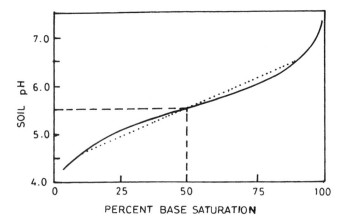

Figure 6.4. Theoretical titration curve for a large number of Florida soils. The dotted line indicates the zone of greatest buffering. The maximum buffering should occur at approximately 50% base saturation. (From Peech, 1941. Soil Sci. 51:473–486. With permission of Williams & Wilkins, Baltimore, MD.)

6.5. BUFFERING CAPACITY

Buffers are systems that maintain their pH within a narrow range; that is, they resist a change in pH on addition of small amounts of acid or base. Mixtures of weak acids and their salts are used to make buffers. For example, acetic acid–sodium acetate, ammonium hydroxide–ammonium chloride and phosphoric acid–sodium phosphate are frequently used buffers. In the acetic acid–sodium acetate buffer, the salt sodium acetate dissociates and gives rise to a large concentration of acetate ions, which suppress the dissociation of acetic acid. If small quantities of an acid such as HCl are added, excess acetate ions in the buffer react with H^+ ions and form acetic acid, and there is no change in pH. Similarly, if small amounts of NaOH are added, hydroxyl ions are neutralized by H^+ ions in solution. When this happens more H^+ ions are released by the dissociation of acetic acid. Thus the buffer resists a change in its pH. Acid soils also react in a similar manner, and the change in their pH requires application of large amounts of a base (lime). Also, a sufficient time interval has to be provided to affect a pH change. Such soils are said to have a large buffering capacity.

The buffering capacity of a soil is related to its cation exchange capacity (CEC) and is therefore related to the clay content and mineralogy and to the amount of soil organic matter present. The larger the amounts of either clay or organic matter, the greater the buffering capacity.

6.6. NATURE OF SOIL ACIDITY

Initially, soil acidity was identified with humic acids. However, it was soon discovered that leaching acid soils with neutral salt solutions produced

Table 6.3 Total Acidity and Exchangeable Al³⁺, Fe³⁺, and H⁺
in the 1 N BaCl₂ Leachates of H-Clays (cmol kg⁻¹)

Soil	Total acidity	Exchangeable Al^{3+}	Exchangeable Fe^{3+}	Exchangeable H^+
Satara F	63.4	44.4	0.5	18.5
Jorhat F	23.3	12.5	0.4	10.4

From Mukherjee et al., 1947. J. Colloid Sci. 1:247–254. With permission of Academic Press, Inc.

extracts that could be titrated for acidity. These leachates contained sizable amounts of Al (Table 6.3), and exchangeable Al was identified as the main cause of soil acidity. Although these results were obtained by several earlier workers such as T.B. Veitch in the United States, C.E. Marshall and R.K. Schofield in the United Kingdom, H. Riehm in Denmark, and V.A. Chernov in the USSR (Jenny, 1961), the subject received considerable attention only later (Lin and Coleman, 1960; Shoemaker et al., 1961; Kamprath, 1970; and Sanchez, 1976). McLean et al. (1965) observed that various forms of soil acidity appear to have the following order of activity: exchangeable or permanent charge H^+ > exchangeable or permanent charge Al^{3+} > hydroxy - Al monomers > hydroxyl Al polymers ≈ organic matter acidity ≈ lattice Al-OH or Si-OH acidity. Brady (1990) suggested the term residual acidity for acidity not determined by active acidity and the term exchangeable or titratable acidity for acidity extracted by KCl. Titratable acidity determined by KCl was much less than that determined by ammonium acetate at pH 4.8, using Shoemaker-McLean-Praftt (SMP) buffer, Woodruff buffer, and Mehlich buffer (Table 6.4) (McLean et al., 1964).

As already brought out in Chapter 4, soils contain a permanent negative charge (permanent acidity) due to isomorphous substitution in clay minerals. They also acquire negative charges with an increase in soil pH (pH-dependent acidity) due to hydroxyl ions on the edges of the clay and those associated with hydroxides and oxyhydroxides of iron and aluminum. Furthermore, data from Ohio soils (Table 6.5) varying in organic matter content show that soil organic matter can largely affect the permanent negative charge and total cation exchange capacity of soils; the effect on pH-dependent CEC was much less. Because the negative charge on soil organic matter is due to -OH and -COOH radicals, the contribution of exchangeable H^+ on soil acidity in organic matter rich soils could be considerable. Data in Table 6.5 also show that liming can considerably reduce pH dependent CEC.

6.7. FACTORS AFFECTING SOIL ACIDITY

A number of factors affect soil acidity. These include (1) the kind of chemical fertilizer used; (2) the amount of basic cations removed by the crop plants; (3) leaching of cations; (4) organic residues and their decomposition;

Table 6.4 Soil Acidity, as Indicated by Various Methods, and Cation-Exchange Properties of Four Representative Ohio Soils

Soil Type	pH	Extr. Al (NH$_4$OAc- pH 4.8) (me/100 g)	Soil Acidity (cmol kg^{-1})					Direct titration		CEC (cmol kg^{-1})	
			CaCO$_3$ incub. pH 6.8	KCl exch.	SMP buffer	Mehlich buffer	Woodruff buffer	pH 6.8	pH 8.1	NH$_4$O Ac pH 7.0	BaCl$_2$ pH 4.1
Bennington si.l	4.5	1.5	9.4	1.6	7.2	9.8	4.4	6.4	12.0	11.0	13.1
Mahoning si.l	4.2	3.7	16.8	4.4	16.0	19.0	10.0	13.0	26.0	13.2	16.5
Ashtabula si.l	4.7	7.8	17.0	4.4	17.2	18.2	11.4	10.0	21.0	17.0	20.0
Trumbull si.l	4.0	5.9	21.8	6.0	19.6	20.6	10.6	17.0	33.0	14.4	19.0

From McLean et al., 1964. Soil. Sci. 97:120. With permission of Williams & Wilkins, Baltimore, MD.

Table 6.5 The Amount of Permanent Charge (KCl-CEC), pH-Dependent CEC and Total CEC with or without Organic Matter in Some Ohio Soils Differing in Organic Matter Content

Soil	Lime added (cmol kg⁻¹)	Soil pH	Organic matter (%)	KCL-CEC (cmol kg⁻¹)		pH-dependent CEC (cmol kg⁻¹)		Total CEC (cmol kg⁻¹)	
				OM	WOM	OM	WOM	OM	WOM
Clermont	4	5.7	1.4	6.4	4.9	3.8	2.4	10.2	7.3
	8	7.2	—	8.2	5.5	1.9	2.9	10.1	8.4
Bennington	0	4.8	2.4	7.3	6.4	10.7	4.6	18.0	11.0
	8	6.2	—	11.1	7.3	5.6	4.3	16.7	11.6
Fries	8	4.9	4.2	9.9	9.8	16.5	9.0	26.4	11.2
	20	6.6	—	16.6	10.6	3.3	6.6	19.5	11.3
Cambridge	0	4.4	9.7	10.9	7.3	31.1	7.8	42.0	11.9
	45	7.0	—	24.4	8.5	8.5	6.0	32.9	9.7

Note: OM, with organic matter; *WOM*, without organic matter.

Adapted from McLean et al. (1965).

**Table 6.6 Effect of Forms of Chemical Fertilizer
on the pH of Surface (0–5 cm) Soil**

Fertilizer	N rate	pH 1:1 (H_2O)		
		Alfisol	**Ultisol**	**Oxisol**
Control	0	6.09	4.92	3.87
AS[a]	100	4.02	3.98	3.51
urea	100	4.24	4.18	3.46
CAN[b]	100	4.64	4.60	3.47

[a] Ammonium sulfate.

[b] Calcium ammonium nitrate.

From Stumpe and Vlek, 1991. Soil. Sci. Soc. Am. J.
55:150. With permission of the Soil Science Society of
America, Madison, WI.

(5) nitrogen transformations; and (6) deposition of nitrate, sulfate, and acid-
forming chemicals by rain.

6.7.1. Chemical Fertilizers

Chemical fertilizers are well known for their acidity and basicity-forming
effects. The basicity-forming effect of Chilean nitrate ($NaNO_3$) was one of the
factors that made it popular in the early days of farming when fertilizers first
appeared. Both nitrogen and phosphorous fertilizers are known for these effects.

In an experiment in the United States at the International Fertilizer Development Center (IFDC) in Alabama, studies were made in plexiglass cylinders
(5 cm internal diameter and 70 cm length) with three soils; an alfisol from
Egdeba, southwestern Nigeria; an ultisol from Onne, southeastern Nigeria;
and an oxisol from Carimagua, eastern Colombia. Following each application
of N at rates of 0, 50, 100 kg ha^{-1} as urea, ammonium sulfate, or calcium
ammonium nitrate, the soils were flushed with distilled water 30 times during
2 years of study with 6-week rest periods after each five cycles. The pH values
recorded after the study are given in Table 6.6. Application of fertilizer N
reduced soil pH in all three soils. In the alfisol, as well as in the ultisol, the
reduction in soil pH was the most with ammonium sulfate and the least with
calcium ammonium nitrate. The differences between the N sources were not
distinct in the oxisol.

Data on changes in soil pH due to fertilizer application from long-term
experiments on acid Indian soils are given in Table 6.7. These data confirm
that continuous application of N results in depletion of soil pH. Application
of lime stopped this depletion in soil pH.

A method for determining the acidity or basicity of fertilizers was developed by Pierre in 1933. The basic assumptions in this method are as follows:

**Table 6.7 Changes in Soil pH as Affected by Intensive Manuring
and Cropping on Acidic Soils in India (1988–1989 Data)**

Treatments	Ranchi (Haplustalf)	Palampur (Hapludalf)
Control	5.3	5.6
Nitrogen	4.4	4.9
NPK + lime	5.8	6.6
LSD (0.05)	0.07	0.2
Initial (1971) value	5.3	5.8

From Nambiar et al. 1992. Annual Report, (ICAR) Indian Agricultural Research Institute, New Delhi, p. 151.

1. The acid-forming effect of fertilizer is caused by all of the contained sulfur and chlorine, one-half of the nitrogen, and one-third of the phosphorous. However, Andrews (1954) maintained that the entire amount of N contributed to soil acidity.
2. The presence of calcium, magnesium, potassium, and sodium in fertilizer raises the pH of the soil.

Using the above assumption let us calculate equivalent acidity for ammonium sulfate $(NH_4)_2SO_4$. When ammonium sulfate is applied to soil, the following reactions will occur:

$$(NH_4)_2SO_4 + 2H_2O \rightarrow H_2SO_4 + 2NH_4OH$$

$$2NH_4OH + 3O_2 \rightarrow 2HNO_3 + 2H_2O$$

$$H_2SO_4 + CaCO_3 \rightarrow CaSO_4 + H_2O + CO_2$$

$$2HNO_3 + CaCO_3 \rightarrow Ca(NO_3)_2 + H_2O + CO_2$$

Thus if all nitrogen is considered to be responsible for causing soil acidity (as proposed by Andrews), each molecule of ammonium sulfate (containing 28 g N) will require two molecules of calcium carbonate (2 × 100 g). Thus 28 g N will react with 200 g CaCO. This is equal to 7.14 kg calcium carbonate (lime) per kg N in fertilizer (Column 3 of Table 6.8). If only half of the nitrogen contributes to soil acidity (as per Pierre), then each molecule of ammonium sulfate will react with 1.5 molecules of calcium carbonate (0.5 molecule for N, 1 molecule for sulfate). Thus 28 g N will react with 150 g $CaCO_3$. This is equal to 5.35 kg calcium carbonate (lime) per kg N (Column 5 of Table 6.8). Amounts of lime needed per 100 kg fertilizer material can be easily calculated by multiplying by the percentage of N (or other nutrient) in the fertilizer material. For ammonium sulfate the values obtained will be 7.14 × 20.5 = 146 (Column 4 of Table 6.8) or 5.35 × 20.5 = 110 (Column 6 of Table 6.8) using

Table 6.8 Equivalent Acidity and Basicity of Chemical Fertilizers

		Lime required to neutralize acidity			
		Andrews		Pierre[a]	
Chemical fertilizer	Percent N	Per kg N	Per 100 kg fertilizer	Per kg N	Per 100 kg fertilizer
1	2	3	4	5	6
Nitrogen fertilizers					
Ammonium sulfate	20.5	7.14	146	5.35	110
Anhydrous ammonia	82.2	3.57	293	1.80	148
Calcium ammonium nitrate	20.5	1.77	36	0	0
Urea	46.6	3.57	166	1.80	84
Urea - ammonia liquor	45.5	3.57	162	1.80	82
Phosphate fertilizers					
Ammonphos A	11.0	6.77	74	5.00	55
Di-ammonphos (18-46-0)	18	—	—	—	—
Ordinary super-phosphate	0	0	0	0	0
Triple super-phosphate	0	0	0	0	0
Potash fertilizers					
Muriate of potash	0	0	0	0	0
Sulfate of potash	0	0	0	0	0

[a] Now official method in the United States.

Andrews or Pierre's assumption. Equivalent acidity or basicity of some chemical fertilizers is given in Table 6.8.

6.7.2. Removal of Basic Cations

All plants take up exchangeable bases during their growth. When they are completely or partly removed from the land, the net result is loss of bases from the soil, and this leads to the development of soil acidity. This has been clearly brought out from data from the long-term experiments conducted in various countries. Crops differ in the quantity of bases removed. For example, leguminous crops remove more calcium and less potassium than cereals.

Pierre and Banwart (1973) suggested that the ratio of excess bases (EB) defined as total cations (Ca^{2+}, Mg^{2+}, K^+, and Na^+) minus total anions (Cl^-, SO_4^{2-}, NO_3^-, and $H_2PO_4^-$) in the plant material to plant nitrogen content as the criterion for determining the acid-forming property of crops. Crop plants with EB:N ratios above 1.0 increase soil acidity, while those below 1.0 decrease acidity. In their study of 149 samples taken at different stages of growth of 26 crop species, only buckwheat, tobacco, and spinach had values slightly above 1.0. Cereal and grasses had average ratios of 0.43 and 0.47, respectively, meaning thereby that only 43 and 47%, respectively, of the nitrogen they received was acid forming.

6.7.3. Nitrogen Transformations

When ammonium or ammonium-forming N fertilizers are applied to soils, little to large amounts of N can be lost by ammonia volatilization (see Chapter 8). This lost N does not contribute to nitrate formation and the resultant acidity. Similarly, the amount of N lost as N_2 or N_2O due to denitrification (see Chapter 8) also does not contribute to soil acidity. Whatever the NH_4 source, nitrification is an acidic process.

6.7.4. Acids Brought in by Rain

A number of gaseous sulfur and nitrogen compounds, especially SO_2 and NO_2, are emitted into the atmosphere through natural processes and/or by man's activities. These gaseous compounds are dissolved in rains, come down to earth, and are added to soil. A study in Alberta (Lau and Das, 1985) showed that about 8.8 kg ha^{-1} of sulphate, 4 kg ha^{-1} of nitrate, and 0.15 kg ha^{-1} of hydrogen were added annually to the soil. Acid rain occurs at many places in the world, and in the United States it exhibits a maximum in the northeastern region and adjacent Canadian provinces. A pH contour map of North America is shown in Figure 6.5. Thus acids are regularly added to soil by precipitation, the amounts being greatest near industrial towns.

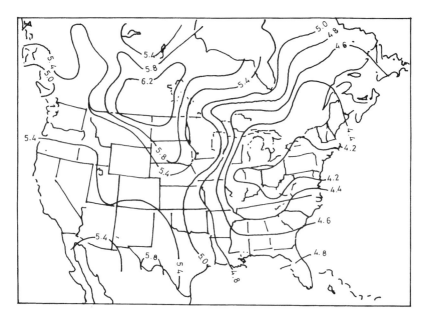

Figure 6.5. Precipitation pH contour map of North America for the period 1972–1982. (From Mohnen and Wilson, 1985. *Acid Deposition—Environmental, Economic and Policy Issues*, **D.A. Adams and W.P. Page, Eds., p. 440. With permission of Plenum Press.)**

6.7.5. Leaching of Bases

Leaching of bases as a factor responsible for soil acidity is well established in pedogenesis.

6.8. SOIL pH AND CROP PRODUCTION

Crops do differ in their soil pH requirements. The optimum soil pH for crop production is generally considered to be between 6.5 and 7.0, and lime applications are made to that effect. Kamprath (1971) and his group of research workers in North Carolina have, however, shown that pH values greater than 6.0 or 6.2 may not only be unnecessary but harmful. On oxisols and ultisols of the warm, humid, southeastern United States they have suggested liming soils sufficient to reduce exchangeable Al^{3+} to less than 10% of the effective CEC (Kamprath, 1970). On the other hand, liming to pH values of 6.5 to 6.8 is recommended for alfisols and mollisols, such as in the midwestern United States (Tisdale et al., 1985).

The chemical and physical properties of oxisols and ultisols in the southern United States are largely controlled by hydroxyaluminum and hydroxyiron coatings on clay fraction. These coatings, on liming, become pH-dependent adsorption sinks for P, Zn, Mg, and other elements. This would explain why a lower pH value is considered suitable for general field crops on ultisols and oxisols than in alfisols and mollisols.

Liming oxisols and ultisols to pH 7.0 reduces phosphorous and micronutrient uptake by crops, reduces water percolation, and subsequently reduces plant growth (Kamprath, 1971). Grove and Sumner (1985) working on an ultisol (Typic Hapludult—Bradson clay loam) in Georgia showed that lime application at some point induced magnesium stress in corn (Figure 6.6). There was also a depression in plant zinc concentration.

Aluminum toxicity is probably the major limiting factor to plant growth and crop production in strongly acidic soils (Foy, 1992). The root system of plants is adversely affected by Al-toxicity (Figure 6.7). In cereals roots appear corolloid with many stubby lateral roots and no fine branching. This limits plants ability to absorb water and nutrients and results in poor growth. Crops and cultivars among a crop differ in tolerance to Al-toxicity, and research is going on to identify such cultivars for increasing crop production in acidic soils.

6.9. LIME REQUIREMENT

Due to the buffering capacity of soils, change in pH takes considerable time. Therefore the ideal procedure for determining lime requirement would be mixing a known amount of soil with different amounts of lime ($CaCO_3$) for a fairly long period (Figure 6.8) and then determining the amount of lime required to obtain the desired pH. This procedure, however, requires a fairly long time, so is not suitable for soil-testing purposes. Advantage has been taken of buffer solution simulation of acidic soils. A number of buffer solutions have been proposed for determining the lime requirement of soils. Some of these are: p-nitrophenol buffered at about pH 7 (Schofield, 1933); barium chloride-triethanolanine at pH 8 (Peech et al., 1947); p-nitrophenol, potassium dichromate, calcium acetate-MgO buffered at pH 7 (Woodruff, 1948); and p-nitrophenol-triethanolamine-potassium chromate-calcium acetate, and calcium chloride buffer adjusted to pH 7.5 with NaOH (Shoemaker et al., 1961); the Schomaker-McLean-Pratt (SMP) buffer has found most favor with United States soil scientists. Lime requirements of some acidic Indian soils, as determined by $CaCO_3$ incubation (for one month), Peech's buffer, SMP buffer, and Schofield buffer, are given in Table 6.9 (Dolui and Saha, 1984). Values determined by Peech and Schofield buffers were very high, while those determined by SMP buffer were closest to results from the $CaCO_3$ incubation procedure. The correlation coefficients (γ) between $CaCO_3$ incubation and the Peech, SMP, and Schofield buffers were 0.917, 0.932, and 0.749 and were significant. The SMP buffer is especially well suited for soils possessing the following properties: lime requirement > 4 cmols kg^{-1} (> 4400 kg lime ha^{-1}; taking the weight of 0 to 15 cm surface soil layer weight at 2.2×10^6 kg ha^{-1}), pH <5.8, organic matter concentration of <10% and appreciable quantities of soluble

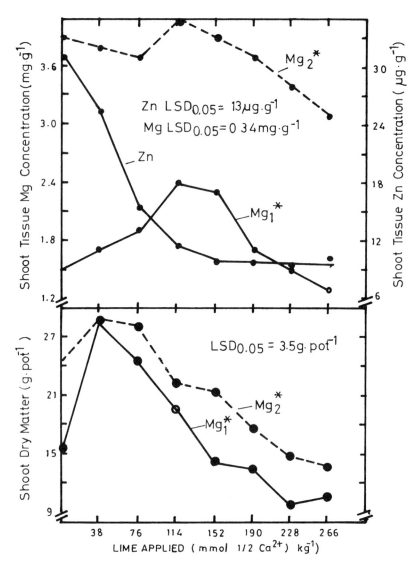

Figure 6.6. Relationships between shoot dry matter productivity, shoot tissue Mg, Zn, concentrations, and lime application rate. *Mg₁, Mg₂ = 2, 10 mmol (1/2 Mg²⁺) kg⁻¹, respectively. (From Grove and Sumner, 1985. Soil Sci. Soc. Am. Proc. 61:428–432. With permission of the Soil Science Society of America, Madison, WI.)

Figure 6.7. Al-toxicity — cotton plants grown on 0, 0.75, 1.5, 3.0, and 6.0 mg Al L⁻¹. (Photo courtesy of Dr. C. Foy, USDA, Maryland.)

(extractable) Al (Tisdale et al., 1985). However, the SMP buffer will give lime requirement values that frequently result in overliming of many low-base-status soils.

On highly leached tropical soils, namely, ultisols and oxisols, there is a considerable amount of pH-dependent charge due to hydroxides of Al and Fe. The lime requirement determined by buffers such as the SMP buffer could be very high. Research with these soils (Sanchez, 1976) has suggested that exchangeable Al^{3+} determination is a better criterion for determining lime requirement for these soils. Exchangeable Al^{3+} can be determined by extracting soils with unbuffered 1 N KCl (Lin and Coleman, 1960). There is very little exchangeable Al^{3+} when soil pH is 5.5 or above (Figure 6.9). However, when Mn toxicity is suspected, liming may be done to raise the pH up to 6.0. The generally suggested lime application on ultisols and oxisols is that which provides sufficient lime to neutralize soil acidity to 20 to 25% Al saturation. Al saturation is calculated by dividing exchangeable Al^{3+} (and exchangeable H^+ if present) by the sum of exchangeable bases plus exchangeable Al^{3+} (and H^+). The relationships between percent Al saturation and soil pH in ultisols and oxisols of Puerto Rico is shown in Figure 6.10.

Liming ultisols and oxisols to pH 5.5 or 6.0 is suggested (Sanchez, 1976) when mangenese toxicity is suspected. Kamprath (1970) suggested that the lime requirement be determined by multiplying exchangeable Al^{3+} (cmol kg⁻¹)

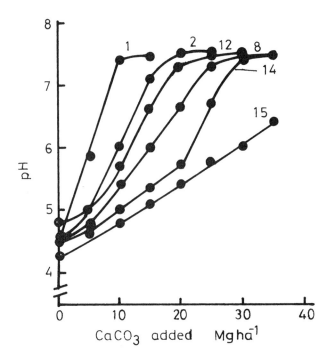

Figure 6.8. Titration curves for some Ohio soils. pH after 17 months of moist incubation with lime. Number on curves indicates different soils. (From Shoemaker et al., 1961. Soil Sci. Soc. Am. Proc. 25:274–277. With permission of Soil Science Society of America, Madison, WI.)

by 1.5 to give the cmol kg^{-1} of Ca or 1.65 cmol kg^{-1} of CaCO$_3$. This amount of lime would neutralize 85 to 90% of exchangeable Al^{3+} in soils containing 2 to 7% organic matter; for soils containing more organic matter, the factor may be raised to 2 to 3. Application of this concept of liming has substantially reduced the rates of lime applied to soils.

Another important component of the lime requirement is the level of exchangeable Al^{3+} that specific crops can tolerate; cotton, sorghum, and alfalfa are susceptible to levels of 10 or 20% Al-saturation, while corn is sensitive to 40 to 60% Al-saturation (Sanchez, 1976) (Figure 6.11). Thus cotton, sorghum, and alfalfa fields need to be limed to nearly zero Al-saturation level, while liming corn fields to 20% Al-saturation is adequate. Crops such as rice (explained earlier), coffee, pineapple, and some pasture species seldom respond to liming even in soils with high Al-saturation. In a recent study on ultisols and inceptisols in Nigeria (Nwachuku and Loganathan, 1991), near maximum corn yield was obtained when liming reduced Al-saturation to 25%. No increase in corn yield was obtained by liming beyond pH 5.5.

**Table 6.9 Lime Requirement (Mg ha⁻¹) by Calcium
Carbonate Incubation and Other Diagnostic Methods**

Location	Incubation	Peech	SMP (pH 6.4)	Schofield
Bagdubi	0.39	5.93	1.72	20.92
Ausgram	0.98	11.86	3.58	20.89
Andibahara	1.28	8.15	1.72	19.55
Kalimpong	1.38	14.08	6.67	23.54
Dimrulia	1.58	8.89	4.20	17.01
Kapgari	1.97	3.70	0.49	14.39
Arambagh	2.32	8.89	3.58	26.08
Khaprail	3.95	17.79	12.84	20.86
Rajganj	5.14	22.23	10.37	24.85
Kharibari	5.93	18.53	9.76	26.16
Kalijhora	6.22	20.01	14.08	20.92
Bhutobori	18.48	33.35	23.10	32.70

From Dolui and Saha, 1984. J. Indian Soc. Soil Sci. 32:158–161.
With permission.

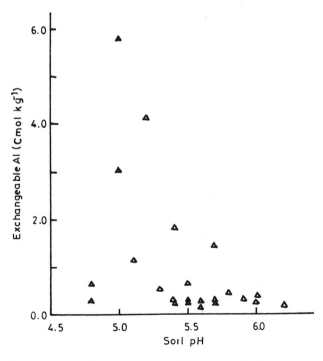

**Figure 6.9. Exchangeable aluminum at different pH values in nine oxisols
and andepts from Panama. (From Mendez-Lay, 1973. M.S.
Thesis, North Carolina University, Raleigh.)**

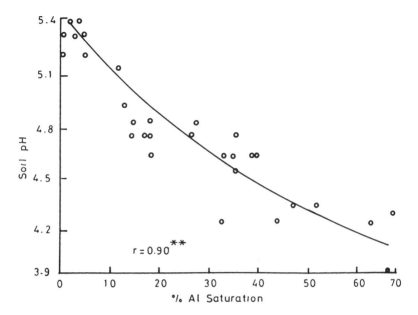

Figure 6.10. Relationship between soil pH and aluminum saturation in eight ultisols and oxisols of Puerto Rico. (From Abruna et al., 1975. *Soil Management in Tropical America*. With permission of North Carolina University Press.)

6.10. LIMING MATERIALS

Limestone.   The most commonly used liming material is limestone or calcium carbonate ($CaCO_3$) and all other liming materials are evaluated relative to their effectiveness compared with calcium carbonate (given a value of 100).

Limestone deposits are found in most countries of the world. It is often mined by open-pit methods (quarrying) using explosives to blast the rock. The broken rock pieces are crushed to sizes of 2.5 cm (1 inch) or less and further ground or pulverized to the desired size. Generally an 8- to 10-mesh* material (more than 75% should pass the screen) is considered desirable (discussed later). The quality of commercial limestone generally used is 90 to 98% calcium carbonate equivalent (CCE), but even inferior materials may be marketed. Lower CCE values are due to its impurities such as clay.

The calcium carbonate equivalent (CCE) of a liming material is determined as shown below giving the examples of limestone ($CaCO_3$) and slaked lime ($Ca(OH)_2$):

---

* A 10-mesh size refers to a sieve where each linear inch is divided into 10 divisions; thus there will be 100 divisions per square inch. The actual size of pores will also depend upon the thickness of the wire used to make the sieve.

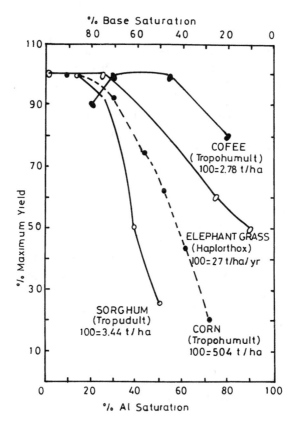

Figure 6.11. **Yield responses to liming in Puerto Rican oxisols and ultisols.
(From Sanchez, 1976. *Properties and Management of Soils in
the Tropics.* With permission of John Wiley & Sons.)**

$$CaCO_3 + 2HCl \rightarrow CaCl_2 + H_2CO_3 \tag{1}$$

$$Ca(OH)_2 + 2HCl \rightarrow CaCl_2 + 2H_2O \tag{2}$$

In equation 1 100 g calcium carbonate or limestone would neutralize
73 g HCl. However, the same amount of HCl is neutralized by only 74 g slaked
lime as shown in equation 2. Thus 74 g slaked lime is equivalent to 100 g
calcium carbonate or limestone. Therefore 100 g slaked lime would equal

$\dfrac{100}{74} \times 100 = 135$ g limestone or calcium carbonate; 135 is the calcium

carbonate equivalent (CCE) of slaked lime.

Calcium Oxide (CaO).   Calcium oxide or (quick) lime is obtained by calcining limestone. Calcium oxide gives very quick results and is therefore used when rapid change in soil pH is required. Because the material is a white powder, which irritates skin and eyes, it is difficult to apply and mix in soil. After application, it forms flakes or granules in the presence of soil water. Due to reaction with atmospheric or soil carbon dioxide, calcium carbonate is formed and is deposited on the surfaces of flakes or granules. In this condition the remaining CaO remains in the soil for a fairly long time. The calcium carbonate equivalent (CCE) for calcium oxide is 179%.

Calcium Hydroxide (Ca[OH]$_2$).   This is slaked lime and is made by reacting calcium oxide with water. It is also known as builders' lime and is used for construction purposes. Calcium hydroxide is generally used for whitewashing houses in India and many other developing nations. With the passage of time calcium carbonate forms on the surface by reaction with atmospheric carbon dioxide, and the color then becomes pale yellow. This is the reason whitewashing is repeated every year. Calcium hydroxide is much easier than calcium oxide to handle and apply. The CCE value for slaked lime is 135%.

Dolomite.   This is calcium magnesium carbonate [Ca Mg(CO$_3$)$_2$]. The CCE value is 109%. Dolomite has the advantage of supplying Mg as a plant nutrient and is especially important in Mg-deficient soils such as some soils developed from glacial till.

Other Materials.   Other liming materials include marl (unconsolidated deposits of calcium carbonate frequently mixed with earth; found near sea coasts), slags (blast furnace slag, basic slag, electric furnace slag), fly ash (from coal burning in thermal power–generating plants), flue dust (from cement-manufacturing plants), sugar lime (from carbonation used in sugar factories), and pulp mill lime. In the United States and other countries, scrubbers (calcium carbonate solutions) are also used to remove SO$_2$ and other gases from smokestacks, and precipitators are used to remove particulate matter. These materials usually have some neutralizing capacity. These and several other materials could be available as industrial wastes at throw-away prices and could be used for liming. A large number of experiments in India were conducted with basic slag as a source of phosphorous with limited success since most Indian basic slags contained only 2 to 3% P$_2$O$_5$, which was too little compared with the grinding and transportation costs. It was never tested as a liming material in India since there was no need; more than 80% of cultivated soils in India have a pH well above 7.5.

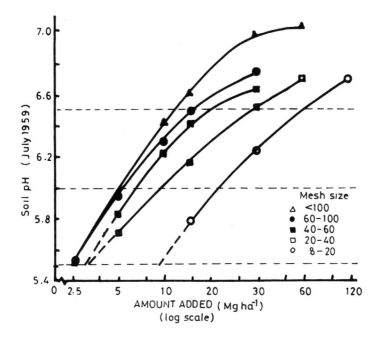

Figure 6.12. Effect of various rates of dolomitic limestone fractions on the pH of Withee silt loam after 2 years of equilibration under field conditions. (From Love et al., 1960. *Trans. 7th Int. Cong. Soil Sci.* **3:293–301. With permission of International Society of Soil Science, Vienna, Austria.)**

6.11. FINENESS OF LIMESTONE

Besides the purity of the limestone, the degree of fineness is also important; the finer the limestone, the faster it reacts. This can be reflected in soil pH changes (Figure 6.12), as well as in crop yields (Figure 6.13). Also less limestone is required to achieve the desired soil pH changes and to improve crop yields if the particle size is finer. However, finer grinding adds to the cost of material. Because lime is generally recommended for application every 3 to 5 years, long-term effects are also important. Generally, lime is considered satisfactory when 75% passes an 8- to 10-mesh screen. When a liming material passes an 8- to 10-mesh screen, it will also have finer particles that will pass finer-mesh sieves. Thus liming material generally has particles of various sizes.

Calcium oxide and calcium hydroxide are powdery materials and do not have problems with particle size.

6.12. BENEFITS OF LIMING

Liming improves soil pH and thus provides soil environments for better plant growth. It prevents toxicity due to excess Al and Mn, and the availability

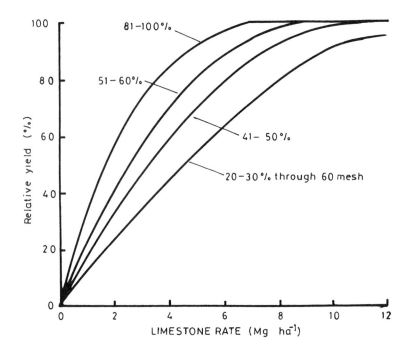

Figure 6.13. The effect of limestone fineness on the response of crops to increased rates of limestone application. These are average data from a number of field experiments. (From Brady, N.C., 1990. *The Nature and Properties of Soils*, 11th ed. With permission of Prentice-Hall, Inc., Upper Saddle River, N.J.)

of P and Mo improves. Microbiological processes such as nitrification and nitrogen fixation also improve. Because of increased microbial activity and plant growth, liming may also indirectly improve the physical conditions of the soil.

REFERENCES

Abruna, F., W. Pearson, and R. Perez-Escolar. 1975. Lime response of corn and beans grown on typical ultisols and oxisols of Puerto Rico, in *Soil Management in Tropical America*, E. Bornemisza and A. Alvardo, Eds., North Carolina State University Press, Raleigh, NC, pp. 261–281.

Andrews, W.B. 1954. *The Response of Crops and Soils to Fertilizers and Manures*, 2nd ed., State College, MS.

Brady, N.C. 1990. *The Nature and Properties of Soils*, 11th ed., Prentice-Hall, Inc. Upper Saddle River, NJ.

Daniels, F. and R.A. Alberty. 1967. *Physical Chemistry*, 3rd ed., John Wiley & Sons, New York, p. 767.

Dolui, A.K. and R. Saha. 1984. A comparison of several methods of lime requirement of some acid soils of Bengal. J. Indian Soc. Soil Sci. 32:158–161.

Foy, C.D. 1992. Soil chemical factors limiting plant growth. Adv. Soil Sci. 19:97–149.

Grove, J.H. and M.E. Sumner. 1985. Lime-induced magnesium stress in corn: impact of magnesium and phosphorous availability. Soil Sci. Soc. Am. J. 49:1192–1996.

Jenny, H. 1961. Reflections on the soil acidity merry-go-round. Soil Sci. Soc. Am. Proc. 61:428–432.

Kamprath, E.J. 1970. Exchangeable Al as a criterion for liming leached mineral soils. Soil Sci. Soc. Am. Proc. 34:252–254.

Kamprath, E.J. 1971. Potential detrimental effects from liming highly weathered soils to neutrality. Soil Crop Sci. Soc. Florida Proc. 31:200–211.

Lau, Y.K. and N.C. Das. 1985. Precipitation quality monitoring in Alberta, in *Impact of Air Toxins on the Quality of Life*, Proc. Ann. Meeting CPANS/PNWIS Sections, Air Pollution Control Association, Pittsburgh, PA, pp. 213–233.

Lin, C. and N.T. Coleman. 1960. The measurement of exchangeable aluminum in soils and clays. Soil Sci. Soc. Am. Proc. 24:444–446.

Love, J.R., R.B. Corey, and C.C. Olsen. 1960. Effect of particle size and rate of application of dolomitic limestone on soil pH and growth of alfalfa. Trans. 7th Int. Cong. Soil Sci. (Madison, WI) 3:293–301.

McLean, E.O., W.R. Hourigan, H.E. Shoemaker, and D.R. Bhumbla. 1964. Aluminum in soils. V. Form of aluminum as a cause of soil acidity and a complication in its measurement. Soil Sci. 97:119–126.

McLean, E.O., D.C. Reicosky, and C. Lakshamanan. 1965. Aluminum in soils. VII. Interrelationships of organic matter, liming, and extractable aluminum with permanent charge and pH dependent cation exchange capacity of surface soils. Soil Sci. Soc. Am. Proc. 29:374–378.

Mendez-Lay, J. 1973. Effects of limes on phosphorus fixation and plant growth in various soils of Panama. M.S. Thesis, Soil Sciences Dept., N.C. State University, Raleigh, 90 pp.

Mohnen, V.A. and J.W. Wilson. 1985. Acid rain in North America: concepts and strategies, in *Acid Deposition—Environmental, Economic and Policy Issues*, D.A. Adams and W.P. Page, Eds., Plenum Press, New York, pp. 439–452.

Mongia, A.D. and A.K. Bandyopadhyay. 1991. Changes in pH of acid soils in different electrolytes. J. Indian Soc. Soil Sci. 39:351–354.

Mukherjee, J.N., B. Chatterjee, and B.M. Banerjee. 1947. Liberation of H^+, Al^{3+} and Fe^{3+} ions from hydrogen clays by neutral salts. J. Colloid Sci. 1:247–254.

Nambiar, K.K.M., P.N. Soni, M.R. Vats, D.K. Sehgal, and K.H. Mehta. 1992. Annual Report. All Indian Coordinated Agronomic Research Project on Long Term Experiments (ICAR), Indian Agric. Research Institute, New Delhi, p. 151.

Nwachuku, D.A. and P. Loganathan. 1991. The effect of liming on maize yield and soil properties in Southern Nigeria. Commun. Soil Sci. Plan Anal. 22:623–639.

Peech, M. 1941. Availability of ions in light sandy soils as affected by soil reaction. Soil Sci. 51:473–486.

Peech, M., L.T. Alexander, L.A. Dean, and J.F. Reed. 1947. Methods of soil analysis for soil fertility investigation. U.S. Dept. Agric. Circ. 757.

Pierre, W.H. and W.L. Banwart. 1973. Excess-base and excess-base/nitrogen ratio of various crop species and parts of plants. Agron. J. 65:91–96.

Ponnamperuma, F.N. 1976. Physicochemical properties of submerged soils in relation to fertility, in *The Fertility of Paddy Soils and Fertilizer Applications for Rice*, Food and Fertilizer Technology Centre for the Asian and Pacific Region, Taipei, Taiwan, pp. 1–27.

Rasnake, M. 1973. Liming acid soils, University of Kentucky, College of Agriculture, Cooperative Extension Services, AGR-19, p. 2.

Sanchez, P.A. 1976. *Properties and Management of Soils in the Tropics*, John Wiley & Sons, New York, p. 618.

Schofield, R.K. 1933. Rapid methods for examining soils. II. The use of p-nitrophenol for assessing lime status. J. Agric. Sci. (Camb.) 23:252–260.

Shoemaker, H.E., E.O. McLean, and P.F. Pratt. 1961. Buffer methods for determining lime requirement of soils with appreciable amounts of extractable aluminum. Soil Sci. Soc. Am. Proc. 25:274–277.

Stumpe, J.M. and P.L.G. Vlek. 1991. Acidification induced by different nitrogen sources in columns of selected tropical soils. Soil Sci. Soc. Am. J. 55:145–151.

Tisdale, S.L., W.L. Nelson, and J.D. Beaton. 1985. *Soil Fertility and Fertilizers*, Macmillan, New York, p. 754.

Woodruff, C.M. 1948. Testing soils for lime requirement by means of a buffered solution and the glass electrode. Soil Sci. 66:53–63.

7   SOIL SALINITY AND SODICITY

Saline/sodic soils contain either elevated levels of Ca or Na or both; those having high Na levels (sodic soils) may have pH values 8.5 or above. When soil pH is as high as 8.5, a number of micronutrients such as Fe, Mn, Cu, and Zn become deficient and crop growth is adversely affected. Also high Na concentrations damage plant roots and affect plant growth adversely. In addition, sodic soils with more than about 20% clay content exhibit poor physical structure as a result of the destruction of aggregation resulting from adsorption of the hydrated Na^+ ions on clay surfaces. These soils crust and cake, have low infiltration rates, and are a poor medium for plant growth. Sodic soils may or may not be high in total soluble salts (saline).

Saline soils are high in concentration of soluble salts and are alkaline if Mg or Na are the dominant cations. However, saline soils may also be high in exchangeable H^+ with pH as low as 4.0, presenting severe soil acidity problems that affect nutrient availability as discussed in Chapter 6. Salinity experiments with cereals conducted in field plots at constant fertilizer application (Francois et al., 1986) demonstrate a 20 to 50% reduction in plant P content. On the other hand, Grattan and Maas (1984) observed P toxicity for soybeans under saline conditions in a solution culture at P levels that were nontoxic under nonsaline conditions. These data indicate the intricacies of soil fertility problems as influenced by salinity. An understanding of the nature of soil salinity/alkalinity can therefore greatly help in soil fertility management on such soils.

7.1. COVERAGE AND SPECIAL FEATURES

Large land masses, particularly in the arid and semiarid regions of the world, have salt deposits in the soil surface during hot summer periods. These salts may move to the subsoil during the rains and may move again to the surface during dry periods. Depending upon the salts present, these soils may have problems of acidity or alkalinity. Also, if Na salts dominate, degraded soil structure may create additional problems. The world distribution of salt-affected areas is given in Table 7.1.

Table 7.1 The World Distribution of Salt-Affected Areas (1000 ha)

Continent	Country	Saline/ solonchaks	Sodic/ solonetz	Total
North	Canada	264	6974	7238
America	United States	5927	2590	8517
Mexico and	Cuba	316	—	316
Central	Mexico	1649		1649
America			—	
South	Argentina	32473	55139	85612
America	Bolivia	5233	716	5949
	Brazil	4141	362	4503
	Chile	5000	3642	8642
	Colombia	907	0	907
	Ecuador	387	—	387
	Paraguay	20008	1894	21902
	Peru	21	—	21
	Venezuela	1240	0	1240
Africa	Afars and Issas	1741	—	1741
	Algeria	3021	129	3150
	Angola	440	86	526
	Botswana	5009	670	5679
	Chad	2417	5850	8267
	Egypt	7360	0	7360
	Ethiopia	10608	425	11033
	Gambia	150	0	150
	Ghana	200	118	318
	Guinea	525	—	525
	Guinea-Bissau	194	—	194
	Kenya	4410	448	4858
	Liberia	362	44	406
	Libyan Arab Jamahiriya	2457	—	2457
	Madagascar	37	1287	1324
	Mali	2770	0	2770
	Mauritania	640	—	640
	Morocco	1148	—	1148
	Namibia	562	1751	2313
	Niger	—	1389	1389
	Nigeria	665	5837	6502
	Rhodesia	—	26	26
	Senegal	765	—	765
	Sierra Leone	307	—	307
	Somalia	1569	4033	5602

Table 7.1 The World Distribution of Salt-Affected Areas (1000 ha) (Continued)

Continent	Country	Saline/ solonchaks	Sodic/ solonetz	Total
	Sudan	2138	2736	4874
	Tunisia	990	—	990
	United Rep. of Cameroon	—	671	671
	United Rep. of Tanzania	2954	583	3537
	Zaire	53	—	53
	Zambia	—	863	863
Southern Asia	Afghanistan	3103	—	3103
	Bangladesh	2479	538	3017
	Burma	634	—	634
	India	23222	574	23796
	Iran	26399	686	27085
	Iraq	6726	—	6726
	Israel	28	—	28
	Jordan	180	—	180
	Kuwait	209	—	209
	Muscat ant Oman	290	—	290
	Pakistan	10456	—	10456
	Qatar	225	—	225
	Sarawak	1538	—	1538
	Saudi Arabia	6002	—	6002
	Sri Lanka	200	—	200
	Syrian Arab Rep.	532	—	532
	United Arab Emirates	1089	—	1089
North and Central Asia	China	36221	437	36658
	Mongolia	4070	—	4070
	USSR	51092	119628	170720
Southeast Asia	Democractic Kampuchea	1291	—	1291
	Indonesia	13213	—	13213
	Malaysia	3040	—	3040
	Socialist Rep. of Vietnam	983	—	983
	Thailand	1456	—	1456
Australasia	Australia	17269	339971	357240
	Fiji	90	—	90
	Solomon Islands	238	—	238

From Massoud, 1977. Proc. Int. Conf. Managing Saline Water. 1976, pp. 432–454. Texas Tech. University Press, Lubbock, TX.

Salt-affected soils are drawing more and more global attention because of population pressure and the consequent ever-increasing food demands in some regions of the world where such soils exist. Heavy food demand has necessitated the development of irrigation in such areas. Without adequate provisions for drainage, irrigation has aggravated the situation of salinity/alkalinity. For food production, these soils are regarded as a class of problem soils that need special remedial measures and management practices. Composition of salts, their distribution in the profile, soil texture and structure, and the species of plants grown determine plant growth on these soils.

The main source of all salts in the soil is the primary minerals of the earth's crust, which are gradually released and made soluble during chemical weathering. In some regions salts are transported away from their source of origin in humid areas through surface and groundwater streams to semi-arid and arid regions. These salts may then move up to the soil surface by capillary movement due to high evaporation rates resulting from the high temperatures of such regions.

The predominant ions at the site of weathering are carbonates, bicarbonates, sulfates, and chlorides of Ca, Mg, K, and Na. Often, as these salts move down the stream, those with low solubility are precipitated, while others undergo further changes through processes of exchange, adsorption, and differential mobility. The net result is that many Ca and Mg salts are removed, leaving high concentration of chloride and sodium ions. In some regions the ocean may be the source of salts, and parent material consists of marine deposits. The Mancos shales occurring in Colorado, Wyoming, and Utah are typical examples of saline marine deposits. Also the low-lying soils along the sea coasts, such as the Kari soils in Kerela (India), get their salts from the ocean. Sometimes, salt is moved inland through the transportation of spray by winds (cyclic salt) or by flooding following hurricanes. These soils are less extensive and have special problems. The discussion in this chapter is restricted to large areas, where the source of salts is surface and groundwaters.

When soils have an excess of chlorides and sulfates, during dry periods these salts are often deposited on the surface giving a white appearance. Such soils have been called "white alkali soils." Truly speaking, they are not alkali soils and should be called saline soils. The presence of high concentrations of carbonates and bicarbonates of Na is generally associated with dispersed humus in the surface soil, which appears black. Hence, the name "black alkali soils" (now known as sodic soils) has been given, because Na is the element responsible for several of their properties. It is desirable to differentiate between sodic and saline soils as the management practices required for them differ markedly. There are three criteria that are used for determining and classifying salinity and sodicity. These are (1) salt concentration, (2) sodium status, and (3) pH. A brief discussion on these follows.

7.2. CRITERIA FOR DETERMINING SALINITY/SODICITY

1. Salt concentration: Salt concentration in soil solution is determined based on the principle of the ability of salt to conduct electricity. This determination is thus made on a soil slurry or saturation extract with a conductivity meter. Electrical conductivity (EC_e or sometimes EC) is measured in decisemens per meter (dS m^{-1}). An earlier unit of measurement was millimhos per centimeter (mmhos cm^{-1}). Since 1 siemen(s) = 1 mho = 1000 mmhos, then 1 dS = 100 mmhos. Thus, 1 dS m^{-1} = 100 mmhos m^{-1} = 1 mmho cm^{-1}. Some useful conversion factors are given in Table 7.2.

Table 7.2 Some Useful Conversion Factors

• Conductivity 1 S cm^{-1} (1 mho/cm) = 1000 mS/cm (1000 mmhos/cm)

1 mS/cm^{-1} (1 mmho/cm) = 1 dS/m = 1000 μS/cm (1000 micromhos/cm)

• Conductivity to mmol (+) per liter:

$$mmol\ (+)/1 = 10 \times EC\ (EC\ in\ dS/m)$$

for irrigation water and soil extracts in the range of 0.1 to 5 dS/m.

• Conductivity to osmotic pressure in bars:

$$OP = 0.36 \times EC\ (EC\ in\ dS/m)$$

for soil extracts in the range of 3 to 30 dS/m.

• Conductivity to mg/1:

$$mg/1 = 0.64\ (EC\ in\ dS/m)$$

or

$$mg/1 = 640 \times EC$$

for waters and soil extracts having conductivity up to 5 dS/m.

• mmol/1 (chemical analysis) to mg/1: multiply mmol/1 for each ion by its molar weight and obtain the sum.

Note: The SI unit of conductivity is siemens (symbol S) per meter. The equivalent non-SI unit is mho and 1 mho = 1 siemens. Thus for those not used to the SI system mmhos/cm can be read for dS/m without any numerical change.

From Abrol et al., 1988. FAO Soils Bull. 39:18. With permission from the Food and Agriculture Organization of the United Nations.

2. Sodium status: Sodium status is determined as the ratio of exchange-able Na^+ to the total cation exchange capacity of the soil; the term used is exchangeable sodium percentage (ESP). Thus:

$$ESP = \frac{\text{Exchangeable Sodium (cmol kg}^{-1}) \times 100}{\text{Cation Exchange Capacity (cmol kg}^{-1})}$$

When the ESP value is 15, the soil pH is 8.5 or above. Higher ESP values increase soil pH to 10.0.

Since determination of ESP is time consuming, a more easily measured characteristic is the sodium absorption ratio (SAR), which gives the ratio of the concentrations of Na^+, Ca^{2+}, and Mg^{2+} in an extract usually obtained by suction from a satured soil. It is calcu-lated as follows:

$$SAR = [Na]^+ \Big/ \left[1/2 \left(Ca + Mg \right)^{2+} \right]^{1/2}$$

where $[Na^+]$, $[Ca^{2+}]$, and $[Mg^{2+}]$ are concentrations of these ions in cmol kg^{-1} soil. The relationship between ESP and SAR is shown in Figure 7.1. The SAR is also used to characterize the sodicity hazard of irrigation water added to soils.

3. pH: A soil is said to be an alkali soil or "sodic soil" when the pH of the saturation extract is above 8.5. For making this determination, soil is placed in a beaker and distilled water is gradually added while stirring the contents of the beaker until a saturated soil paste is made. A suction filter is then used to obtain a sufficient amount of the extract (often the same extract is used for SAR determination).

Soil water content in the field normally fluctuates between the permanent-wilting percentage (lower end) and field capacity (upper end); the upper end is approximately two times the lower end. Measurements in soil indicate that, over a considerable textural range, soil water content at saturation percentage (SP) is approximately four times that at 1.5 MPa (permanent wilting percentage) (Table 7.3) (USDA, 1954). Thus the soluble salt concentration measured in saturation extract tends to be nearly one-half the concentration of the soil solution at field capacity and about one-fourth of the concentration at perma-nent-wilting percentage. The salt dilution effect that occurs due to high water-retention capacity of fine-textured soils is thus accounted for. EC_e is therefore considered a reasonable approximation of the salt concentration that the grow-ing plants encounter in soils.

7.3. CLASSIFICATION

Salt-affect soils can be broadly classified as below:

1. Saline soils: These are the soils that contain sufficient amounts of neutral soluble salts (chlorides and sulfates of Na, Ca, and Mg) to

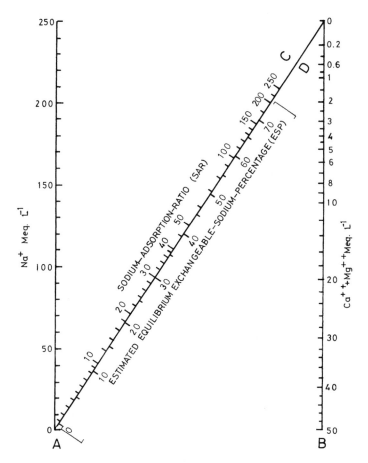

Figure 7.1. Nomogram for determining the SAR value of a saturation extract and for estimating the corresponding ESP value of soil at equilibrium with the extract (From USDA, 1954).

adversely affect the growth of most crop plants. EC_e of such soils is more than 4 dSm^{-1} and SAR is less than 13 to 15 (ESP less than 15). pH of saturation extract of these soils is less than 8.5 (Table 7.4).

2. Sodic soils: These soils (also known as alkali soils) contain sufficient sodium salts to give a saturation extract pH above 8.5. EC_e of these soils is less than 4 dSm^{-1}, and SAR is more than 13 to 15 (ESP more than 15) (Table 7.4). Sodium concentration in these soils could be very high, and the soil pH may rise to 10.0. Soils with SAR values between 5 and 13 to 15 may exhibit various aspects of sodicity, such as alkaline pH, moderate dispersion and crusting, and reduced infiltration.

3. Saline-sodic soils: These soils have mixed characteristics of both saline and sodic soils. While soil pH is less than 8.5, the EC_e may be

Table 7.3 Relation of Saturation Percentage (SP) to 1.5 MPa Water Content as Influenced by Soil Texture

Soil group	Soil samples (number)	1.5 MPa			SP			SP/1.5 MPa		
		Min.	Max.	Av.	Min.	Max.	Av.	Min.	Max.	Av.	SD
Coarse	10	3.4	6.5	5.0	16.0	43.1	31.8	4.68	8.45	6.37	1.15
Medium	23	6.6	14.2	10.8	26.4	60.0	42.5	3.15	5.15	3.95	.48
Fine	11	16.1	21.0	18.5	41.8	78.5	59.5	2.03	4.26	3.20	.60
Organic	18	27.6	51.3	37.9	81.0	255	142	2.53	4.97	3.66	.75

Adapted from USDA, 1954. SD, standard deviation.

Table 7.4 Important Characteristics of Saline, Sodic, and Saline-Sodic Soils

Soil	Saturation extract pH	Electrical conductivity (dS m^{-1})	Sodium adsorption
Saline	<8.5	>4	<13–15
Saline-sodic	<8.5	>4	>13–15
Sodic	>8.5	<4	>13–15

more than 4 dSm^{-1} and SAR more than 13 to 15 (ESP more than 15) (Table 7.4).

7.4. RECLAMATION AND MANAGEMENT OF SALINE SOILS

Saline soils are sometimes recognized by the presence of a white salt crust on the surface during hot summer months. However, gypsiforous soils may also have a white crust, but the restricted solubility of gypsum limits EC$_e$ to about 2.8 dSm^{-1}. Also, as mentioned earlier, some black mellow soils may be saline if salts have hydrolyzed and precipitated humic materials in the surface. During the crop growth period saline soils are generally characterized by spotty growth of crop plants, often with a blue-green tinge. In the field there may be barren spots and plant growth is generally stunted. Moderate salinity can often go undetected because it causes no apparent injuries. Succulent leaves with a darker, blue-green color could be an indication, but the final judgment can be made after soil analysis. Plants in salt-affected soils often have symptoms similar to those for stress (drought) conditions, although plants may not wilt because the osmotic potential of the soil solution usually changes gradually and plants adjust their internal salt content to maintain turgor and thus avoid wilting.

The reclamation of saline soils centers around removal of excess salt from these soils. Methods commonly adopted are scraping, flushing, leaching, and drainage.

7.4.1. Scraping

This refers to the mechanical removal of salts using the available tools. Disposal of scraped salts poses a problem. Thus this method has limited applicability.

7.4.2. Flushing

Washing away the salts by flushing water over the surface can be and is sometimes used to desalinize soils, but has limited application because only a small fraction of accumulated salt can be flushed away; a large part moves down the profile with water.

7.4.3. Leaching

Leaching salts out of the active root zone of crops is the most effective way to reclaim saline soils. For this purpose the first and foremost requisite is to have a reliable estimate of the quantity of water required to accomplish leaching of salts. The major factors determining the amount of water needed for leaching are (1) the initial salt content of the soil; (2) the desired level of salt content for good growth of crop plants; (3) the depth to which reclamation is required; (4) soil characteristics such as texture, permeability, etc.; and (5) the crop and its variety to be grown. Where groundwater tables are within a few m of the soil surface, leaching without drainage will have little lasting effect on soil salinity.

A useful rule of thumb is that a unit depth of water will remove nearly 80% of salts from a unit soil depth (Abrol et al., 1988). Thus 1 m-ha water will remove 80% of salts from the top 1 m of soil from one hectare of land. However, soil properties, particularly texture, are important and for reliable estimates it is desirable to conduct salt-leaching tests on a limited area and prepare leaching curves. Results of such a test in Iraq (Dieleman, 1963) are shown in Figure 7.2. These data show that for reducing salt content to 20% of the initial amount (removal of about 80% of the salts), the depth of leaching water required per unit depth of soil was less than 0.5 on Dujailah (silt loam, loam), about 0.75 at Dalmaj (clay to silty clay), and about 1.6 at Annanah (silty clay).

Regarding methods of leaching, sprinkling is better than flooding. Because of a slower wetting rate under sprinkling, the zone of complete leaching at the end of irrigation extends more deeply into the profile than under flood irrigation. Results comparing sprinkler vs. flooding confirm that the salts removed per unit quantity of water leached can be increased appreciably by leaching at soil water contents of less than saturation (Figure 7.3). Also when flooding is used as a method of leaching salts, more salts move upward and accumulate in the soil surface on evaporation (as shown by dashed line in Figure 7.3). Nielsen et al. (1966) showed that 25 cm of sprinkled water reduced the salinity of the upper 60 cm of soil to the same degree as 75 cm of ponded water.

7.4.4. Drainage

One of the most important requirements of the management of the saline soils is that the desired salt concentration in the root zone achieved by leaching is maintained for long periods. To achieve this, evaporation from groundwater must be prevented by keeping the groundwater table below the depth that will cause rapid soil salinization. Provision of adequate drainage is the only way to control the groundwater table. In addition to surface drainage, adequate subsurface drainage is essential. This is normally achieved by the use of open

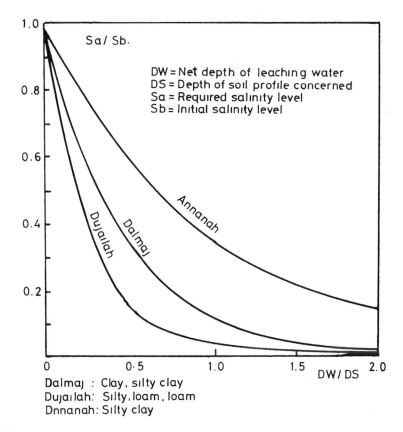

Dalmaj : Clay, silty clay
Dujailah: Silty, loam, loam
Dnnanah: Silty clay

Figure 7.2. Typical leaching curves for soils in Iraq (From Dieleman, 1963. International Institute for Land Reclamation and Improvement, Wageningen. With permission from Food and Agriculture Organization of the United Nations.)

drain ditches or with buried tiles. For dryland saline seeps, salinity buildup from seepage of salt-laden water is prevented by cropping the recharge area above the seep with deep-rooted perennial crops such as grasses or alfalfa (Brown et al., 1982).

When irrigation is available and used for crop production, careful planning can help considerably to overcome the salinity problem. When furrow irrigation is practiced, most salts accumulate on the top of the ridge. Planting seeds on the sides of the ridges can help to overcome the salinity problem and permit satisfactory germination. Thus where soil and farming practices permit, furrow planting may help in obtaining better crop stands and yields.

7.5. CROP PRODUCTION ON SALINE SOILS

Plant tolerance to salinity is usually appraised in one of three ways: (1) the ability of a plant to survive on saline soils; (2) the absolute plant growth or

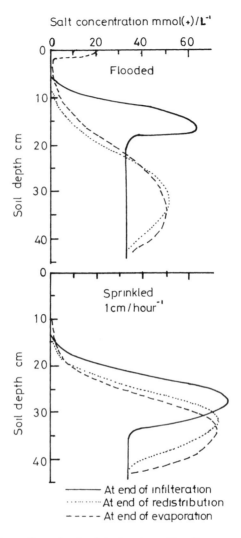

Figure 7.3. Effect of method of irrigation (flooding vs. sprinkler) and water redistribution following irrigation and evaporation on the salt concentration profiles (From Bresler and Hanks, 1969. Soil Sci. Soc. Am. Proc. 33:827–832. With permission of SSSA, Madison, WI.)

yield; and (3) the relative growth on saline soil as compared with that on nonsaline soils (Maas, 1986). Regarding the ability of plants to survive on saline soils, it may be pointed out that salinity affects plants at all stages of development, but may vary from one growth stage to another. For example, sugarbeet is more sensitive during germination, while rice, barley, wheat, corn, sorghum, and cowpea are more sensitive during early seedling growth. The

Table 7.5 Relative Salt Tolerance of Various Crops at Emergence and During Growth to Maturity

| Crop | | Electrical conductivity of saturated soil extract | |
Common name	Botanical name	50% Yield (dS m^{-1})	50% Emergence (dS m^{-1})
Barley	*Hordeum vulgare*	18	16–24
Cotton	*Gossypium hirsutum*	17	15
Sugarbeet	*Beta vulgaris*	15	6–12
Sorghum	*Sorghum bicolor*	15	13
Safflower	*Carthamus tinctorius*	14	12
Wheat	*Tritcum aestivum*	13	14–16
Beet, red	*Beta vulgaris*	9.6	13.8
Cowpea	*Vigna unguiculata*	9.1	16
Alfalfa	*Medicago sativa*	8.9	8–13
Tomato	*Lycopersicon lycopersicum*	7.6	7.6
Cabbage	*Brassica oleracea capitata*	7.0	13
Corn	*Zea mays*	5.9	21–24
Lettuce	*Lactuca sativa*	5.2	11
Onion	*Allium Cepa*	4.3	5.6–7.5
Rice	*Oryza sativa*	3.6	18
Bean	*Phaseolus vulgaris*	3.6	8.0

From Maas, 1986. App. Agric. Res. 1:12–26. With permission of Springer-Verlag, Berlin, Germany.

relative salt tolerance of some crops at emergence, as well as in relation to yield, is given in Table 7.5.

Maas (1986) pointed out that yield response curves provide two essential parameters sufficient for expressing salt tolerance: (1) threshold — the maximum allowable salinity without yield reduction below that for nonsaline conditions; and (2) slope — the percent yield decrease per unit increase in salinity beyond the threshold. The division for classifying crop tolerance to salinity can then be made as shown in Figure 7.4. Based on these criteria, a list of some food, fiber, forage, and vegetable crops is given in Table 7.6.

7.6. RECLAMATION AND MANAGEMENT OF SODIC SOILS

By definition sodic soils are those where sodium salts dominate, exchangeable sodium percentage (ESP) is above 15 (SAR 13 to 15), and pH is above 8.5. After a thorough survey of published data, Abrol et al. (1980) observed that for sodic soils most often an ESP of 15 to 20 is associated with a saturation paste pH of 8.2. Therefore this pH value would be more realistic for diagnostic purposes; a saturation paste pH of 8.5 is generally associated with higher

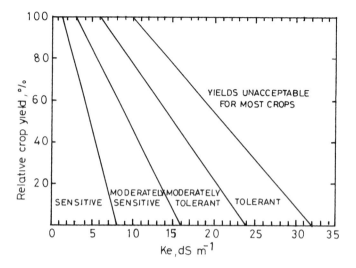

Figure 7.4. Divisions for classifying crop tolerance to salinity. Ke is EC$_e$, i.e., electrical conductivity of saturation extract (From Maas, 1986. App. Agric. Res. 1:12–26. With permission of Springer-Verlag, Berlin, Germany.)

values of ESP. Diagnosis at an earlier stage of sodicity is desirable. If soil pH is determined in a 1:2 soil to solution ratio, the diagnostic pH limit for sodic soils should be 9.0 instead of 8.2 as suggested for saturation paste pH (Abrol et al., 1988).

Excess exchangeable sodium in soils adversely affects plant growth in the following ways:

1. Excess exchangeable sodium has a marked adverse effect on the soil's physical properties; the soil tends to become deflocculated, and soil permeability is considerably reduced.
2. Excess exchangeable Na$^+$ increases soil pH, which (a) lowers the solubility of Ca and Mg carbonates, and these nutrients may become deficient; (b) increases the availability of Mo and Bo, which may reach toxic levels; and (c) decreases the availability of Zn, and this deficiency may limit plant growth.
3. Excess exchangeable sodium results in dispersion of surface soils, reduced aggregation and infiltration rate, and increased soil resistance and crusting. A general relationship between ESP and sodicity hazard is shown in Table 7.7.

Reclamation of sodic soils therefore aims at reducing the exchangeable Na to the extent that it does not degrade soil physical properties or interfere with the plant growth. For this purpose a number of soil amendments are used,

Table 7.6 Salt Tolerance of Some Food, Fiber, Forage, and Vegetable Crops

Crop	EC$_e$ of saturation extract	
	Threshold (dS m^{-1})	Slope (%/dS m^{-1})
Tolerant		
Barley (*Hordeum vulgare*)	8.0	5.0
Sugarbeet (*Beta vulgaris*)	7.0	5.9
Wheat-durum (*Triticum turgidum*)	5.9	3.8
Cotton (*Gossypium hirsutum*)	7.7	5.2
Bermuda grass (*Cynodon dactylon*)	6.9	6.4
Alkali grass (*Puccinellia airoides*)	—	—
Kallar grass (*Diplachne fusca*)	—	—
Salt grass (*Districhlis stricta*)	—	—
Wheat grass (*Agropyron cristatum*), fairway crested	7.5	6.9
Asparagus (*Asparagus officinalis*)	4.1	2.0
Moderately tolerant		
Sorghum (*Sorghum bicolor*)	6.8	1.6
Soybean (*Glycine max*)	5.0	20
Fescue, tall (*Festuca elatior*)	3.9	5.3
Hardingrass (*Phalaris tuberosa*)	4.6	7.6
Wheat grass (*Agropyron sibricum*), standard crested	3.5	4.0
Moderately sensitive		
Broad bean (*Vicia faba*)	1.6	9.6
Corn (*Zea mays*)	1.7	12
Peanut (*Arachis hypogaea*)	3.2	29
Sugarcane (*Saccarum officinarum*)	1.7	5.9
Alfalfa (*Medicago sativa*)	2.0	7.3
Orchard grass (*Dactylis glomerata*)	1.5	6.2
Clovers, ladino, red strawberry (*Trifolium repens, T. pratense, T. fragiferum*)	1.5	12
Broccoli (*Brassica oleracea botrytis*)	2.8	9.2
Cabbage (*Brassica oleracea capitata*)	1.8	9.7
Celery (*Apium graveolens*)	1.8	6.2
Sweet corn (*Zea mays*)	1.7	12

**Table 7.6 Salt Tolerance of Some Food, Fiber, Forage,
and Vegetable Crops (Continued)**

	EC$_e$ of saturation extract	
Crop	Threshold (dS m^{-1})	Slope (%/dS m^{-1})
Cucumber (*Cucumis sativus*)	2.5	13
Lettuce (*Lactuca sativa*)	1.3	13
Pepper (*Capsicum annum*)	1.5	14
Potato (*Solanum tuberosum*)	1.7	12
Radish (*Raphanus sativus*)	1.2	13
Spinach (*Spinacia oleracea*)	2.0	7.6
Sweet potato (*Ipomea batatas*)	1.5	11
Tomato (*Lycopersicon lycopersicum*)	2.5	9.9
Turnip (*Brassica rapa*)	0.9	9.0
Sensitive		
Bean (*Phaseolus vulgaris*)	1.0	19
Carrot (*Daucus carota*)	1.0	14
Onion (*Allium cepa*)	1.2	16

Adapted from Maas, 1986.

**Table 7.7 Exchangeable Sodium Percentage (ESP)
and Sodicity Hazard**

ESP	Sodicity hazard
<15	None to slight
15–30	Light to moderate
30–50	Moderate to high
50–70	High to very high
>70	Extremely high

From Abrol et al., 1988. FAO Soils Bull. 39:18. With permission from the Food and Agriculture Organization of the United Nations.

which include gypsum, calcium chloride, elemental sulfur, sulfuric acid, iron sulfate, aluminum sulfate, and pyrite. Even hydrochloric acid, available as a byproduct of the caustic soda industry, has been suggested as an amendment (Ahmad et al., 1986).

7.6.1. Gypsum

Gypsum is a mineral that occurs extensively as a natural deposit in semi-arid and arid regions. For use as an amendment it needs to be ground to a reasonable particle size (to pass about a 2-mm mesh). The chemical formula is $CaSO_4 \cdot 2H_2O$. Gypsum has a low solubility (0.25% in water) and therefore does not easily leach out. When applied and incorporated into soil (15 cm of surface soil is considered enough; incorporation into deeper layers may be less advantageous as part of it may be utilized in neutralizing soluble carbonates at the expense of replacement of exchangeable Na), it reacts with sodium salt, as well as replaces exchangeable Na, which is then leached out as sodium sulfate. The following reactions occur in soil:

a. $Na_2CO_3 + CaSO_4 \rightarrow CaCO_3 + Na_2SO_4 \text{(leachable)}$

b. $\frac{Na}{Na} [\text{clay micelle}] + CaSO_4 \rightarrow Ca [\text{clay micelle}] + Na_2SO_4 \text{(leachable)}$

Both the above reactions are reversible, so adequate leaching arrangements have to be made to leach out sodium sulfate.

Sandoval et al. (1972) demonstrated that for sodic (solinetzic) soils containing a gypsum layer within 0.7 m of the surface, deep plowing would bring the gypsum to the surface, reducing ESP considerably and permitting profitable crop production. Several million ha of such soils exist in the Great Plains, and thousands of ha have been deep plowed in Alberta, Canada.

7.6.2. Sulfur

When elemental powdered sulfur (yellow) is applied in soil, it is oxidized by sulfur oxidizing bacteria (*Thiobacillus thiooxidans*) to SO_3^{2-}, which, when dissolved in water, produces sulfuric acid. Sulfuric acid then reacts with calcium carbonate (which is generally present in such soils) and forms calcium sulfate. The following reactions occur:

a. $2S + 3O_2 \rightarrow 2SO_3$

b. $SO_3 + H_2O \rightarrow H_2SO_4$

c. $H_2SO_4 + CaCO_3 \rightarrow CaSO_4 + H_2O + CO_2$

d. $\frac{Na}{Na} [\text{clay micelle}] + CaSO_4 \rightarrow Ca [\text{clay micelle}] + Na_2SO_4 \text{(leachable)}$

7.6.3. Pyrite

Pyrite (FeS_2) is a waste material of the steel industry and can therefore be available to farmers at a fairly low cost. This material has been widely used in India. Naturally occurring pyrite minerals also occur. Pyrite oxidation (see Chapter 11 on sulfur) produces H_2SO_4, which reacts with $CaCO_3$ in soil as explained above and produces $CaSO_4$.

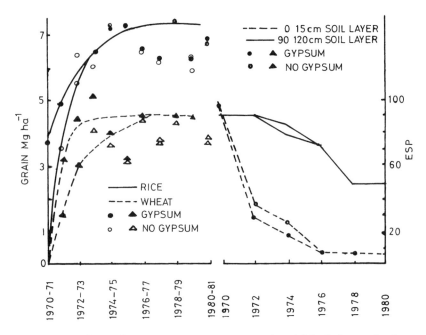

Figure 7.5. Effect of gypsum treatment on grain yield of rice and wheat and exchangeable sodium percentage (ESP) in a sodic soil from 1970 to 1980. (Drawn from the data by Singh and Abrol, 1988).

As regards their effects on soil properties and yield of rice and wheat, pyrite was found to be about one-fourth as effective as gypsum (Verma and Abrol, 1980a,b).

7.7. CROP PRODUCTION ON SODIC SOILS

Crops differ considerably in their sensitivity to sodicity. In general, grain legumes (beans) are very sensitive; cereals (other than rice and barley), cotton, sugarcane, and forage legumes such as *Trifolium alexandrinum* and *Melilotus parviflora* are less sensitive; and grasses, however, such as Bermuda grass (*Cynodon dactylon*), Rhodes grass (*Chloris gayana*), Para grass (*Brachiaria mutica*), Karnal grass (*Diploachne fusca*), barley, rice, alfalfa (*Medicago sativa*), green manuring legume dhaincha (*Sesbania aculeata*), and sugarbeet are fairly tolerant.

The tolerance of rice is primarily due to its ability to grow in standing water. Adding large masses of water lowers the concentration of sodium in the soil solution and thereby lowers pH to near neutrality (see Figure 6.3). The adverse effects of sodicity in lowering soil permeability are also to the advantage of the rice crop. Long-term field studies in India have shown that when rice was included in the crop sequence, there was a gradual reduction in sodicity so that in a period of about 10 years the upper 1.2 meters of soil was nearly free of the sodicity problem (Singh and Abrol, 1988) (Figure 7.5).

Amunson and Lund (1985) measured selected properties of a naturally saline and sodic soil in San Joaquin Valley, California, in pedons that had been reclaimed a minimum of 0, 5, 8, 15, or 25 years. More than 7.2×10^4 kg ha^{-1} of soluble salt was leached from the upper 1 m of the soil profile after 5 years, with a little or no additional removal with time. A steady-state soil solution composition was reached after approximately 15 years, as indicated by EC values of the saturation extracts. The authors also found that the smectite minerals were unstable in the reclaimed soils and were transformed to kaolinite, which thus resulted in the reduced dispersibility of the clay fraction and which may influence water infiltration in these soils.

REFERENCES

Abrol, I.P., R. Chhabra, and R.K. Gupta. 1980. A fresh look at the diagnostic criteria for sodic soils. Proc. Int. Symp. Salt Affected Soils, Central Soil Salinity Research Institute, Karnal, India, February 18–21, 1980, pp. 142–147.

Abrol, I.P., J.S.P. Yadav, and F.I. Masoud. 1988. Salt affected soils and their management. FAO. Soils Bull. 39:131.

Ahmad, M., B.H. Niazi, Rahmatullah, and G.R. Sandhu. 1986. Reclamation of calcareous saline sodic soils by gypsum and HCl under rice cultivation. Trop. Agric. (Trinidad) 63:271–272.

Amundson, R.G. and L.J. Lund. 1985. Changes in the chemical and physical properties of reclaimed saline-sodic soil in the San Joaquin Valley of California. Soil Sci. 140:213–222.

Bresler, E. and Hanks, R.J. 1969. Numerical method for estimating simultaneous flow of water and salt in unsaturated soils. Soil Sci. Soc. Am. Proc. 33:827–832.

Brown, P.L., A.D. Halvorson, F.H. Siddoway, H.F. Mayland, and M.R. Miller. 1982. Saline seep diagnosis, control and reclamation. USDA Conservation Res. Rep. No. 30, 22 pp.

Dieleman, P.J., Ed. 1963. Reclamation of salt affects soils in Iraq. International Institute for Land Reclamation and Improvement, Wageningen, Pub. No. 11, p. 175.

Francois, L.E., E.V. Maas, T.J. Donovan, and V. L. Youngs. 1986. Effect of salinity in grain yield and quality, vegetative growth, and germination of semi-dwarf and durum wheat. Agron. J. 78:1053–1058.

Grattan, S.R., and E.V. Maas. 1984. Interactive effects of salinity and substrate phosphate on soybean. Agron. J. 76:668–676.

Maas, E.V. 1986. Salt tolerance of plants. App. Agric. Res. 1:12–26.

Massoud, F.I. 1977. Basic principles for prognosis and monitoring of salinity and sodicity. Proc. Int. Conf. Managing Saline Water for Irrigation. Texas Tech. Univ., Lubbock, August 16–20, 1976, pp. 432–454.

Nielson, D.R., J.W. Bigger, and J.N. Luthin. 1966. Desalinization of soils under controlled unsaturated conditions. Sixth Congress Int. Comm. on Irrigation and Drainage, New Delhi, India.

Sandoval, F.M., J.J. Bond, and G.A. Reichman. 1972. Deep plowing and amendment effect on a sodic clay pan soil. Trans. ASAE 15:681–687.

Singh, S.B. and I.P. Abrol. 1988. Long-term effect of gypsum application and rice-wheat cropping on changes in soil properties and crop yield. J. Indian Soc. Soil Sci. 36:316–319.

USDA. 1954. Diagnosis and improvement of saline and alkali soils. U.S. Dept of Agriculture, Washington, DC. Agric. Handbook No. 60, p. 160.

Verma, K.S. and I.P. Abrol. 1980a. Effect of gypsum and pyrites on soil properties on a highly sodic soil. Indian J. Agric. Sci. 50:844–851.

Verma, K.S., and I.P. Abrol. 1980b. Effect of gypsum and pyrite on yield and chemical composition of rice and wheat grown in a highly sodic soil. Indian J. Agric. Sci. 50:935-942.

8   NITROGEN

We live immersed in an atmospheric sea of 79% nitrogen (N), yet N is the nutrient that most often limits biological productivity.

Nitrogen occurs in many forms in agricultural ecosystems because it can exist in a number of valence states within an ecosystem. The valence state in which nitrogen exists depends primarily on the ambient environment at micro-sites within the soil. The transformations and flow of nitrogen from one valence state to another constitute the basics of the nitrogen cycle (Figure 8.1). For example, atmospheric N_2 (0 charge) gas is converted by lightening to various oxides and finally to nitrate (+5 charge), which falls with the rain and is taken up by growing plants. Also N_2 gas can be converted to ammonia (–3 charge) by microbial fixation, with the NH_3 being used in various biochemical reactions within the plant. When plant residues decompose, much of the N they contain will undergo several microbial conversions and will eventually end up back as nitrates. Also under anaerobic conditions, nitrates can be reduced to various oxides and ultimately to N_2 gas again. Nitrogen in inputs such as fertilizers and manures is also subjected to these same microbial transformations. As indicated by Figure 8.1, the N cycle is very complex and is influenced by many factors.

A global inventory of N in the biosphere shows that N is distributed in terrestrial, oceanic, and atmospheric components in the ratio 1:70:11818 (Table 8.1). Thus the bulk of the biospheric N is in the atmosphere. The atmospheric column on an hectare of land will contain approximately 8.4×10^4 Mg ha^{-1} N. Yet for growing most cereals and nonlegume forage crops, one has to apply large amounts of manure and/or fertilizer N. Nitrogen in soil originates from plant and animal residues, from fixation by legumi-nous plants and trees, and from components of rain such as nitrates (Figure 8.1). Total N content in the top 15 to 20 cm of surface soils ranges from 0.01% (or even less in desert soils) to more than 2.5% in peats. N content in the subsurface of any soil is generally less than in the surface layer since most organic residues are deposited on the soil surface. Under suitable conditions some of the organic nitrogen mineralizes into inorganic form and eventually may be present in ammonium (NH_4^+) and nitrate (NO_3^-) form.

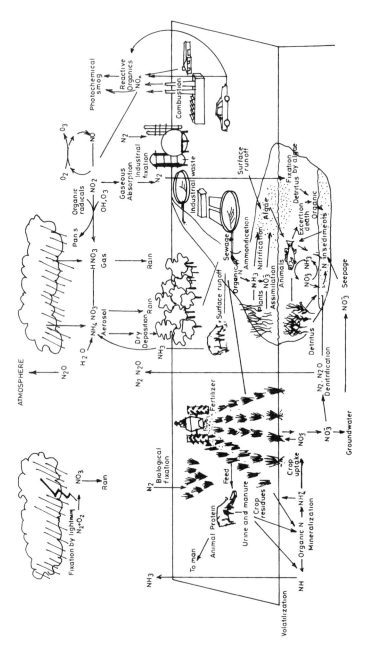

Figure 8.1. Schematic representation of the nitrogen cycle, emphasizing human activities that affect fluxes of nitrogen. (From NAS, 1978.)

Table 8.1 Global Inventories of Nitrogen in the Biosphere (Million Mg N)

Terrestrial

Plant biomass	$1.1–1.4 \times 10^4$
Animal biomass	2×10^2
Litter	$1.9–3.3 \times 10^3$
Soil organic matter	3×10^5
Soil insoluble inorganic	1.6×10^4
Soil soluble inorganic	3×10^3
(Soil microorganism included in total soil organic matter)	5×10^2

Oceanic

Plant biomass	3×10^2
Animal biomass	1.7×10^2
Dissolved organic matter	5.3×10^5
Particulate organic matter	$0.3–2.4 \times 10^4$
N_2 dissolved	2.2×10^7
N_2O dissolved	2×10^2
NO_3^- dissolved	5.7×10^5
NO_2^- dissolved	5×10^2
NH_4^+ dissolved	7×10^3

Atmospheric

N_2	3.9×10^9
N_2O	1.3×10^3
NH_3	0.9
NH_4^+	1.8
NO_x	1–4
NO_3	0.5
Organic N	1

From Winteringham, F.P.W. 1980. *Soil N as Fertilizer or Pollutant.* With permission of IAEA.

However, inorganic N in soil at any moment is only a small fraction of total soil N. The bulk of soil N in a surface soil is present in the organic form. In fields where chemical fertilizer is not applied, nonleguminous crop plants obtain most of their nitrogen from soil organic N after its mobilization. In subsoils (especially those with illitic clays) much of the nitrogen present is trapped in the clay lattice as nonexchangeable ammonium and is largely unavailable for plant uptake.

8.1. SOIL ORGANIC N

Soil organic N consists of proteins (20 to 40%), amino sugars such as the hexosamines (5 to 10%), purine and pyrimidine derivatives (1% or less), and complex unidentified compounds formed by reaction of ammonium with lignin, polymerization of quinones with nitrogen compounds, and condensation of sugars and amines. Some of these have already been discussed in the chapter on organic matter. Part of the organic N is also present as clay-humus complexes, which are resistant to decomposition. This would also explain why only a very small part of immobilized fertilizer N becomes available to the growing crop plants.

8.2. MINERALIZATION OF SOIL ORGANIC NITROGEN

Mineralization of soil organic nitrogen is the microbial process by which organic forms of N in soils are converted to inorganic forms (ammonium, nitrite, and nitrate). Mineralization takes place in three step-by-step reactions, namely, aminization, ammonification, and nitrification. Of these three reactions, the first two are carried out by heterotrophic microorganisms, while the third one is carried out by autotrophic bacteria. Heterotrophs derive their energy from oxidation of organic carbon compounds, while autotrophs obtain their energy from specific inorganic salts and their carbon from bicarbonate salts in the soil. Organic N in soils is ultimately derived from decomposition of plant material returned to the soil. This organic N may be present in both relatively labile forms (crop residues and microbial biomass) and in organic compounds more resistant to decomposition (lignoproteins, various types of humates, and condensed cyclic molecules).

8.2.1. Aminization

The heterotrophs, including bacteria, fungi, and actinomycetes, break down complex organic molecules releasing amines and amino acids; this process is known as aminization. Bacteria and actinomycetes often dominate in neutral and alkaline conditions, while fungi are more active under acid conditions. Most N undergoing aminization during a growing season originates from degradation of proteins and amino acids in decomposing crop residues and microbial cells, with lesser amounts originating from decomposition of the more resistant sources such as lignoproteins and humates.

8.2.2. Ammonification

Ammonification consists of the biological processes by which organic forms of soil nitrogen are converted to ammonia or ammonium ions. The final reaction in these processes is the hydrolysis of amino groups. The amines and amino acids released in aminization are reacted upon by other heterotrophs, which release N in the inorganic NH_4^+ form. Both aerobic and anaerobic microorganisms are capable of carrying out this reaction. Also a very diverse population of bacteria, fungi, and actinomycetes is capable of releasing ammonium. The ammonium released may be

1. Lost by ammonia volatilization
2. Utilized by plants
3. Absorbed on the exchange complex of clay minerals
4. Fixed in the crystal lattice of 2:1 expanding clay minerals
5. Immobilized by soil microorganisms
6. Nitrified

Ammonification can occur in both aerobic and anaerobic environments, although rates are generally more rapid in the aerobic environment.

8.2.3. Nitrification

This is a two-step process. In the first step ammonium is converted to nitrite $(NO_2)^-$, and in the second step nitrite is converted to nitrate $(NO_3)^-$. A group of obligate autotrophic bacteria known as *Nitrosomonas* is responsible for the first step, that is, conversion of ammonium to nitrite. Conversion of nitrite to nitrate is carried out by another group of obligate autotrophic bacteria known as *Nitrobacter*. It should be mentioned that, although *Nitrosomonas* and *Nitrobacter* are the most important organisms responsible for the reactions mentioned above, a few heterotrophs can also carry out these reactions, usually at much lower rates. Nitrates so formed may be

1. Taken up by plants.
2. Lost by leaching — creating health hazards by increasing nitrate concentration in underground water.
3. Under anaerobic conditions lost by denitrification — creating atmospheric pollution problems. Nitrous oxide (N_2O), one of the products of denitrification, is involved in depletion of the ozone layer.
4. Immobilized by soil microorganisms.

8.3. FACTORS AFFECTING NITRIFICATION

With the exception of rice most of our agricultural production is from well-drained soils, which favor nitrification. Nitrates therefore dominate as the form in which inorganic N is present in soil. A major exception is land continuously under perennial grasses where ammonium may be present in larger amounts. Therefore most crop plants have developed to utilize nitrate $(NO_3)^-$ as a major source of N. In recent years, however, considerable evidence has accumulated to show that even upland plants grow better when a mixture of ammonium NH_4^+ and nitrate NO_3-N are present.

It is obvious that the factors that have a pronounced effect on the activity of nitrifying bacteria will affect nitrification. Some of these factors are soil water content, aeration, pH, temperature, supply of ammonium, and population of nitrifying organisms.

8.3.1. Soil Water and Aeration

Nitrification is brisk in soils having adequate soil water and air; in most soils this is near field capacity.* Data on the effect of soil water potential on nitrification, as reported by Mahli and McGill (1982), are given in Table 8.2. At a water potential of 0 K Pa, that is, saturation, there is little or no air in soil pores, and nitrification completely ceases (100% water-filled pore space). In well-aerated soils nitrification increases as the water content increases from 1.5 M Pa (megapascal) potential (permanent wilting point) to −33 kPa (or −0.033 M Pa) (near field capacity), nitrification being greatest near field capacity. Again, data in Table 8.2 also indicate that at the same soil water potential, nitrification rate increased as soil clay content decreased (from silty clay loam to sandy loam). The latter soil had lesser water content and therefore better aeration. These relationships between soil water content and rates of aerobic microbial activity (such as nitrification) were shown graphically by Linn and Doran (1984) where soil water content was expressed as percent water-filled pore space (percent saturation). They found that ammonification and nitrification for the soils studied was maximum near 60% water-filled pore space, which corresponds closely to field capacity (Figure 8.2). Ammonification results from activity of many microbial species, both fungal and bacterial. However, nitrification is specific for a small group of bacteria. These bacteria are obligate aerobes and are more sensitive to water stress than some of the soil fungi. Consequently, as a soil becomes dry, often soil ammonium accumulates because ammonification proceeds more rapidly than nitrification.

8.3.2. Soil pH

Morrill and Dawson (1967) studied nitrification in 116 soils of the United States ranging in pH from 4.4 to 8.8 and observed four different patterns, which are as follows:

1. In some acid soils having pH < 5.39, NH_4^+ oxidized slowly to NO_3^- without the appearance of NO_2^-.

---

* Water content in soil when, after saturation, it is allowed to drain under gravitational pull until percolation ceases; this state may be reached at a soil water potential of −33 K Pa (or −0.033 M Pa) in medium- to heavy-textured soils to 0 to 10 K Pa (or −0.01 M Pa) in light, sandy soils. Also in dry soils ammonium and sometimes nitrite may accumulate, presumably because *Nitrobacter* is more sensitive to water stress than other microbes.

**Table 8.2 Effect of Soil Water Potential on Nitrification in
Three Canadian Soils**

Soil water potential (KPa)	Period of incubation (days)	Rate of nitrification (μg N g^{-1} day^{-1})		
		Cooking Lake	**Falun**	**Malmo**
0	8	0.0 d[a]	0.0 d	0.0
−33[b]	8	3.6 a	3.4 a	3.2 a
−700	11	2.0 b	2.1 b	2.0 b
−1500	17	1.3 c	1.3 c	1.2 c

Note: Cooking Lake is gray luvisol sandy loam; Falun is dark gray
chernozemic loam; and Malmo is black chernozemic clay loam.

[a] In each column, the values are significantly different (95% level of
probability) when not followed by the same letter.

[b] Water content at −33 KPa in Cooking Lake, Falun, and Malmo was 15.5,
20.7, and 39.7%, respectively.

From Malhi and McGill, 1982. Soil Biol. Biochem. 14:393–399. With
permission of Pergamon Press.

2. In some acid soils having pH $\leq$ 5.39, there was accumulation of
   NH_4^+ with very little oxidation to NO_2^- and NO_3^-.
3. In soils having pH 5.01 to 6.38, NH_4^+ and NO_2^- are rapidly oxidized
   to NO_3^-.
4. In soils having pH 6.93 to 7.85, NH_4^+ oxidized to NO_2^-, which
   accumulated for extended periods before being oxidized to NO_3^-.

They also observed that on liming acid soil patterns "1" and "2" identified
above changed to pattern "3." In another study (Dancer et al., 1973) observed
a linear relationship between the soil pH range of 4.7 to 6.4 and nitrification
(Figure 8.3). Effects of liming to obtain a desired pH confirmed this relationship (Figure 8.4).

Studies involving still wider pH ranges suggest a pH optimum for nitrate
formation between pH 6.0 and 9.4. Nitrification can continue even when pH
is greater than 9.4. Nitrite oxidation may be inhibited more than ammonium
oxidation at high pH values, sometimes resulting in accumulation of nitrites.

Nitrification of both soil and fertilizer nitrogen releases H$^+$ ions as shown
in the reaction below:

$$2\,NH_4^+ + 3\,O_2 \rightarrow 2\,NO_2^- + 2\,H_2O + 4\,H^+.$$

Nitrification is thus an acid-producing process per se, and continuous use
of ammonium or ammonium-producing fertilizers or even heavy dressings of
organic manures result in lowering the soil pH. (See also Chapter 6 on soil
acidity.)

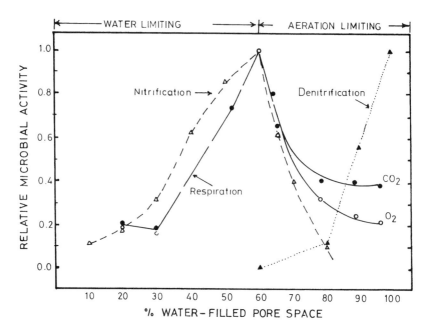

Figure 8.2. The relationship between water-filled pore space and relative amount of microbial nitrification (after Greaves and Carter, 1920), denitrification (after Nommik, 1956), and respiration (O_2 uptake, $\bigcirc$–$\bigcirc$, CO_2 production, ●–●) (Linn and Doran study). Data for nitrification originally expressed as percentage water-holding capacity. (From Linn and Doran, 1984. Soil Sci. Soc. Am. J. 40:1267–1272. With permission of SSSA.)

8.3.3. Soil Temperature

Soil temperature as a factor controlling nitrification has been a matter of concern in cooler climates, since the availability of native soil N depends very much on nitrification. Soils in cooler climates have, in general, much more organic matter than soils in warmer climates. Therefore the contribution of native soil N toward crop production could be substantial in cooler regions, with nitrification values of over 200 kg N ha[-1] being reported for some growing seasons.

The optimum temperature for nitrification in temperate soils is between 25 and 35°C (Sabey et al., 1959). However, Malhi and McGill (1982) from Canada reported an optimum of 20°C, while Schloesing and Muntz (1879) reported the optimum temperature for nitrification to be 37°C, with the process continuing to 46 to 50°C and ceasing at 55°C. The effect of temperature on nitrification varies with the climate, and this seems to be well supported when comparing data of Malhi and McGill (1982), Myers (1975), and Sabey et al. (1959) (Figure 8.5). Thus there is a climatic selection of species of nitrifiers,

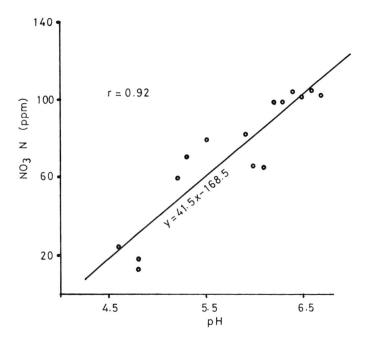

Figure 8.3. Relation between soil pH and NO$_3^-$-N accumulation in soils treated with 100 ppm of NH$_4^+$-N as (NH$_4$)$_2$SO$_4$ and incubated for 15 days at 23°C. (From Dancer et al., 1973. Soil Sci. Am. Proc. 31:67–69. With permission of SSSA.)

with those from cooler climates having lower temperature optima and less heat tolerance than species from warmer regions.

Nitrification has been shown to proceed even at 0°C (Eggleton, 1935). Thus nitrifying bacteria appear to be resistant to freezing at least in natural environments, as nitrification may be rapid when frozen soils thaw. Nitrates may also be redistributed in soil profile by internal drainage after thaw.

8.3.4. Supply of Ammonium

Since ammonium is the substrate that is nitrified, an adequate supply is essential for nitrification. Factors such as ammonia volatilization loss reduce the amount of ammonium and thereby the amount of nitrates formed. Similarly, rapid immobilization of ammonium by soil microorganisms, especially when organic residues with wide C:N ratio are incorporated, will also reduce nitrification.

Very high ammonium concentrations may retard nitrification; the maximum tolerable amounts could be as high as 800 μg g^{-1} soil. The depressing effect of high concentrations of ammonium may be attributed to toxic levels of ammonia, salt effects, or to lowering of pH when ammonium sulfate is added as a source of N. Because of the toxic effect of high NH$_4^+$ concentrations, application of

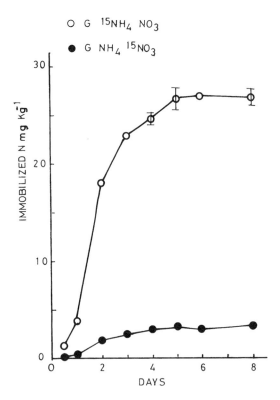

Figure 8.7. Immobilization of N in a loamy soil (soil I) incubated at 10°C with 500 mg C-glucose kg⁻¹ soil. 100 mg N kg⁻¹ soil applied as $^{15}NH_4 \cdot NO_3$ or $NH_4{}^{15}NO_3$. Values are the mean of two replications. (From Recous and Mary, 1990. Soil Biol. Biochem. 22:913–922. With permission from Elsevier Science Ltd.)

and fixed ammonium has been reported (Sparks et al., 1979; Jensen et al., 1989). Jensen et al. (1989) separated clay and silt fractions of soil by ultrasonic technique and showed that in four Danish soils fixed ammonium in the clay fraction varied from 255 to 430 μg N g⁻¹, while that in the silt fraction varied from 72 to 166 μg N g⁻¹. Similarly, for a Canadian soil with 37% clay, Kowalenko and Ross (1980) reported values of 335 μg g⁻¹ for clay and 8.3 μg g⁻¹ for silt.

In addition to clay minerals, soil organic matter can also fix $NH_4{}^+$ in nonexchangeable forms. Also NH_4-organic matter complexes are extremely resistant to microbial decomposition. Aromatic compounds and their unsaturated alicyclic counterparts are primarily responsible for NH_4-fixation by soil organic matter. Mechanisms responsible for ammonium fixation by soil organic matter are not yet well understood.

In addition to clay type and content, important factors affecting ammonium fixation in soil are (1) amount of ammonium added; (2) soil water content, drying, and wetting; (3) presence of other ions (especially K^+); and (4) freezing and thawing.

The amount of ammonium fixed in the edges of clay lattices generally increases with increasing amounts of ammonium added.

Soil water content affects ammonium fixation because clay particles remain expanded in moist soils and fixation is reduced. On drying, the inter-layer spaces in the clay and minerals contract, trapping ammonium ions that have diffused into the edges of the lattice. In a clayey soil at 60% of maximum water-holding capacity, NH_4 fixation was one-fourth that for the dry soil (Osborne, 1976).

Regarding the presence of other ions, a study by Chen et al. (1989) showed that saturation of Beaumont and Lake Charles soils with Ca reduced fixation of added ammonium and resulted in release of fixed ammonium (Figure 8.8); ^{15}N data showed that this release was from the chemically added fraction and not from the indigenous (interior) fixed ammonium. Typically, K^+ and NH_4^+ ions in the external solution cause the lattices of 2:1 clay to collapse, restricting diffusion into and out of the edges of clay particles. On the other hand, larger cations (especially Na^+ because of its hydration) expand the lattices and allow more diffusion and exchange to occur.

Regarding anions, the presence of phosphate may increase ammonium fixation. Mandal and Mukhopadhyay (1984) reported 980 to 1134 mg kg^{-1} soil NH_4^+-N fixation with DAP as compared with 910 to 1097 mg kg^{-1} soil with ammonium sulfate in three soils of West Bengal, India.

Since freezing removes water from the system, it may influence the amount of ammonium fixed. For example, Walsh and Murdock (1960) showed that a treatment for 5 days at $-15°C$ increased NH_4^+ fixation in a number of podzolic soils from 0.08 to 0.14 cmol kg^{-1} of soil.

8.6. NH_4^+ VERSUS NO_3^- NUTRITION OF PLANTS

Theoretically, NH_4^+ should be the preferred form of N since it does not need to be reduced before incorporation into organic materials. However, in most well-drained soils suitable for crop production, oxidation of NH_4^+ to NO_3^- is fairly rapid, and therefore most plants growing under well-drained conditions have developed to grow better with NO_3^-. In recent years there has been increased interest in NH_4^+ versus NO_3^- nutrition of crops, and the results of a number of studies indicate better growth of plants and higher yield with a mixture of NH_4^+ and NO_3^- rather than a single ionic form (Hageman, 1984). The availability of nitrification inhibitors (chemicals that inhibit or retard nitrification) has permitted maintaining higher concentrations of ammonium for longer periods in fields (Joseph and Prasad, 1993). For wheat and many other crops, NH_4^+ to NO_3^- ratios of 50:50 and 75:25 appear to be near optimum. With coarse-textured soils, especially when they are slightly alkaline, an enhanced NH_4^+ regime may be advantageous for the growth of corn. Thus for

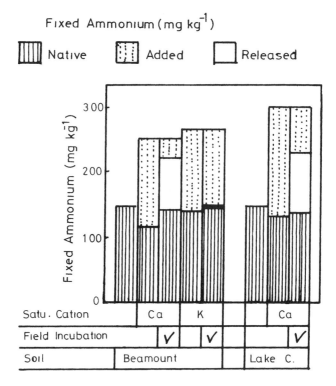

Figure 8.8. Native and added fixed NH₄⁺ as influenced by chemical treatment and field incubation in Beaumont and Lake Charles
soils. (From Chen et al., 1989. Soil Sci. Am. J. 53:1039. With
permission of SSSA.)

crops growing on well-drained soils, a mixed availability of NH_4^+ and NO_3^-
nitrogen is considered ideal.

Rice, which grows under submerged soil conditions, must feed mostly on
NH_4^+ because NO_3^- is not stable under those conditions. Nitrates under submerged conditions are lost by denitrification (discussed later in this chapter).
The term amoniphilic (ammonia-liking) plants has been suggested for plants
such as rice that feed on NH_4^+ (Prasad et al., 1983). When urea is used as
fertilizer, rice plants may also absorb some directly as molecular urea (Mitsui
and Kurihara, 1962; Saraswathi et al., 1991).

8.7. BIOLOGICAL NITROGEN FIXATION

Nitrogen-fixing organisms can be broadly classified into those that are
free-living and those living in association with plants. Nutman (1971) has
made a fairly detailed listing of N-fixers of both groups. Total global biological
nitrogen fixation by all the different types of organisms is estimated to be

Table 8.3 Global Rates of Nitrogen Fixation

Process	Amounts (million Mg yr^{-1})
Terrestrial biofixation	
Legumes	36
Rice paddies	4
Other crops	5
Grasslands[a]	45
Forests[a]	40
Others	10
Total	140
Oceanic biofixation	20–120
Industrial fixation (including fertilizers)	89
Terrestrial combustion processes	19

[a] Would also include legumes.

From Winteringham, 1980. *Soil N as Fertilizer or Pollutant*. With permission of IAEA.

approximately 175 million Mg yr^{-1}, of which legumes alone account for almost half (80×10^6 Mg). Estimates of global fixation and losses of N by different sources are given in Table 8.3, while N-fixation in kg^{-1} ha^{-1} year by selected organisms or systems are given in Table 8.4. A list of nitrogen-fixing organisms and associations is given in Table 8.5.

Although the overall process $N_2 \rightarrow 2\ NH_3$ is exergonic (–104 kJ/mole), high energy input (940 kJ/mole) is required to rupture the $N \equiv N$ bond. Thus, whatever the source, the manufacture of nitrogen fertilizer is highly energy dependent. Recent interest in biological N_2-fixation and biofertilizers (Verma, 1993) is due to the fact that biological fixation uses solar energy, the most inexhaustible energy source, for producing ammonia from nitrogen for use in the protoplasm of organisms.

8.7.1. Nitrogen Fixation by Legumes

Rhizobium species of bacteria living in a symbiotic relationship in root nodules of legumes are capable of converting atmospheric N_2 gas to NH_3. NH_3 is then utilized by the legume to form amino acids and proteins. The rhizobia obtain the energy needed for their growth and for the fixation of N ultimately from products of photosynthesis formed in the legume. By some estimates, this can be as much as 35% of the energy fixed by photosynthesis. For this reason, seldom is biomass production by an N-fixing legume as great as that of cereal crops, when no other factors limit growth.

**Table 8.4 Biological Nitrogen Fixation by
Different Organisms/Systems**

Organism or system	N-fixed $(kg\ ha^{-1}\ yr^{-1})$
Legumes	
Forage	57–700
Grain	17–270
Nodulated nonlegumes	
Alnus	40–300
Hippophae	2–179
Ceanothus	60
Coriaria	150
Plant algal associations	
Gunnera	12–21
Azollas	31
Lichens	39–84
Free-living microorganisms	
Blue green algae	25
Azotobacter	0.3
Clostridium pasteurianum	0.1–0.5

From Nutman (1971) and Evans and Barber (1977).

Although the causative organism responsible for nitrogen fixation in pure culture was isolated by Beijerinck in 1888 (Waksman, 1952), growing legumes to maintain and build up soil fertility has been practiced since ancient times. It is now well established that the enzyme nitrogenase is involved in nitrogen fixation both in free-living organisms and those living in associations. The nitrogenase enzyme complex in combination with the necessary reactants catalyzes the reduction of nitrogen to ammonia. This enzyme complex consists of two protein components. The first is a larger molybdenum-iron protein having a molecular weight of 2 to 2.7×10^5 and containing one or two atoms of molybdenum, 17 to 36 iron atoms, and 14 to 28 acid labile sulfur atoms per protein molecule. The second is an iron-protein molecule having a molecular weight of 55 to 67×10^3 and containing four atoms each of iron and acid-labile sulfur per protein molecule (Evans and Barber, 1977).

There is no doubt a specificity exists between *Rhizobium* strain and the legume, and compatibility between the two is essential for successful nodulation. This necessitates using specific cultures for different legumes; when growing a new legume species on a soil, it is necessary that the appropriate *Rhizobium* culture be applied. The benefit of seed inoculation with specific *Rhizobial* cultures has been proved the world over.

There is considerable variation in nitrogen fixation by different legumes. Perennial legumes cut for forage generally fix more than 100 kg N ha^{-1},

Table 8.5 Nitrogen-Fixing Organisms and Associations

Kind of organisms	Examples
Free-living organisms	
Photosynthetic bacteria	*Rhodospirillum, Chromatium, Chlorobium*
Aerobic bacteria	*Azotobacter, Beijerinckia, Derxia,* *Spirillum, Thiobacillus*
Facultative bacteria	*Klebsiella, Bacillus, Vibrio* (methane oxidizers)
Obligate anaerobic bacteria	*Clostridium, Methanobacterium,* *Desulfovibrio*
Actinomycetes	*Mycobacterium, Nocardia, Actinomyces*
Blue-green algae	
Heterocystous	*Anabaena, Aphanizomenon, Gloeotrichia,* *Nostoc, Calothrix, Shizothrix*
Nonheterocystous	*Gloeocapsa, Oscillatoria, Plectonema*
Nonobligatory Associations[a]	
Animal associations (bacteria)	Termites, sea urchins, cattle sheep

Aquatic macrophytes
   (blue-green algae)
	Azolla (*Anabaena*): *Sargassum*
	   (*Dichothrix*, blue-green algae); *Codium*,
	   a macro-green alga, *Azotobacter*

Leaf surfaces of angiosperms (bacteria)	Various tropical plants
Leaf nodules (bacteria)	*Psychotria*, a tropical plant (*Klebsiella*)
Lichens (blue-green algae and fungi)	*Peltigera aphthosa* (*Nostoc*)
Root nodules of tropical gymnosperms (blue-green algae)	*Podocarpus, Macrozamia, Cycas*
Obligatory symbioses	

Nonleguminous angiosperms
   (root nodules)
	Alder tree —*Alnus* sp., *Myrica, Elaeagnus,*
	   Coriaria, Dryas

Legume root nodules
   (*Rhizobium* spp.)
	Beans, peas, alfalfa, clover, soybeans,
	   lentils

[a] Nitrogen-fixing agent in parentheses.

Modified from Burns and Hardy (1975).

sometimes as much as 400 to 600 kg N ha^{-1}. Both temperate and tropical species are equally effective. As compared with forage legumes, grain legumes fix appreciably less (Table 8.6) because of their shorter growing season and because the nitrogen left in the haulm is usually not recorded. Intercropping systems in which a grain legume is grown between widely spaced cereal crop rows (such as maize, sorghum, millet, cotton, and sugarcane) are being developed. These intercropping systems permit the production of grain legumes as a bonus crop, thereby providing bonus nitrogen fixation. In some of these

Table 8.6 Nitrogen Fixed by Different Legumes

Legumes	kg N ha^{-1} yr^{-1}
Forage Legumes	
Temperate	
Clovers	23–620
Lucerne	164–300
Tropical	
Stylo	30–196
Tick clover	700
Grain Legumes	
Temperate	
Vetch and tick beans	57–190
Peas	46
Lupins	128
Tropical	
Lentil	35–77
Pigeonpea	41–150
Cowpea	73–354
Soybean	17–206
Cluster beans	37–196
Groundnut	33–111
Chickpea	41–270
Mung beans	224

From Nutman, 1971. Sci. Prog. Oxford 59:55–74. With permission of Science Reviews Ltd., Middlesex, England.

experiments additional yields of the main crop have been reported, suggesting some transfer of nitrogen from the companion legume crop. Also, multiple and relay-cropping systems have been developed with a legume crop as a component.

The practice of growing legumes to a certain vegetative stage and then incorporating them into the soil is known as "green manuring" and is another way of utilizing biological N fixation *in situ*. A large number of crops and species have been used as green manure including sunnhemp (*Crotolaria juncea* L.), dhaincha (*Sesbania aculeata* [wild] Poir), Phillipasera (*Phaseodu trilobus* Ait), mung bean (*Vigna aureus* L.), cowpea (*Vigna aculeata*), cluster beans (*Cyamopsis tetragonoloba* L.), senji (*Melilotus alba* Medik.), and khesari (*Lathyrus satisvus*). In some regions green matter from leguminous and nonleguminous plants may be added as green manure. The tree species and other plants used include *Thespesia populanea* L., *Cassia Auriculata* L., *Pongamia glabra* Vent., *Azadirachta indica* A. Juss., *Calotropis gigantea, Jatropha gossypifolia* L., *Jatropha glandulfora* Roxb., *Glircidia meculata* (H.B. and K.) *Tephrosia purpurea* L., *Tephrosia candida* (Roxb) DC., *Cassia tora* L., and *Ipomea carnea* auct. (non-Jacq.).

Likewise, in temperate regions a number of legume species are used as green manures. For cool season plantings, especially if growth is to be terminated within 40 to 70 days, species such as field peas (*Pisum sativum* L.), faba bean (*Vicia faba* L.), hairy vetch (*Vicia villosa* Roth subsp. *villosa*), black medic (*Medicago lupulina* L.), lentils (*Lens culinaris*, Medikus), and crimson clover (*Trifolium incarnatum* L.) are often used. For warmer season plantings some of the above species plus soybean (*Glycine max* L. Merr.), flatpea (*Lathyrus syvestris* L.), and tinga pea (*Lathyrus lingitanus* L.) are possible selections. In hot seasons many of the species listed above for the tropics can be used. If the potential growth period is longer, then a number of perennials and biannual species are potential candidates, such as red clover (*Trifolium pratense* L.), sweet clover (*Melilotus alba* Medikus and *officinalis* Lam.), lespedeza (*Lespedeza stipulacae* Maxim), alfalfa (*Medicago sativa* L.), and others. If grown as a winter cover crop, hairy vetch and the nonlegume rye (*Secale cereale* L.) usually have sufficient winter-hardiness to survive. For milder areas (southern United States), crimson clover is also a good winter cover crop.

Data from a large number of green-manure field experiments show a grain yield increase for rice equivalent to 50 to 120 kg N fertilizer ha^{-1} or even more depending upon the green-manure biomass incorporated and its total nitrogen content (Goswami et al., 1988; John et al., 1989a; Singh et al., 1991). Use of green manures in temperate regions was summarized recently by Power (1987).

Little work has been done for estimating the amounts of N fixed by naturally occurring legumes, which are widely distributed as herbs in grasslands, bushes, and trees in savannah and trees and creepers in tropical forests. Orchard and Darb (1956) reported that a plantation of *Acacia mollisima* fixed as much as 270 kg N ha^{-1} yr^{-1}. By fixing atmospheric nitrogen, legumes provide stability to natural ecosystems.

8.7.2. Nitrogen Fixation by Blue-Green Algae (Cyanobacteria)

Blue-green algae may contribute 25 to 30 kg N ha^{-1} per cropping season, depending on ecological conditions (Venkataraman, 1979). However, Watanabe et al. (1977) measured C_2H_2 reduction before and after removing cyanobacterial growth from the rice field and concluded that the nitrogenase activity could account for a daily input of approximately 0.5 kg N ha^{-1}.

8.7.3. Azolla-Anabaena Systems

The Art of Feeding the People, a book on agricultural techniques written in 540 A.D. by Jia Ssu Hsieh, describes the cultivation and use of *Azolla* in rice fields in China (Chu, 1979), indicating that *Azolla* has been used in rice fields in China since ancient times. International interest in Azolla-anabaena systems for meeting a part of the nitrogen requirement of rice is rather recent. The ecosystem requirements for *Azolla* include free-floating water, sufficient light, and temperature around 15°C. The most favorable temperature for the growth of *A. pinnata* is 16 to 17°C; much of the *Azolla* may die when the temperature rises to 20 to 24°C (Becking, 1979). The inability of *Azolla* to grow under high temperatures is a major barrier in using *Azolla-Anabaena* systems for supplying nitrogen to rice in many developing countries in Asia and Africa. However, when the temperature during the two-month period before rice planting does not rise above 30°C, it should be possible to use *Azolla* for this purpose.

The major nutrient requirement for *Azolla* is phosphate (10 to 25 kg P_2O_5 ha^{-1}). The addition of potassium also helps in the growth of *Azolla*. *Azolla* grows best in soils having pH between 5.5 and 7 (Singh, 1979), although some growth is reported even at pH 10. Acidic soils (pH 3.0 to 3.5) do not permit the growth of *Azolla*, and the inoculum dies.

Although laboratory estimates of N_2-fixation by *Azolla* could be as high as 335 to 670 kg N ha^{-1} (Becking, 1979), field data normally suggest an N_2-fixation equaling a basal N fertilization of 30 to 40 kg N ha^{-1}.

8.7.4. Nonphotosynthetic Bacteria

The nitrogen-fixing power of nonphotosynthetic bacteria has been known for decades, and there is a fairly long list of species capable of this process (LaRue, 1977). The amount of nitrogen fixed by these bacteria is generally relatively small, and therefore there has been little interest in this topic. However, with the energy crisis in recent years, there is renewed interest in this avenue of nitrogen fixation.

8.7.5. Nodule-Forming Nonlegumes

Most nodulating nonleguminous plants with the capacity of fixing nitrogen are woody, and therefore their economic application is generally restricted to forestry and the general ecology of a region. Nevertheless, they do play an important role in overall human welfare. These associations have been recently discussed in detail by Becking (1977).

8.7.6. Nonsymbiotic N_2-Fixation

In addition to symbiotic N_2-fixation, nitrogen is also fixed in soil by non-symbiotic bacteria *Azotobacter* spp. (aerobic) and *Clostridium* spp. (anaerobic). Available estimates suggest that 10 to 15 kg N ha^{-1} yr^{-1} is fixed by nonsymbiotic N_2-fixing bacteria. Cultures of these organisms have been successfully used in

Soviet agriculture, but the United States experience has not been very encouraging. Nevertheless, there is an interest in the use of cultures of *Azotobacter* spp. in developing countries (Thomas, 1993).

Other bacterial species (*Beijerinckia*) that fix atmospheric N inhabit the surfaces of a number of tropical plants and are in tropical soils (Becking, 1961). Dobereiner et al. (1972) reported nitrogen fixation by a *Paspalum notatum - Azotobacter paspali* association. In such an associative system no nodules are produced, but in some cases bacteria live underneath a mucilaginous sheath on the root surface. Dobereiner et al. (1972) suggested fixation of about 90 kg N ha^{-1} for a *P. notatum – A. pasali* association. Later they reported the occurrence of a nitrogen fixing *Spirillum lipoferum* on the roots of the grass *Digitaria decumbens* (Dobereiner and Day, 1976). The amount of N fixed by associative systems is considered lesser than that for *Rhizobia*.

8.8. NITROGEN FERTILIZERS OR INDUSTRIAL NITROGEN FIXATION

Essentially all nitrogen fertilizers are produced from ammonia gas. Exceptions are small amounts of naturally occurring guano (bird droppings from South America) and sodium nitrate deposits. Over 90 million megagrams (Mg) of nitrogen are commercially fixed each year worldwide for use as fertilizers.

There are three methods by which atmospheric elemental nitrogen can be converted into a chemical form that can be directly used as a chemical fertilizer or can be used for making a chemical fertilizer. These methods or processes are (1) the cynamide process, (2) the arc process of nitric acid synthesis, and (3) the Haber-Bosch ammonia synthesis process. A brief discussion on these follows.

8.8.1. Cyanamide Process

This process was developed by Frank and Caro in Germany in 1898 and involves passing purified nitrogen gas over calcium carbide kept at 1100°C.

$$CaC_2 + N_2 \xrightarrow{\text{heat}} CaCN_2 + C$$

This process is used only on a very limited scale in Germany and other European countries, often for purposes other than making chemical fertilizers. Calcium cyanamide, when used as a fertilizer, has some toxic effects on crop plants, and application is generally recommended 4 to 6 weeks before sowing of a crop. Calcium cyanamide has also been used as a herbicide.

8.8.2. Arc Process

This process involves passing elemental nitrogen and oxygen through an arc that is expanded in an electromagnet to increase the contact with gases.

In nature, lightning also accomplishes this process and therefore is a source of nitrates in rain. The reactions are

$$N_2 + O_2 \rightarrow 2\ NO$$

$$2\ NO + O_2 \rightarrow 2\ NO_2$$

$$3\ NO_2 + H_2O \rightarrow 2\ NHO_3 + NO$$

This process can be used where electrical power is inexpensive.

8.8.3. Haber-Bosch Process (Ammonia Synthesis)

Ammonia synthesis by the Haber-Bosch process is one of the few, most significant, scientific discoveries of the early twentieth century, and it led to Haber receiving a Nobel prize for this invention. Ammonia synthesis is based on the reaction of N_2 and H_2 in the presence of a catalyst, the main component of which is magnetite (Fe_3O_4), and at temperatures up to 1200°C. The pressure required varies from 200 to 1000 atm. The basic reaction is

$$3\ H_2 + N_2 \rightarrow 2\ NH_3$$

While N_2 is obtained from the elemental atmospheric N, the source of H_2 is natural gas, naptha, fuel oil, and coal; the first three are products of the petroleum industry. Natural gas is by far the most important feed stock used for ammonia production. Where electrical power is inexpensive, H_2 can also be obtained by the electrolysis of water.

The anhydrous ammonia produced by this process can be used as such as a fertilizer (comprising 41.8% of total N consumed in the United States in 1990), or it can be reacted with nitric, sulfuric, and phosphoric acids to make ammonium nitrate, ammonium sulfate, and ammonium phosphate, all of which are important fertilizers. In the United States anhydrous ammonia is often delivered to farm communities by underground pipelines, resulting in a very inexpensive fertilizer material costing half as much as most other sources. Furthermore, anhydrous ammonia when reacted with carbon dioxide yields urea, which is an important N source in many Asian countries, as well as elsewhere in the world. The reactions are

$$2\ NH_3 + CO_2 \rightarrow NH_2COONH_4$$

$$NH_2COONH_4 \rightarrow NH_2CONH_2 + H_2O$$

(Urea)

Urea and ammonium nitrate can also be dissolved to make nitrogen fertilizer solutions, notably UAN, which is also used widely. UAN is often the fertilizer

Table 8.7 General Composition of Some Common N and NP Fertilizers

Fertilizer	Percent				
	N	P_2O_5	CaO	MgO	S
N Fertilizers					
Ammonium sulfate	21	—	—	—	24
Anhydrous ammonia	82	—	—	—	—
Ammonium chloride[a]	25–26	—	—	—	—
Ammonium nitrate-sulfate	30	—	—	—	5–7
Ammonium nitrate with lime (ANL)/CAN[b]	20.5	—	10	7	—
Calcium nitrate	15	—	34	—	—
Calcium cyanamide	22	—	54	—	0.2
Sodium nitrate	16	—	—	—	—
Urea	45–46	—	—	—	—
Urea-sulfate	30–40	—	—	—	6–11
Urea-sulfur	30–40	—	—	—	10–20
Urea-ammonium nitrate (solution)	28–32	—	—	—	—
NP Fertilizers					
Ammoniated ordinary superphosphate	4	16	23	0.5	10
Monoammonium phosphate	11	48–55	2	0.5	1–3
Diammonium phosphate	18–21	46–54	—	—	—
Ammonium phosphate-sulfate	13–16	20–39	—	—	3–14
Ammonium polyphosphate solution	10–11	34–37	—	—	—
Urea-ammonium phosphate	21–38	13–42	—	—	—
Urea-phosphate	17	43–44	—	—	—

[a] 66% chloride.

[b] Calcium ammonium nitrate in India.

From Tisdale et al., 1993. *Soil Fertility and Fertilizers*, 5th ed., p. 156. With permission of Prentice-Hall, Inc., Upper Saddle River, NJ.

of choice to mix with liquid P and K sources for use as a preplant starter fertilizer, and also for injection into sprinkler systems for use in fertigation.

A list of commonly used nitrogen fertilizers is provided in Table 8.7.

8.9. EFFICIENT NITROGEN MANAGEMENT

8.9.1. Recovery of Fertilizer Nitrogen

Recovery of fertilizer nitrogen is calculated as the fraction of fertilizer N applied that is removed in the harvested crop. Recovery values often vary from

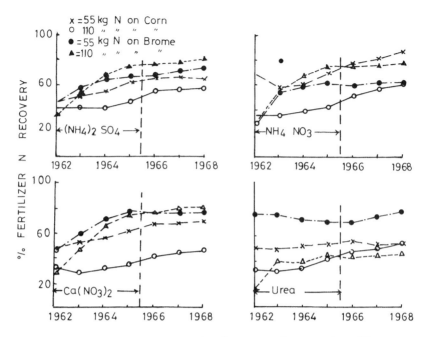

Figure 8.9. Fertilizer N recovery in plant tops from various N sources. (From Power et al., 1973. Agron. J. 65:765–768. With permission of SSSA.)

20 to 80% depending upon crop, amount of N, source of N, soil, frequency and amount of precipitation, and method used to calculate recovery. It is generally determined by the difference method using the expression:

$$\% \text{ recovery} = (Nf - Nck)/Nap \times 100$$

when Nf and Nck refer to N uptake in kg ha^{-1} in fertilized and check (no fertilizer) plots, respectively, and Nap is the amount of nitrogen applied in kg ha^{-1}. Since the amount of N applied may affect the quantity of native N released by soil (priming effect), this kind of recovery is referred to as apparent recovery. In a field experiment at Mandan, North Dakota (Power et al., 1973), N was applied to corn or bromegrass (*Bromus inermis* L.) at 55 or 110 kg N ha^{-1} for 4 consecutive years (1962 to 1965), and barley was grown for the following 3 years (1966 to 1968) without fertilizer. Data on apparent recovery of N are shown in Figure 8.9. Fertilizer N recovery was greater with bromegrass than with corn, due to a better root system and greater biomass for bromegrass. Also, recovery was greater with 55 kg N ha^{-1} than for 110 kg N ha^{-1}. Final recoveries, including residual N taken up by barley, ranged from 45 to 55% at 110 kg N ha^{-1} applied to corn to about 88% at 110 kg N ha^{-1} applied to bromegrass. At 110 kg N ha^{-1} ammonium sulfate and ammonium nitrate com-

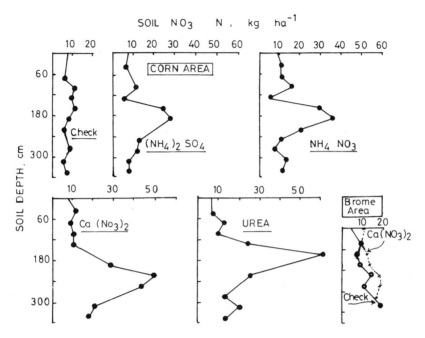

Figure 8.10. **Soil nitrate-N distribution with depth after residual growth and N-uptake responses were complete (0 and 111 kg N/ha). (From Power et al., 1973. Agron. J. 65:765–768. With permission of SSSA.)**

pared well and gave higher recovery than calcium nitrate and urea; this may have been due to more ammonia volatilization and/or leaching of nitrates for the latter two fertilizers (Figure 8.10). Similarly, the recovery of N applied to winter wheat was reported as 26 to 45% (Muchova and Apltauer, 1983; Janzen, et al., 1990) and 37 to 52% for spring wheat (Tandon, 1980). In recent years a large number of experiments have been conducted with ^{15}N-enriched or -depleted fertilizer material, and data on ^{15}N uptake permit the determination of direct recovery of the ^{15}N applied. On coarse-textured alkaline (pH 7.7 to 7.9) soils of Punjab, India, ^{15}N recovery by spring wheat was 65% with potassium nitrate and 33 to 44% with urea (Katyal et al., 1987). In a recent study in Nebraska (Francis et al., 1993) corn was grown on a Wood River silt loam and received 75, 150, 225, and 300 kg N ha[-1] as ammonium nitrate. At maturity only 21.2 to 54.3% of N present in the corn came from fertilizer, and only 24.4. to 35.3% of applied N was recovered by the crop (probably more was taken up, but some was lost by NH_3 volatilization before maturity) (Table 8.8). Generally, the difference method gives recovery values somewhat greater than the isotopic method. This may be due to the isotopic method reducing uptake of nonisotopic soil nitrogen and to failure of the isotopic method to evaluate residual effects of treatment in previous years.

Table 8.8 Total Plant N, Percentage of Plant N Derived from Fertilizer (Ndff), and Percentage of Fertilizer N Recovered by Corn at Maturity on Wood River Silt Loam in Nebraska

Fertilizer N kg ha^{-1}	Total N uptake by plant tops (kg ha^{-1})	Ndff (%)	Fertilizer N recovered (%)
75	87.4	21.2	24.4
150	122	35.0	28.5
225	164	39.8	28.9
300	194	54.5	35.3

From Francis et al. 1993. Agron. J. 85:659–663. With permission of ASA.

Table 8.9 ^{15}N Balance in Rice-Wheat, Corn-Wheat-Mungbean, and Wheat-Mungbean-Corn Rotations (% of Applied N)

kg N ha^{-1} applied to 1st crop	Recovered by			Left in soil	Unaccounted N
	1st crop	2nd crop			
	Rice	Wheat			
60	35.4	4.1		16.7	43.8
120	31.2	4.6		25.6	38.6
	Corn	Wheat	Mungbean		
120	20.8	7.0	0.8	40.1	31.3
	Wheat	Mungbean	Corn		
90	45.3	2.5	1.5	48.2	2.5

From Goswami et al. (1988) and Subbiah et al. (1985).

The efficiency of fertilizer N in tropical and subtropical environments is very low (Prasad and DeDatta 1979). Data are available from a number of experiments using ^{15}N. In a field experiment at the International Rice Research Institute, Manila, Philippines, 29 to 40% of applied urea N was taken up by rice grain and straw (John et al., 1989a). At the Indian Agricultural Research Institute, New Delhi, India, George and Prasad (1989) found that the recovery of applied urea N by rice was 31.0, 26.7, and 25.9% at 50, 100, and 150 kg N ha^{-1}, respectively. Goswami et al. (1988) reported that in the rice-wheat rotation (growing rice July to November and growing wheat November to April) recovery of 60 kg N ha^{-1} applied to rice was 35.4% by rice and 4.1% by the succeeding wheat (Table 8.9). Subbiah et al. (1985) reported that, when 120 kg N ha^{-1} was applied to maize in a single dressing, only 20.8% could be recovered. Recovery could be raised to 38.3% by application in three split doses. Wheat succeeding maize recovered only 7% of that applied to maize. Succeeding crops of green gram (mung beans), corn, and wheat in the same field over the next 4 years recovered only negligible amounts (Table 8.9). However, when 90 kg N ha^{-1} was applied to wheat,

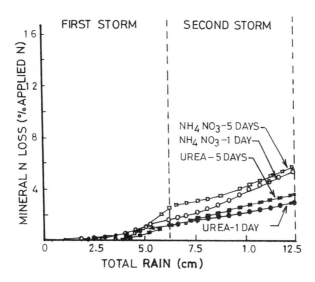

Figure 8.11. The effect of treatments on the amount of mineral nitrogen loss by runoff from fallow plots during five inches of simulated rain. Pelleted ammonium nitrate and urea applied to the surface of all plots at a rate of 44.8 kg N ha^{-1} either 1 day or 5 days before rainfall. (From Moe et al., 1968. Soil Sci. 105:428–433. With permission of Williams & Wilkins.)

recovery was 46.3% by wheat, 2.5% by a succeeding green gram, and 1.5% by the maize that followed green gram. Thus efficiency of applied N was greater for wheat grown during the cold, dry period than for maize or rice grown during the warm, rainy season. Heavy rain and irrigation may have been responsible for N loss from the soil.

Data on N balance sheets (Table 8.9) show that about 16 to 48% of applied N may be retained or immobilized in the soil and 2.5 to 43.8% of applied N remained unaccounted (lost by various N loss mechanisms operating in soil). N loss mechanisms in soil therefore need to be carefully understood.

8.9.2. Nitrogen Loss Mechanisms

If not taken up by the crop plants or immobilized in soil, fertilizer N applied to crop fields may be lost by surface runoff, ammonia volatilization, nitrate leaching, denitrified as N_2 or N_2O, or lost as ammonia from foliage (Figure 8.1).

8.9.2.1. Surface Runoff

When heavy rains follow an application of fertilizer N, part may be lost by surface runoff, particularly with urea and nitrates, which are very soluble and

**Table 8.10 Estimated Rates of Urea Hydrolysis
in Soils and Overlying Paddy Water**

Time interval (hr)	Rate of hydrolysis	
	Soil (mol s^{-1})	Water (mol s^{-1})
0–12	1×10^{-5}	2×10^{-7}
12–24	6×10^{-5}	2×10^{-6}
24–48	8×10^{-5}	4×10^{-6}

From Hongprayoon et al. 1991. Soil Sci. Soc. Am. J. 55:1130–1134. With permission of SSSA.

not retained by soil particles. Data from a simulated rain storm applying 12.5 cm of rain on a Lanesville silt loam in Indiana, having a 13% slope and a fragipan layer at 60 to 95 cm depth, are shown in Figure 8.11 (Moe et al., 1968). Losses were greater (5% of applied N) from ammonium nitrate than from urea (2 to 2.5%).

8.9.2.2. Urea Hydrolysis

When urea is used as a fertilizer and applied to a moist soil, it is first hydrolyzed to ammonium carbamate and then to ammonium carbonate by the enzyme "urease," which is generally present in most soils. Ammonium carbonate readily degrades to NH_3 and O_2 gases. The pH of the soil in the vicinity of a urea prill may rise to 8.0 or above due to the formation of ammonium carbonate. This can result in loss of N by ammonia volatilization. Urea hydrolyzes fairly rapidly in soils.

The main site of urea hydrolysis in rice paddies is in the soil and not the floodwater (Delaune and Patrick, 1970). After 30 hours of incubation, 3 and 64% of the added urea was hydrolyzed in the floodwater and soil, respectively. In soil the rate of urea hydrolysis ranged from 1 to 8×10^{-5} mol s^{-1}, while hydrolysis rate in water varied from 2×10^{-7} to 4×10^{-6} mol s^{-1}, depending upon time interval (Hongprayoon et al., 1991) (Table 8.10). Depletion of O_2 in submerged soils retards urea hydrolysis. The order of hydrolysis rate is oxidized soil > reduced soil > floodwater.

Use of urease inhibitors such as phosphorodiamidate (PPD) and N (N-butyl) thiophosphoric triamide (NBPT) have been proposed as a strategy to reduce volatilization loss by retarding the hydrolysis of urea (Vlek et al., 1980). Because both urease inhibitors keep floodwater ammonia concentration low, they result in lesser ammonia volatilization (Buresh et al., 1988a). However, these urease inhibitors may have some phytotoxicity.

Amendment of urea with PPD or NBPT resulted in increased rice grain yield on a silty clay soil but not on a clay soil (Buresh et al., 1988b). Similar effects occurred for no-till corn (Schlegel et al., 1986).

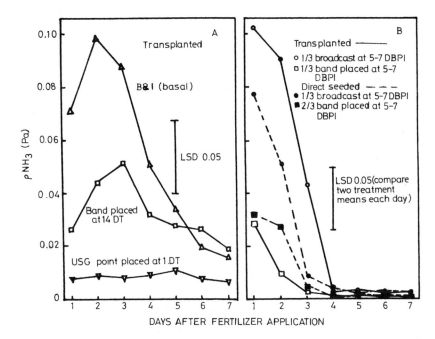

Figure 8.12. Partial pressure of NH$_3$ (pNH$_3$) in paddy floodwater at 1400 h, as affected by (A) method of basal N-fertilizer application and (B) N application 5–7 days before pancile initiation (*DBPI*), Munoz, Nueva Ecija, Philippines, 1987 dry season. *B&I*, broadcast and incorporated; *DT*, days after transplanting; *LSD*, least significant difference, *USG*, urea supergranules. (From Schnier et al., 1990. Biol. Fertil. Soils 10:89–96. With permission of Springer-Verlag.)

8.9.2.3. Ammonia Volatilization from Soils

Losses of N by ammonia volatilization after fertilizer application to crops range from 0 to 50% of the N applied. The factors that control ammonia volatilization loss include fertilizer form, method of application, soil pH, soil water content, cation exchange capacity of soil, wind velocity, air temperature, crop, and stage of crop growth. Ammonia loss under upland conditions is greatest from ammonium or ammonium-producing fertilizers applied on calcareous soils, where ammonium salts react with calcium carbonate to form precipitates of low solubility. Ammonia volatilization losses are very high when prilled urea is broadcast on a wet surface soil or on floodwater in rice paddies (Table 8.11) and can be considerably reduced by band or point placement of fertilizer urea or by use of coated urea or urea supergranules (USG — a modified urea form where the size of each granule may be a few mm to 1 cm in diameter) (Schnier et al., 1990) (Figure 8.12). Data in Figure 8.12

Table 8.11 Effect of Fertilizer and Water Management on Ammonia Volatilization Loss (% of Applied N) from Flooded Rice Fields

Urea applied kg N ha^{-1}	Method of application	Mabitac silty clay	Calavun clay	Aguilar silty loam	San Marcelino loamy sand
53	B_5[a]	48	27	14	26
53	B_5 & I	42	12	10	30
53	B_o & I	11	11	4	14
80	B_5	56	23	10	36
80	B_5 & I	43	12	10	26
80	B_o & I	7	16	5	10
LSD (p 0.05)		8	11	5	10

[a] B_5 — basal dressing broadcast into 5 cm deep floodwater 10 days after transplanting. I — incorporated into soil by rotary harrow; B_o — basal dressing broadcast onto soil without surface water and incorporated into soil by rotary harrow; water returned to a depth of 5 cm 2 days after urea application.

From Freney et al. 1990. Biol. Fert. Soils 9:31–36. With permission of Springer-Verlag.

also show higher ammonia concentration in floodwater when urea was broadcast and incorporated than when it was band placed. Ammonia volatilization losses from rice fields range from 7 to 48% of the quantity of fertilizer N applied, depending upon the dose of N, its method of application, and soil characteristics (Freney et al., 1990). Losses were greater when urea was applied in 5 cm of standing water than when applied to a saturated soil. Incorporation of urea into the soil reduced ammonia volatilization loss.

Fertilizer applications to grazed pastures, which probably are one of the main sources of atmospheric NH_3, can result in large losses of ammonia, particularly if the applied urea falls on a urine patch. Losses by ammonia volatilization from urine-affected and unaffected soils were 27% and 7% of the applied N, respectively (Black et al., 1984).

Legume green manures, often advocated as an N source in soil-conserving production systems, may be particularly vulnerable to volatile N loss. In greenhouse studies at Alberta, Canada, (Janzen and McGinn, 1991) as much as 14% of the N applied as field-grown lentil (*Lens culineris*) residues was volatilized within 14 days of application.

8.9.2.4. Nitrogen Losses from Plants

The total amount of N in the aboveground biomass (grain plus stover) reaches a maximum well before maturity, often followed by a subsequent decline. In perennial species much of this N is probably translocated and stored in roots as a reserve to initiate growth the next season. However, for annual species (including most grain crops) there is little evidence of appreciable translocation to roots. Postanthesis N losses from wheat ranged from 5.9 to 80 kg N ha^{-1} (Daigger et al., 1976; Harper et al., 1987; Papkosta and Gagianas,

1991) due to ammonia volatilization from the aboveground biomass. In a recent study in Nebraska (Francis et al., 1993) postanthesis fertilizer N losses ranged from 45 to 81 kg N ha^{-1} for irrigated corn. Although not measured directly, volatile NH_3 losses from aboveground plant material appeared to be the most plausible explanation for this plant loss of N. Patron et al. (1988) at Fort Collins, Colorado, measured actual ammonia loss and reported that spring wheat lost ammonia at a relatively low and fairly constant rate (60 to 120 mg NH_3-N m^{-2} s^{-1}) during presenescence (before milk ripe stage), but at rates of 200 to 300 ng NH_3-N m^{-2} s^{-1} during final plant senescence. Presumably in well fertilized crops, translocation of N from vegetative to reproductive tissue may be inefficient, allowing NH_3 to escape through the stomata in the respiration stream.

From a field study at Cuttack, India, and a pot-culture study at Fort Collins, Colorado, Mosier et al. (1990) suggested that young rice plants may facilitate the efflux of N_2 and N_2O from the soil to the atmosphere.

Thus the failure to include labeled N losses from aboveground plant biomass in N balance studies leads to overestimation of N losses from soil by direct ammonia volatilization, denitrification, and leaching.

8.9.2.5. Denitrification

Denitrification is the process by which nitrates are reduced to N_2 gas or various gaseous oxides of N. In soils most denitrification results from action of certain anaerobic and especially facultative microbes. Chemical denitrification can also occur, especially in acid soils. In microbial denitrification, nitrates serve as the oxygen source for the organisms. Soluble organic carbon compounds are used for microbial growth. For denitrification to occur, there are four requirements — anaerobic environment, presence of nitrates, presence of soluble carbons, and presence of denitrifying organisms.

When soils become submerged as in rice paddies or under heavy rainfall or irrigation under upland conditions, oxygen is excluded and anaerobic decomposition takes place. Plowing under a green manure crop can sometimes result in such a flush of microbial activity that soil O_2 supply is temporarily depleted to the extent that denitrification can occur. Some anaerobic organisms that belong to the genera *Pseudomonas, Bacillus, Chromobacteria,* and *Thiobacillus* (Table 8.12) can obtain their oxygen from nitrates and nitrites and release N_2 and N_2O. The most probable biochemical pathway is shown below:

$$2\ HNO_3 \xrightarrow[-2\ H_2O]{+4\ H} 2\ HNO_2 \xrightarrow[-2\ H_2O]{+2\ H} 2\ NO \xrightarrow[-H_2O]{+2\ H} N_2O \xrightarrow[-H_2]{+2\ H} N_2$$

Because of the release of N_2O (a greenhouse gas) and its involvement in ozone depletion, the process of denitrification has received considerable attention.

Table 8.12 Taxonomy of Denitrifying Microorganisms[a]

Organism	Comments
Achromobacter liquefaciens	Oxidizes CH_4 with NO_3^-
Alcaligenes sp.	Oxidizes CH_4 with NO_3^-
Bacillus	Many species known to denitrify
Chromobacterium	
Corynebacterium nephridii	Produces only NO and N_2O
Flursarium spp.	Two species of this fungus reduce NO_2^- (but not NO_3^-) to N_2O
Halobacterium	—
Hydrogenomonas spp. (= *Alkalingenes* sp.)	—
Hyphomicrobium sp.	Oxidizes methanol with NO_3^-
Micrococcus denitrificans	Chemolithotroph; oxidizes H_2
Moraxella	—
Propionibacterium	—
Pseudomonas spp.	Many well-known denitrifying species
Spirillum	—
Thiobacillus spp.	Chemolithotrophs; oxidize S and $S_2O_3^{2-}$ with NO_3^-
Veillonella alcalescens	Strict anaerobe; both assimilatory and dissimilatory NO_3^- reduction
Xanthomonas	—

[a] Not all species within a genus may be capable of denitrification.

From Brezonik. 1977. Progress in Water Tech. 8:373–392. With kind permission from Elsevier Science Ltd., Kidlington, U.K.

The ratio of N_2 to N_2O and other oxides produced by denitrification depends on many environmental factors. Generally, however, the more anaerobic the environment, the greater the N_2 production.

Estimates of N losses by denitrification vary from 3 to 62% of applied N in arable soils. Losses were greatest from rice paddies. It may be mentioned that these values are arrived at by subtracting ammonia volatilization losses from the unaccounted ^{15}N. The actual recovery of $(N_2 + N_2O)$-^{15}N was only 0.1% of the applied urea in the study of DeDatta et al. (1991) and only 0.51% of the applied N in the study by John et al. (1989b) under field conditions. Consequently, the tremendous dilution of the isotope that occurs creates a large error in their measurements. Field measurement techniques of N_2 and N_2O therefore need to be improved. Also, N losses from the aboveground portion of the plant must be accounted for.

While denitrification in soil is primarily controlled by three primary or proximal factors (oxygen, nitrate, and carbon), these primary factors are affected by several physical and biological factors (distal factors). Thus it usually becomes necessary to focus on distal rather than proximal factors as

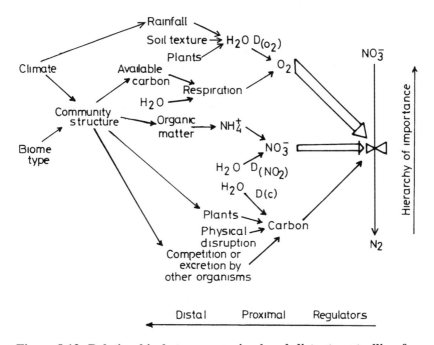

Figure 8.13. **Relationship between proximal and distant controlling factors of denitrification. (From Tiedje, 1987. *Environmental Microbiology of Anaerobes*, Zehnder, A.J.B., Ed. With permission of John Wiley & Sons.)**

controllers of denitrification (Groftman et al., 1988). The relationship between proximal and distal factors controlling denitrification is shown in Figure 8.13, while Figure 8.14 shows how the controlling factors change as the scale of investigation increases. Effects of primary factors on denitrification are briefly discussed. As already pointed out, an anaerobic condition is the basic requirement for denitrification to proceed. Therefore as oxygen concentration decreases, denitrification increases (Table 8.13). Because water and air comprise the soil pore volume, oxygen content decreases as water content increases and vice versa. The effect of increasing % water filled space on denitrification is seen in Figure 8.2.

A source of energy (carbon) is essential for denitrification. Consequently, denitrification generally increases with an increase in water-soluble, as well as with mineralizable, soil organic C.

In addition to oxygen, nitrate, and carbon, soil pH also influences the rate of denitrification. Since bacterial activity is generally low below pH 5, denitrification losses are very low in soils having pH 4 or less (Bremner and Shaw, 1958). Denitrification is most rapid at pH values 6 to 8.

Soil temperatures most suited to denitrification are 25 to 60°C; it is slow at lower temperatures and may be inhibited at temperatures above 60°C (Tisdale et al., 1985). An example of the effect of some distal factors (soil texture

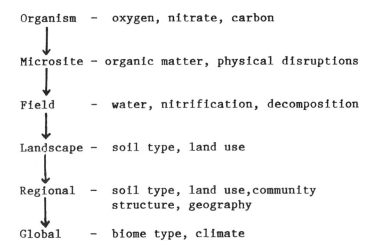

Organism — oxygen, nitrate, carbon

Microsite — organic matter, physical disruptions

Field   — water, nitrification, decomposition

Landscape — soil type, land use

Regional — soil type, land use,community
             structure, geography

Global   — biome type, climate

Figure 8.14. Factors controlling denitrification at different levels of investigation. (From Tiedje, 1987. *Environmental Microbiology of Anaerobes*, **Zehnder, A.J.B., Ed. With permission of John Wiley & Sons.)**

Table 8.13 Nitrate Uptake by a Plant (*Hordeum vulgare*) and Reduction by a Denitrifier (*Pseudomonas fluorescans*) at Different Oxygen Concentrations

Oxygen level (% v/v)	Denitrification (μg NO_3^--N mg^{-1} biomass h^{-1})	Plant uptake (μg NO_3^--N mg^{-1} biomass h^{-1})
20	0	1.85
10	0	1.15
5	0	0.52
1	0	0
0.05	0	0
0.01	0.15	0
0	3.5	0

From Wilson, R.J.R., Editor. 1988. *Advances in Nitrogen Cycling in Agricultural Ecosystems*. With permission of C.A.B. International.

and drainage) is given by data in Table 8.14. Denitrification loss was greatest in heavy (clay loam) soil. Also the increase in denitrification due to impoverished drainage was greater in clay loam than in loam soils and was least in sandy loam soil.

8.9.2.6. Leaching

Amount and intensity of rainfall, quantity and frequency of irrigation, evaporation rate, temperature effects, soil properties (particularly, texture and

Table 8.14 Annual N Loss Due to Denitrification as Affected by Denitrification, Soil Drainage, and Soil Texture for Forested Soils in Michigan

Soil	Denitrification N loss (kg N ha^{-1} yr^{-1})
Sand loam	
Well drained	0.6
Somewhat poorly drained	0.8
Poorly drained	0.5
Loam	
Well drained	10
Somewhat poorly drained	11
Poorly drained	24
Clay loam	
Well drained	18
Somewhat poorly drained	17
Poorly drained	40

From Wilson, R.J.R., Ed. 1988. *Advances in Nitrogen Cycling in Agricultural Ecosystems*. With permission of C.A.B. International.

structure), the type of land use, cropping and tillage practices, and the amount and form of N applied interact in complex ways to determine the amount of nitrates leached through the effective root zone, vadose zone, and eventually into the underground water.

Nitrate leaching can be accelerated by preferential flow of water through worm or root channels and natural fissures and cracks in soils. This is especially evident in vertisols. A number of computer simulation models are now available to predict nitrate leaching rates. The Nitrate Leaching and Economic Analysis Package (NLEAP) is an excellent computer program for identifying situations or practices where leaching could be excessive (Shaffer et al., 1991).

Leaching of nitrate is likely to be greatest under bare fallow. A 10-year comparison of nitrate profiles under pasture, fallow wheat, and continuous fallow in western Canada (Rennie et al., 1976) revealed leaching losses of 0 (pasture), 500 (fallow wheat), and 1082 (continuous fallow) kg N ha^{-1} to 3.6 m depth. Vegetation reduces nitrate leaching; perennials such as trees and grass are more effective than annual crops. The average leaching loss of N was 4 kg N ha^{-1} yr^{-1} from grassland and 23 kg N ha^{-1} yr^{-1} from cultivated soils in a United Kingdom experiment (Cooke, 1976).

In a 10-year study in Minnesota (Nelson and MacGregor, 1973) on a Webster loam, about 50% of applied fertilizer N was taken up by corn, and at higher fertilizer rates quite substantial amounts of nitrate were present after harvest in the 5.5-m soil profile (Figure 8.15). These data also show that removal of N by the crop changed little when the rate of N increased from

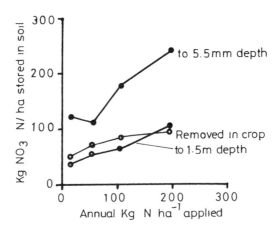

Figure 8.15. Relationship among fertilizer-N rate and amount of nitrate-N stored in soil 1.5 m and 5.5. m and nitrogen removed in crop. (From Nelson and MacGregor, 1973. Soil Sci. Soc. Am. Proc. 37:583–586. With permission of SSSA.)

100 to 200 kg N ha⁻¹. However, the increase in soil nitrate content (5.5-m depth) was considerable. These data clearly indicate that N leaching is greatly increased when fertilizer N rate exceeds crop needs. Thus, to control nitrate leaching into the groundwater, it is essential to develop reliable methods for predicting crop fertilizer needs.

The effects of land use and N fertilization on nitrate concentration in groundwater based on Western European experiences are summarized in Table 8.15. These data show that arable lands under vegetable cultivation contribute the most amount of nitrate to groundwater, and grasslands contribute the least. However, excessive leaching can even occur from grasslands when fertilizer N is applied at rates that exceed the need of the grass.

Gilliam (1991), based on an analysis of water samples over a 20-yr period in North Carolina, observed that drainage conditions prevailing in soil profiles affected nitrate concentration in soil water. In the lower coastal plain region, soils are poorly drained and have high organic matter content and high water tables. In these soils there is sufficient organic matter to provide an energy source for microorganisms so that denitrification occurs and nitrate concentrations are reduced (Figure 8.16*A*). On the other hand, in moderately well-drained soils little denitrification occurs and nitrate accumulation in groundwater is possible (Figure 8.16*B*). In these soils no nitrate was observed below 4 m because of the presence of an impermeable horizon. Water reaching these layers flows laterally to lower elevations, where it frequently enters a stream. Gilliam observed that these nitrates could be either utilized by plants along the way or lost by denitrification. No nitrate problem was recorded in the Upper Coastal Plains or Piedmont soils (Figure

**Table 8.15 Measured Site and Land-Use-Specific Nitrate N
Input into the Groundwater
(Mean Concentration of the Annual Groundwater Recharge)**

Soil	Land use (crop rotation, N fertilizer)	Mean nitrate concentration (mg N L^{-1})
Sand	Arable land (cereal-sugarbeet/potatoes-cereal, $\approx$ 120 kg N ha^{-1} yr^{-1})	25–30
	Arable land (cereal-winter catch crops –sugarbeet/potatoes-cereal, $\approx$ 120 kg N ha^{-1})	14–16
	Grassland (meadow, $\approx$ 250 kg N ha^{-1} yr^{-1})	3–7
	Grassland (intensively grazed pasture, $\approx$ 250 kg N ha^{-1} yr^{-1}; = 2 livestock units ha^{-1}, $\approx$ 180 grazing days)	14–20
	Field cropping of vegetables, including special crops such as asparagus, tobacco ($\approx$ 300–600 kg N ha^{-1} yr^{-1})	34–70
	Woodland (coniferous tree stands)	2.5
	Woodland (alder tree stands)	10
Loess	Arable land (cereal-sugarbeet-cereal $\approx$ 150 kg N ha^{-1} yr^{-1})	7–14

Adapted from Strebel et al. (1989).

8.16*C*). In those situations Gilliam concluded that nitrate leaching to ground-water was possible only with high rates of fertilizer or manure application (Figure 8.16*D*).

Nitrate accumulation in groundwater can result not only from excess N fertilization, but can also occur when excessive rates of animal manures or other organic wastes are applied to soils. In addition, there are a number of geological strata (especially marine shales) that contain excessive nitrates and/or exchangeable ammonium, which, given the right conditions, can result in nitrates leaching into groundwater.

8.10. INCREASING N USE EFFICIENCY

Since fertilizer N efficiency is determined by the biomass yield and nitrogen uptake by a crop, all factors that affect biomass yield and nitrogen concentration in tissue will affect N use efficiency. These factors can be broadly classified into five groups, namely, soil factors, crop factors, environmental factors, agronomic practices, and fertilizer management. Some of these factors are enumerated below:

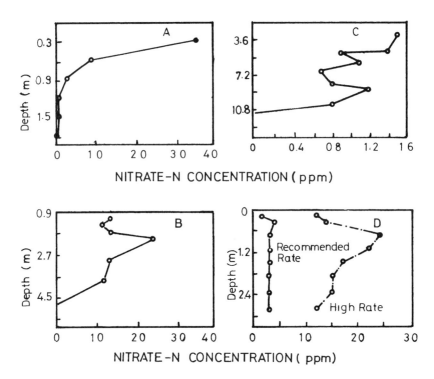

Figure 8.16. Typical nitrate-N profiles in North Carolina soils: A. Soil water NO₃-N in a cultivated, poorly drained, high water table, coastal plain soil. B. Soil water NO₃-N in a cultivated, moderately well drained, coastal plain soil with aquatard between 3 and 3.9 m. C. Soil nitrate-N profile of a cultivated Piedmont soil. D. Soil nitrate-N profile of an area of coastal Bermuda on a coastal plain soil receiving different rates of N. (Adapted from Gilliam, 1991. Better Crops 75:6–8. With permission.)

Soil factors
   1. Initial fertility—very fertile soils give lower fertilizer N response.
   2. Texture and structure—more leaching occurs in light, sandy soils.
   3. pH, salinity/alkalinity—greater ammonia volatilization occurs in calcareous and alkaline soils.
   4. Topography—surface runoff losses are greatest from upper topographic regions.
   5. Drainage—poor drainage can lead to large denitrification losses.
Crop factors
   1. Choice of crop (where more than one crop can be grown) and its yield potential (N uptake potential).

2. Choice of crop variety, its growth period, yield potential, and efficiency of nutrient utilization.
3. Resistance of the variety to diseases and pests, drought, and other stress factors.
4. Resistance of the crop variety to soil problems such as water logging, salinity, alkalinity, nutrient toxicities, etc.

Environmental factors

1. Rainfall and its distribution—large rains can lead to severe runoff and leaching losses.
2. Sunny days and sunshine hours, day length, heat units.
3. Occurrence and duration of frost, low night temperature.
4. Occurrence of thunderstorms, hail, tornadoes, wind damage.

Agronomic practices

1. Timely sowing—delays can reduce yield considerably.
2. Adequate plant population.
3. Water management—water-efficient irrigation practices in irrigated areas, water conservation practices for dryland conditions.
4. Adequate weed control.

Fertilizer management

1. Rate of application matched to crop needs.
2. Method of application to reduce nutrient losses.
3. Time of application matched to crop nutrient uptake pattern.
4. Source of N—modified urea materials, NH_4^+/NO_3^- ratio, liquids.
5. Fertilizer amendments—nitrification and urease inhibitors, coatings.

The discussion on all the factors mentioned above is beyond the scope of this chapter and is restricted to fertilizer N management.

8.10.1. Rate of Nitrogen Application

Fertilizer nitrogen efficiency, in most crops, is generally greater at lower fertilization levels than at higher levels and decreases considerably when N rates increase beyond the optimum. For example, data in Table 8.16 show that the increase in fertilizer N uptake by corn was greatly reduced beyond 100 kg N ha^{-1}. As a consequence, much more nitrate N was lost at rates above 100 kg N ha^{-1} than below. Numerous data supporting this conclusion exist for most crops from all parts of the world.

8.10.2. Method of N Application

Since part of the N applied in ammonium or ammonium-producing fertilizers (urea) is lost through ammonia volatilization, particularly on calcareous and alkaline soils and in rice paddies, fertilizer placement at a few centimeters

Table 8.16 Nitrogen Balance for Microplots Treated with
^{15}N-Enriched Fertilizer

Fertilizer (kg N ha^{-1})	Crop uptake (% of initial N applied[a])	Soil 0–2.4 m (% of initial N applied[a])	Unaccounted for fertilizer (% of initial N applied[a])
50	48 ± 1.1 (5.0)[b]	23.7 ± 7.5	28.3 ± 7.4
100	49.8 ± 3.1 (5.7)	20.0 ± 5.0	30.2 ± 7.0
150	53.1 ± 2.9 (5.9)	19.8 ± 8.1	27.1 ± 6.5

[a] Mean ± the standard deviation.

[b] Values in parentheses are included in the total and represent an estimate of the contribution of roots to total removed by plants based on roots comprising 15% of total plant dry weight at physiological maturity.

From Francis et al. 1993. Agron. J. 85:659–663. With permission of ASA.

depth is considered desirable and profitable. An example of the advantage of placement of N in rice fields has been shown in Figure 8.12.

Soil injection is essential when anhydrous ammonia is used as a fertilizer. In a study in the United States (Stanley and Smith, 1956) loss of anhydrous ammonia was negligible when injected at a depth of 22.5 cm (Figure 8.17) irrespective of soil water content up to 18% (w/w). However, when injected at a depth of 7.5 cm, about 12% of the anhydrous ammonia was lost in a dry soil (2% w/w); loss was reduced to a little over 1% when soil water content increased to about 16%. Soil water dissolves ammonia and changes it to ammonium, which can be retained on the exchange complex of the soil.

8.10.3. Time of N Application

Because N is susceptible to loss from soil, it is desirable to apply N in small doses during the growth of the crop. Often fertilizer is applied in two or three split doses; the amounts at each application may be equally or differentially divided depending upon the rate of N, duration and need of the crop at different growth stages, and soil texture. Data from a field experiment on spring wheat conducted on light textured soils of Punjab (India) are shown in Figure 8.18. On such soils a single basal application was desirable at 40 kg N ha^{-1}, two split doses when 80 to 120 kg N ha^{-1} is applied (the most frequently recommended rate of N for wheat in India), and three split doses when 160 to 200 kg N ha^{-1} is required.

Although split application of fertilizer N is most efficient, in regions such as the corn belt of the United States where N fertilizers are relatively cheap, many producers find it more profitable to apply a little extra N to counteract losses but to apply all the fertilizer in one preplant operation. However, under

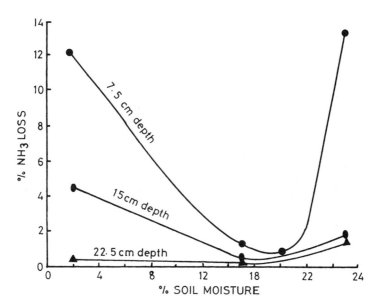

Figure 8.17. **Losses of ammonia from a Putnam loam soil as influenced by depth of application and soil moisture. Anhydrous ammonia applied at the rate of 220 kg N ha⁻¹ (at 100-cm spacings). (From Stanley and Smith, 1956. Soil Sci. Soc. Am. J. 20:557–561. With permission of SSSA.)**

certain conditions, this practice can result in excessive nitrate leaching and pollution of groundwater.

8.10.4. Nitrification Inhibitors

Since leaching and denitrification losses of fertilizer N take place only after ammonium or ammonium-producing fertilizers are nitrified, retarding the nitrification rate has been considered one way to reduce N losses and thereby increase N efficiency. By this means, fertilizer N is maintained primarily in the ammonium (nonleaching) form for several weeks until plant growth and N uptake rates increase to the extent that much of the nitrate formed is used immediately and nitrates do not accumulate in the soil. Nitrification inhibitors (NIs) emerged as a group of agrochemicals with the development of N-serve or nitrapyrin (NP) (2-chloro-6 [trichloromethyl] pyridine). Since that time a large number of chemicals have been reported as NIs, but only seven have been produced commercially; these are N-serve or NP; AM (2-amino-4-chloro-6 methyl pyrinidine); DCD (Dicyandiamide); ST (2-sulfanilamido thiazole); thiourea; Dwell or Ferrozole or Etridiazole (5-ethyoxy-3-trichloromethyl-1,2,4,-thiadizole); and MBT (2 mercaptobenzothiazole). One natural product, namely, neem (*Azadiraclita indica* Juss) seed extract (powder or cake), is also reported

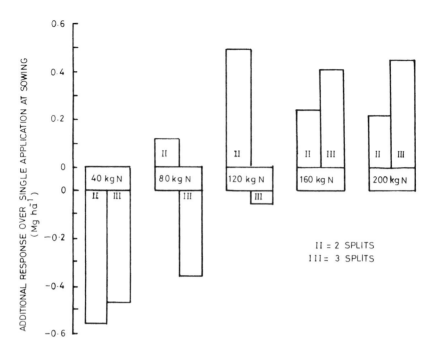

Figure 8.18. The effect of two or three split doses of N at increasing levels of N applied to wheat in a light textured soil. Means of 2 years. (From Sandhu and Gill, 1971. Indian J. Agric. Sci. 41:815–823.)

to have nitrification-inhibiting properties (Reddy and Prasad, 1975; Thomas and Prasad, 1982). Success with the use of NIs under field conditions has been mixed; while some researchers found increased crop yield and N use efficiency with NIs, others did not (Prasad and Power, 1995). In field experiments with rice conducted at the Indian Agricultural Research Institute, the NIs, N-Serve, AM, ST, DCD, and neem cake showed a positive response (Reddy and Prasad, 1977; Sudhakara and Prasad, 1986). Prasad et al. (1981) reported that averaged over 6 years neem cake–coated urea (NCU) gave 0.4 Mg ha[-1] more rice grain than prilled urea (PU). Furthermore, the residual effects resulted in 0.5 Mg ha[-1] more grain in succeeding wheat. Similarly, field tests carried out in Japan (Nishihara and Tsunyoshi, 1968) showed an increase in rice yield with AM. Wells et al. (1989) summarized results with DCD on rice from Arkansas, California, Louisiana, Mississippi, and Texas and reported that DCD increased yields in both drill seeding and water-seeded, continuously flooding, rice culture. Use of DCD was advantageous only if the flood was delayed for more than 14 days after urea application.

With corn, the results of field experiments in the eastern corn belt of the United States showed that in 70% of the trials in Indiana there was an increase in yield with NIs (Nelson and Huber, 1980). Results of experiments with DCD nitrapyrin in the north central United States indicated that the greatest benefit for NI use was obtained on coarse, textured soil. Results from experiments with NIs in the southeastern United States suggest limited benefits to corn due to high soil temperatures. Townsend and McRae (1980) from Nova Scotia, Canada, also observed that except on light, sandy soils, no yield advantage was gained with NP. In grain sorghum NP or Dwell applied with urea, anhydrous ammonia or urea ammonium nitrate solution did not increase yield nor improve efficiency of fertilizer N applied to grain sorghum during a period of 4 years, even with supplementary irrigation to promote leaching and/or denitrification (Westerman et al., 1981). Cotton is sensitive to DCD, and there was a tendency for cotton yields to decline with DCD.

8.10.5. Slow-Release N Fertilizers

One way of reducing N losses and increasing fertilizer use efficiency is to use fertilizer materials that release plant-available N slowly in the soil so that plants can utilize most before it is lost. Slow-release nitrogen fertilizers offer alternatives to the practices of split application and placement of N for increasing the efficiency of nitrogen applied to summer cereals (Prasad, 1982). These materials have drawn attention of agronomists, soil scientists, and fertilizer technologists the world over.

These materials can be broadly classified into two groups: (1) chemical compounds with inherently slow rates of dissolution, for example, urea-form, oxamide isobutylidene diurea (IBDU), and (2) urea and other nitrogen fertilizers provided with a coating that acts as a moisture barrier. Materials such as sulfur, shellac, plastics, and gypsum have been used.

Sulfur-Coated Urea (SCU). Of the slow-release fertilizers SCU, developed by the Tennessee Valley Authority, has been most widely tested. Due to slow release of N from SCU and to the localized acidifying effect of the sulfur, ammonia volatilization losses are usually reduced by using this material. SCU has been tested mostly for the rice crop. Engelstad et al. (1972), summarizing the results before 1972, observed that the response to SCU was generally good and superior to the response to urea in a single application and often exceeded yields for split applications of prilled urea applied at the same N rate.

Isobutylidene Diurea (IBDU). IBDU as a source of N has been the subject of considerable study in Japan, where it was developed (Hamamoto, 1986). This material gave 20% more rice than ammonium sulfate at equivalent N rates. It has also been evaluated in the United States (Hughes, 1976) and India (Rajale and Prasad, 1975). Release of N from IBDU depends upon soil pH and on fertilizer particle size. Hughes (1976) reported that N release from

IBDU was much more rapid in acid than in alkaline soils. He observed that 75% of N from IBDU was released after 10 weeks with a particle size of 0.6 to 0.7 mm, 58% in 21 weeks with 1.0 to 1.2 mm, and 50% in 32 weeks with 1.7 to 2.0-mm size particles.

Plastic-Coated Urea.   In recent years, urea and other fertilizer materials have been encapsulated in decomposable polyvinylchloride or other plastics that are slowly decomposable to control release rate.

8.10.6. Urea Supergranules (USG)

Urea supergranules were developed based on the mudball technique followed in China for increasing the efficiency of fertilizer N. In the mudball technique, a small amount of fertilizer is placed in the center of a mudball by making a hole in the mudball and closing the hole after fertilizer injection. The mudballs are then dried and kept ready for use. Mudballs are placed deep in rice fields at an appropriate time. This is certainly a laborious and time-consuming process. Urea supergranules are aimed at overcoming this. Urea supergranules and briquettes can be made of different shapes and sizes. The 1-g size has generally been most tested. The deep placement of USG by hand or machine is an agronomically efficient, economically sound, and environmentally safe practice in the traditional transplanted rice growing areas, especially those that have nonpermeable soils. Studies over a period of 10 years at IARI in New Delhi (Prasad et al., 1989) showed that USG had a yield advantage over urea of 0.4 Mg ha^{-1} (10.7%). A large number of trials at research centers, as well as on farmer fields, have shown USG to be more efficient than prilled urea for rice on medium and heavy soils only (Kumar et al., 1989). In light, sandy soils the USGs have not been superior due to increased leaching (Katyal et al., 1988). For best results USG should be placed 4 to 5 cm below the soil surface (Singh et al., 1989). To some extent this requirement restricts acceptance because a low-cost applicator is not yet available.

8.11. NITROGEN AVAILABILITY INDICES

A nitrogen availability index is a quantitative value or a certain soil property (or group of properties) that presumably correlates with the amount of nitrogen made available to a crop during the growing season. Such indices may be used to predict nitrogen fertilization rate or other management practices. This topic was reviewed by Keeney (1982).

Total soil N and soil organic matter were the first indices used, and they are still used by some as an index of available soil N. Generally, however, total soil N by the Kjeldahl method is too costly and time consuming for routine soil testing. Nevertheless, soil organic matter can be determined quite rapidly and it is still widely used. In addition to total N and organic matter, other popular indices are chemical extractants or hydrolysis agents such as 1 or 6 N $(NH_4)_2SO_4$ (Gallagher and Bartholomew, 1964), $Ca(OH_2)$ (Prasad, 1965), 0.1 N $Ba(OH)_2$ (Jenkinson, 1968), alkaline $KMnO_4$ (Stanford, 1978), 0.01 N $CaCl_2$ (Standford, 1968), and 1 N NaOH (Cornforth and Walmsley, 1971). An electroultrafiltration (EUF) procedure (a form of electrophoresis) is used in Germany to determine the fertilizer requirement for sugar beets (Nemeth, 1979). A number of biological tests have been proposed such as NO_3-N liberated in 14 to 78 days of incubation (Gallagher and Batholomew, 1964); N mineralized in 14 to 112 days of incubation (Stanford and Smith, 1972); inorganic N released by autoclaving (Standford and DeMar, 1970); and anaerobic mineralization (Cornforth and Walmsley, 1971). None of the above soil tests has been universally accepted. The current thrust in the United States is on measurement of residual soil profile NO_3-N (Stanford and Smith, 1972). Soil samples are taken soon after harvest or just before planting to a depth of 60 to 120 cm. This test is known as the preplant nitrate test (PPNT) (Fox and Piekielek, 1983). This method has met with success in North Dakota, Oklahoma, and Nebraska. The Nebraska interpretation employs an equation to calculate the fertilizer N requirement as a function of the yield goal, the crop N content, and a constant 56 kg ha^{-1} soil N addition.

In addition to or in place of PPNT, there is considerable effort to use PSNT (Pre-sidedress soil nitrate test) (Meisinger et al., 1992; Fox et al., 1992). Sampling of soil (30 cm depth) before sidedressing and analyzing for NO_3-N is considered to provide a more reliable index of available soil N and is used for making a decision on the amount of N to be sidedressed. The major drawback is the short time (about 2 to 3 weeks) available for collecting the soil samples, analyzing for NO_3-N, deciding the rate of N to be applied, and then actually applying the fertilizer N. A more rapid approach is to determine the degree of N stress by measuring the chlorophyll content of corn leaves using a handheld chlorophyll meter (Schepers et al., 1992) or by remote sensing.

8.12. NITROGEN-DEFICIENCY SYMPTOMS

Nitrogen-deficiency symptoms are the most prevalent and easiest to identify. Young crop plants exhibit yellowish green foliage and stunted growth. In older plants there is yellowing or firing on the lower leaves, usually starting at leaf tips and margins (Figure 8.19). Ears are small, and the protein content in grain is low. Nitrogen deficiency is a major barrier to yield all over the world, particularly in developing countries. An effective integrated approach employing organic manures, biofertilizers, chemical fertilizers, nitrification inhibitors, and coated and long-persisting nitrogen fertilizers is the key to sustainable agriculture. Specific plans would differ from country to country and region to region because of differences in climates, soils, crops, labor supply, knowledge base, financial resources, and other reasons.

Figure 8.19. Nitrogen deficiency in corn (maize). Firing (or yellowing) of the midrib of older, lower levels of the plant (left), close-up of firing (right). (From *Corn Field Manual*, J.R. Simplot Company Minerals & Chemical Division, Pocatello, ID, ©1984. With permission.) See Plate 1 following p. 170.

REFERENCES

Becking, J.H. 1961. Studies on nitrogen fixing bacteria of the germ *Beijerinekia*. I. Biological and ecological distribution in soils. Plant Soil 14:49–81.

Becking, J.H. 1977. Dinitrogen fixation in higher plants other than legumes, in *A Treatise on Dinitrogen Fixation*, R.W.F. Hardy and W.S. Silver, Eds., John Wiley & Sons, New York, pp. 185–276.

Becking, J.H. 1979. *In Nitrogen and Rice*, International Rice Research Institute, Los Banos, Philippines.

Black, A.S., R.R. Sherlock, N.P. Smith, K.C. Camerson, and K.M. Goh. 1984. Effect of previous urine application on ammonia volatilization from 3 nitrogen fertilizers. N.Z. J. Agric. Res. 27:413–416.

Black, A.S. and S.A. Waring. 1972. Ammonium fixation and availability in some cereal producing soils in Queensland. Aust. J. Soil Res. 10:197–207.

Bremner, J.M. and K. Shaw. 1958. Denitrification in soil. II. Factors affecting denitrification. J. Agric. Sci. Camb. 51:40–52.

Brezonik, P.L. 1977. Denitrification in natural waters. Progress in Water Technology 8:373–392.

Buresh, R.J., S.K. DeDatta, J.L. Padilla, and M.I. Samson. 1988a. Effect of two urease inhibitors on floodwater ammonia following urea application to lowland rice. Soil Sci. Soc. Am. J. 52:856–861.

Buresh, R.J., S.K. DeDatta, J.L. Padilla, and M.I. Samson. 1988b. Field evaluation of two urease inhibitors with transplanted lowland rice. Agron. J. 80:763–768.

Burns, R.C. and R.W.F. Hardy. 1975. *Nitrogen Fixation in Bacteria and Higher Plants*, Springer-Verlag, New York.

Chen, C., F.T. Turner, and J.B. Dixon. 1989. Ammonium fixation by high-charge smectite in selected Texas Gulf Coast soils. Soil Sci. Soc. Am. J. 53:1035–1040.

Chu, L.C. 1979. Use of azolla in rice production in Elina, in *Nitrogen and Rice*, International Rice Research Institute, Los Banos, Philippines, pp. 375–418.

Cooke, G.W. 1976. A review of the effects of agriculture on the chemical composition and quality of surface and underground waters, in *Agriculture and Water Quality*, MAFF Tech. Bull. 32., Her Majesty's Stationery Office, London, pp. 5–57.

Cornforth, J.W. and D. Walmsley. 1971. Methods of measuring available nutrients in West Indian soils. Plant Soil 35:389–399.

Daigger, L.A., D.H. Sander, and G.A. Peterson. 1976. Nitrogen content of winter wheat during growth and maturation. Agron. J. 68:815–818.

Dancer, W.S., L.A. Peterson, and G. Chesters. 1973. Ammonification and nitrification of N as influenced by soil pH and previous N treatments. Soil Sci. Soc. Am. Proc. 37:67–69.

DeDatta, S.K., R.J. Buresh, M.I. Samson, W.N. Obcemea, and J.G. Real. 1991. Direct measurement of ammonia and denitrification fluxes from urea applied to rice. Soil Sci. Soc. Am. J. 55:543–548.

Delaune, R.D. and W.H. Patrick, Jr. 1970. Urea conversion to ammonia in waterlogged soils. Soil Sci. Soc. Am. Proc. 34:603–607.

Dobereiner, J., J.M. Day, and P.J. Dart. 1972. Nitrogenese activity and oxygen sensitivity of the *Paspalum notatum – Azotobacter paspali* association. J. Gen. Microbiol 71:103–116.

Dobereiner, J. and J.M. Day. 1976. Associative symbiosis in tropical grasses: characterization of microorganisms and dinitrogen fixing sites, in W.E. Newton and C.J. Nyman, Eds., Proc. 1st Int. Symp. Nitrogen Fixation, Washington State University Press, Pullman, pp. 518–538.

Eggleton, W.G.E. 1935. The nitrification of ammonia in the field and in laboratory experiments. Annals Appl. Biol. 22:419–430.

Engelstad, O.P., J.G. Getsinger, and P.G. Stangel. 1972. Tailoring of fertilizers for rice. Tennessee Valley Authority Bull. I-52, p. 56.

Evans, H.J. and L.E. Barber. 1977. Biological nitrogen fixation for food and fiber production. Science 197:332–339.

Fox, R.H. and W.P. Piekielek. 1983. Response of corn to nitrogen fertilizer and the prediction of soil nitrogen availability with chemical tests in Pennsylvania. Penn. Agric. Expt. Stn. Bull. 843.

Fox, R.H., J.J. Meisinger, J.T. Sims, and W.P. Piekielek. 1992. Predicting N fertilizer needs for corn in humid regions: advances in the mid-Atlantic states, in *Predicting N Fertilizer Needs for Corn in Humid Regions*, B.R. Bock and K.R. Kelley, Eds., National Fertilizer and Environmental Research Center, TVA, Muscle Shoals, AL, Bull Y-226, pp. 43–56.

Francis, D.D., J.S. Schepers, and M.F. Vigil. 1993. Post-anthesis nitrogen loss from corn. Agron. J. 85:659–663.

Freney, J.R., A.C.F. Trevitt, S.K. DeDatta,W.N. Obcemea, and J.G. Real. 1990. The interdependence of ammonia volatilization and denitrification as nitrogen loss processes in flooded rice fields in the Philippines. Biol. Fert. Soils 9:31–36.

Gallagher, P.W. and W.V. Bartholomew. 1964. Comparison of nitrate production and other procedures in determining nitrogen availability in southeastern coastal plain soils. Agron. J. 56:179–184.

George, M. and R. Prasad. 1989. Studies on the effect of fertilizer N utilization by rice using ^{15}N technique in rice based multiple cropping systems. Res. Dev. Agric. 6:115–118.

Gilliam, J.W. 1991. Fertilizer nitrates not causing problems in North Carolina groundwater. Better Crops. 75:6–8.

Goswani, N.N., R. Prasad, M.C. Sarkar, and S. Singh. 1988. Studies on the effect of green manuring in nitrogen economy in a rice-wheat rotation using ^{15}N technique. J. Agric. Sci. Camb. 11:413–417.

Greaves, J.E. and E.G. Carter. 1920. Influence of moisture on the bacterial activities of the soil. Soil Sci. 10:361–387.

Groffman, P.M. and J.M. Tiedje. 1989. Denitrification in north temperate forest soils: Spatial and temporal patterns at the landscape and seasonal scales. Soil Biol. Biochem. 21:613–620.

Groffman, P.M., J.M. Tiedje, G.P. Robertson, and S. Christensen. 1988. Denitrification at different temporal and geographical scales: proximal and distal controls, in *Advances in Nitrogen Cycling in Agricultural Ecosystems*, J.R. Wilson, Ed., C.A.B. International, Wallingford, UK, pp. 174–192.

Hageman, R.H. 1984. Ammonium vs nitrate nutrition of higher plants, in *Nitrogen in Crop Production*, R.D. Hauck, Ed., Am. Soc. Agron., Madison, WI, pp. 67–85.

Hamamoto, M. 1986. Isobutylidene diurea as a slow acting nitrogen fertilizer and the studies in this field in Japan. Proc. Fert. Soc. London 90:1–79.

Hardy, R.W.F. and M.D. Marvelka. 1975. Nitrogen fixation research: a key to world food. Science. 188:633–643.

Harper, L.A., R.R. Sharpe, G.W. Langdale, and J.E. Giddens. 1987. Nitrogen cycling in a wheat crop: soil, plant and aerial transport. Agron. J. 79:965–973.

Hongprayoon, C., C.W. Lindau, W.H. Patrick, Jr., D.R. Bouldin, and K.R. Reddy. 1991. Urea transformations in flooded soil columns. I. Experimental results. Soil Sci. Soc. Am. J. 55:1130–1134.

Hughes, T.P. 1976. Nitrogen release from isobutylidene diurea: soil pH and feritlizer particle size effects. Agron. J. 68:103–106.

Janzen, H.H. and S.M. McGinn. 1991. Volatile loss of nitrogen during decomposition of legume green manure. Soil Biol. Biochem. 23:291–297.

Janzen, H.H., C.W. Lindwall, and C.J. Roppel. 1990. Relative efficiency of point-injection and surface application for fertilization of winter wheat. Can. J. Soil Sci. 70:189–201.

Jenkinson, D.S. 1968. Chemical tests for potentially available nitrogen in soils. J. Soil Fd. Agric. 19:168–178.

Jensen, E.S., B.T. Christensen, and L.H. Sorensen. 1989. Mineral-fixed ammonium in clay- and salt-size fractions of soils incubated with ^{15}N ammonium sulphate for five years. Biol. Fertil. Soils 8:298–302.

John, P.S., R.J. Buresh, R.K. Pandey, R. Prasad, and T.T. Chua. 1989a. Nitrogen-15 balances for urea and neem-coated urea applied to lowland rice following two cowpea cropping systems. Pl. Soil 120:233–241.

John, P.S., R.J. Buresh, R. Prasad, and R.K. Pandey. 1989b. Nitrogen gas (N_2 + N_2O) flux from urea applied to lowland rice as affected by green manure. Pl. Soil 119:7–13.

Joseph, P.A. and R. Prasad. 1993. The effect of dicyandiamide and neem cake on the nitrification of urea-derived ammonium under field conditions. Biol. Fertil. Soils 15:149–152.

Katyal, J.C., B. Singh, P.L.G. Vlek, and R.J. Buresh. 1987. Efficient nitrogen use as affected by urea application and irrigation sequence. Soil Sci. Soc. Am. J. 51:366–370.

Katyal, J.C., B. Singh, and P.L.G. Vlek. 1988. Effect of granule size and placement geometry on the efficiency of urea supergranules for wetland rice grown on a permeable soil. Fert. Res. 15:193–201.

Keeney, D.R. 1982. Nitrogen-availability indices, in *Methods of Soil Analysis, Part 2: Chemical and Microbial Properties*, 2nd ed., A.L. Page, Ed., Am. Soc. Agron, Madison, WI, pp. 711–733.

Kowalenko, C.G. and G.T. Ross. 1980. Studies on the dynamics of recently clay-fixed NH_4^+ using ^{15}N. Can. J. Soil Sci. 60:61–70.

Kumar, V., G.C. Shrotriya, and S.V. Kaore. 1990. Scope of urea supergranules for rice in India. Fertil. News 35(7):23–29.

LaRue, T.A. 1977. The bacteria, in *A Treatise on Dinitrogen Fixation*, R.W.F. Hardy and W.S. Silver, Eds., John Wiley & Sons, New York, pp. 19–62.

Linn, D.M. and J.W. Doran. 1984. Effect of water-filled pore space on carbon dioxide and nitrous oxide production in tilled and non-tilled soils. Soil Sci. Soc. Am. J. 48:1267–1272.

Malhi, S.S. and W.B. McGill. 1982. Nitrification in three Alberta soils: effect of temperature, moisture and substrate concentration. Soil Biol. Biochem. 14:393–399.

Mandal, B. and A.K. Mukhopadhyay. 1984. Ammonium fixation in soils from application of NH_4^+-producing fertilizers. J. Indian Soc. Soil Sci. 32:486–487.

Meisinger, J.J., F.R. Magdoff, and J.S. Schepers. 1992. Predicting N fertilizer needs for corn in humid regions: underlying principles, in *Predicting N Fertilizer Needs for Corn in Humid Regions*, B.R. Bock and K.R. Kelley, Eds., National Fertilizer and Environmental Research Center, TVA, Muscle Shoals, AL. Bull Y-226, pp. 7–27.

Mitsui, S. and K. Kurihara. 1962. The intake and utilization of carbon by plant roots for ^{14}C-labelled urea. IV. Absorption of intact urea molecule and its metabolism in plants. Soil Sci. Pl. Nutr. 8:9–15.

Moe, P.G., J.V. Mannering, and C.B. Johnson. 1968. A comparison of nitrogen losses from urea and ammonium nitrate in surface runoff water. Soil Sci. 105:428–433.

Morril, L.G. and J.E. Dawson. 1967. Patterns observed for the oxidation of ammonium to nitrate by soil organisms. Soil Sci. Soc. Am. Proc. 31:757–760.

Mosier, A.R., S.K. Mohanty, A. Bhadrachalam, and S.P. Chakravorti. 1990. Evolution of dinitrogen and nitrous oxide from the soil to the atmosphere through rice plants. Biol. Fertil. Soils. 9:61–67.

Mouchova, H. and J. Apltauer. 1983. Effects of nitrification inhibitor N-Serve on the utilization of fall applied urea by wheat. Fert. Res. 4:165–180.

Myers, R.J.K. 1975. Temperature effects on ammonification and nitrification in a tropical soil. Soil Biol. Biochem. 7:83–87.

NAS. 1978. *Nitrates: An Environmental Assessment*, National Academy of Science, Washington, DC.

Nelson, D.W. and D.M. Huber. 1980. Performance of nitrification inhibitors in the Mideast (east), in *Nitrification Inhibitors — Potentials and Limitations*, J.J. Meisinger, G.W. Randall and M.L. Vitosh, Eds., Spec. Pub. 37, Am. Soc. Agron., Madison, WI, pp. 75–88.

Nelson, W.W. and J.M. MacGregor. 1973. Twelve years of continuous corn fertilization with ammonium nitrate or urea nitrogen. Soil Sci. Soc. Am. Proc. 37:583–586.

Nemeth, K. 1979. The availability of nutrients in soils as determined by electroultrafiltration (EUF). Adv. Agron. 31:155–187.

Nishihara, T. and T. Tsunyoshi. 1968. The effect of some nitrification inhibitors on the availability of basic fertilizer nitrogen by rice plants on dry paddy fields. Bull. Fac. Agric. Kagoshima Univ., pp. 131–141.

Nommick, H. 1957. Fixation and defixation of ammonium in soils. Acta Agric. Scand. 7:395–436.

Nutman, P.S. 1971. Perspectives in biological nitrogen fixation. Sci. Prog. (Oxford) 59:55–74.

Orchard, E.R. and G.D. Darb. 1956. Fertility changes under continued wattle culture with special reference to nitrogen fixation and base status of the soil, 6th Int Cong. Soil Sci. Trans. 4:305–310.

Osborne, G.J. 1976. The significance of inter-calary (CHECK) ammonium in representative surface and subsoils from southern New South Wales. Aus. J. Soil Res. 14:381–388.

Papkosta, D.K. and A.A. Gagianas. 1991. Nitrogen and dry matter accumulation, remobilization, and loss for Mediterranean wheat during grain filling. Agron. J. 83:864–870.

Patron, W.J., J.A. Morgan, J.M. Altenhofer, and L.A. Harper. 1988. Ammonia volatilization from spring wheat plants. Agron. J. 80:419–425.

Power, J.F., Ed. 1987. *The Role of Legumes in Conservation Tillage Systems*, Soil Conserv. Soc. Am., Ankeny, IA, p. 153.

Power, J.F., J. Alessi, G.A. Reichman, and D.L. Grunes. 1973. Recovery, residual effects, and fate of nitrogen fertilizer sources in a semi-arid region. Agron. J. 65:765–768.

Powlson, D.D., G. Pruden, A.E. Johnson, and D.S. Jenkinson. 1986. The nitrogen cycle in the broadwalk wheat experiment: recovery and losses of ^{15}N-labelled fertilizer applied in spring and inputs of nitrogen from the atmosphere. J. Agric. Sci. Camb. 107:591–609.

Prasad, R. 1965. Determination of potentially available nitrogen in soils — a rapid procedure. Plant Soil 23:261–263.

Prasad, R. 1982. The use of nitrification inhibitors and slow-release nitrogen fertilizers for manipulation of growth and yield of rice, in *Chemical Manipulation of Crop Growth and Development*, J.S. McLaren, Ed., Butterworths Scientific, London, pp. 451–464.

Prasad, R. and J.F. Power. 1995. Nitrification inhibitor for agriculture, health and environment. Adv. Agron. 54:233–281.

Prasad, R. and S.K. DeDatta. 1979. Increasing fertilizer nitrogen efficiency in wetland rice, in *Nitrogen and Rice*, International Rice Research Institute, Manila, Philippines, pp. 465–484.

Prasad, R., G.B. Rajale, and B.J. Lakhdive. 1971. Nitrification retarders and slow-release nitrogen fertilizers. Adv. Agron. 23:337–383.

Prasad, R., J. Thomas, V.V.S.R. Gupta, and S. Singh. 1983. Ammoniphilic plants for reducing water pollution. Env. Conserv. 10(3):260–261.

Prasad, R., S.N. Sharma, M. Prasad, and R.N.S. Reddy. 1981. Efficient utilization of nitrogen in rice-wheat rotation. Indian Soc. Agron. Natl. Symp. Crop Management to Meet the New Challenges, Hissar, pp. 37–43.

Prasad, R., S.N. Sharma, S. Singh, and M. Prasad. 1989. Relative efficiency of prilled urea and urea supergranules for rice, in *Soil Fertility and Fertilizer Use*, Vol. 3, G.C. Shrotriya and S.V. Kaore, Eds., IFFCO, New Delhi. pp. 38–46.

Rajale, G.B. and R. Prasad. 1975. Nitrogen and water management for rice. Il Riso 24:117–125.

Recous, S. and B. Mary. 1990. Microbial immobilization of ammonium and nitrate in cultivated soils. Soil Biol. Biochem. 22:913–922.

Reddy, R.N.S. and R. Prasad. 1975. Studies on the mineralization of urea, coated urea and nitrification inhibitor treated urea in the soil. J. Soil Sci. 26:304–312.

Reddy, R.N.S. and R. Prasad. 1977. Effect of variety, rates and sources of nitrogen on growth characters, yield components and yield of rice. Il. Riso 26:217–223.

Rennie, D.A., G.J. Racz, and D.K. McBeath. 1976. Nitrogen losses. Proc. Western Canada Nitrogen Symp., Alberta, Edmonton, pp. 325–353.

Sabey, B.R., L.R. Frederick, and W.V. Bartholomew. 1959. The formation of nitrate from ammonium in soils. III. Influence of temperature and initial population of nitrifying organisms on the maximum rate and delay period. Soil Sci. Soc. Am. Proc. 23:463–465.

Sandhu, H.S. and G.S. Gill. 1971. Compaction response of C306 and kalyan varieties of wheat to nitrogen application. Indian J. Agric. Sci. 41:815–823.

Saraswathi, P., P.V. Balachandran, and P.A. Wahid. 1991. Inhibition of urea hydrolysis in flooded soils and its significance in the molecular absorption of urea by rice. Soil Biol. Biochem. 23:125–129.

Schaffer, M.J., A.D. Halvorson, and F.J. Pierce. 1991. Nitrate leaching and economic analysis package (NLEAP): model description and application, in *Managing Nitrogen for Groundwater Quality and Farm Profitability*, R.F. Follett, D.R. Keeney and R.M. Cruse, Eds., Soil Sci. Soc. Am., Madison, WI, pp. 285–322.

Schepers, J.S., T.M. Blackmer, and D.D. Francis. 1992. Predicting N fertilizer needs for corn in humid regions: using chlorophyll meters, in *Predicting N Fertilizer Needs for Corn in Humid Regions*, B.R. Bock and K.R. Kelley, Eds., National Fertilizer and Environmental Research Center, TVA, Muscle Shoals, AL, Bull Y-226, pp. 105–114.

Schlegel, A.J., D.W. Nelson, and L.E. Sommers. 1986. Field evaluation of urease inhibitors for corn production. Agron. J. 78:1007–1012.

Schloesing, M.T. and A. Muntz. 1879. Recherches sur la nitrification. Comptes Rendus 89:1074–1077.

Schnier, H.F., M. Dingkuhn, S.K. DeDatta, E.P. Marqueses, and J.E. Faronilo. 1990. Nitrogen-15 balance in transplanted and direct-seeded flooded rice as affected by different methods of urea application. Biol. Fertil. Soils 10:89–96.

Shaffer, M.J., A.D. Halvorson, and F.J. Pierce. 1991. Nitrate leaching and economic analysis package (NLEAP): model description and application, in *Managing Nitrogen for Groundwater Quality and Farm Profitability*, R.F. Follett, D.R. Keeney and R.M. Cruse, Eds., Soil Sci. Soc. Am., Madison, WI, pp. 285–322.

Singh, P.K. 1979. Use of azolla in rice production in India, in *Nitrogen and Rice*, International Rice Research Institute, Los Banos, Philippines, pp. 467–418.

Singh, S., R. Prasad, and S.N. Sharma. 1989. Growth and yield of rice affected by spacing, time and depth of placement of urea briquettes. Fert. Res. 19:99–101.

Singh, Y., C.S. Khind, and B. Singh. 1991. Efficient management of leguminous green manures in wetland rice. Adv. Agron. 45:135–189.

Smith, B.E., R.R. Eady, R.N.F. Thorneley, M.G. Yates, and J.R. Postgate. 1977. in *Recent Developments in Nitrogen Fixation*, W.E. Newton, J.R. Postgate and C. Rodriquez-Barruco, Eds., Academic Press, New York.

Sparks, D.L., R.L. Blevins, H.H. Barley, and R.I. Barnhisel. 1979. Relationship of ammonium nitrogen distribution to mineralogy in a Hapludalf soil. Soil Sci. Soc. Am. J. 43:786–789.

Stanford, G. 1968. Extractable organic nitrogen and nitrogen mineralization in soils. Soil Sci. 106:345–351.

Stanford, G. 1978. Evaluation of ammonium release by alkaline permanganate as an index of soil nitrogen availability. Soil Sci. 126:244–253.

Stanford, G. and S.J. Smith. 1972. Nitrogen mineralization potentials of soils. Soil Sci. Soc. Am. Proc. 36:465–472.

Stanford, G. and W.H. DeMar. 1970. Extraction of soil organic nitrogen by autoclaving in water. II. Diffusible ammonia, an index of soil nitrogen availability. Soil Sci. 109:190–196.

Stanford, G., J.N. Carter, D.T. Westermann, and J.J. Meisinger. 1977. Residual nitrate and mineralizable soil N in relation to N uptake by irrigated sugarbeets. Agron. J. 69:303–308.

Stanley, F.A. and G.E. Smith. 1956. Effect of soil moisture and depth of application on retention of anhydrous ammonia. Soil Sci. Soc. Am. J. 20:557–561.

Strebel, O., W.H.M. Duynisveld, and J. Bottcher. 1989. Nitrate pollution of groundwater in Western Europe. Agric. Ecosystem. Environ. 26:186–214.

Subbiah, B.V., M.S. Sachdev, R.P. Arora, and Y.K. Sud. 1985. Efficiency of fertilizer use in multiple cropping system — studies with isitope techniques. Fertil. News 30(2):45–48.

Sudhakara, K. and R. Prasad. 1986. Relative efficiency of prilled urea, urea supergranules (USG) and USG coated with neem cakes or DCD or direct seeded rice. J. Agric Sci. Camb. 106:185–190.

Tandon, H.L.S. 1980. Soil fertility and fertilizer use research on wheat in India: a review. Fertil. News 25(10):45–78.

Thomas, G.V. 1993. Biological nitrogen fixation by asymbiotic and nonleguminous symbiotic system, in *Organics in Soil Health and Crop Production*, R.K. Thampan, Ed., Peekay Tree Crops. Dev. Foundation, Cochia, India, pp. 104–124.

Thomas, J. and R. Prasad. 1982. Studies on mineralization of neem and sulfur coated and N-Serve treated urea. Fertil. News 27(10):38–43.

Tiedje, J.M. 1987. Ecology of denitrification and dissimilatory nitrate reduction to ammonium, in *Environmental Microbiology of Anaerobes*, A.J.B. Zehnder, Ed., John Wiley & Sons, New York, pp. 179–243.

Tisdale, S.L., W.L. Nelson, and J.D. Braton. 1985. *Soil Fertility and Fertilizers*, 4th ed., Macmillan, New York, p. 754.

Townsend, L.R. and K.B. McRae. 1980. The effect of nitrification inhibitor nitrapyrin on yield and in nitrogen fractions in soil and tissue of corn grown in Annapolis Valley of Nova Scotia. Can. J. Pl. Sci. 66:337–347.

Venkataraman, G.S. 1979. Algal inoculation in rice fields, in *Nitrogen and Rice*, International Rice Research Institute, Los Banos, Philippines, pp. 311–321.

Verma, L.N. 1993. Biofertilizers in agriculture, in *Organics in Soil Health and Crop Production*, R.K. Thompson, Ed., Peekay Tree Crops Dev. Foundation, Cochia, India, pp. 151–183.

Vlek, P.L.G., J.M. Stumpe, and B.H. Byrnes. 1980. Urease activity and inhibition in flooded soil systems. Fert. Res. 1:191–202.

Waksman, S.A. 1952. *Soil Microbiology*, John Wiley & Sons, New York, p. 356.

Walsh, L.M. and J.T. Murdock. 1960. Native fixed ammonium and fixation of applied ammonium in several Wisconsin soils. Soil Sci. 89:183–193.

Watanabe, I., K.K. Lee, B.V. Alimango, M. Sato, D.C. del Rosario, and M.R. deGuzman. 1977. Biological nitrogen fixation in paddy field studied by in situ acetylene reduction assays. IRRI. Res. Pap. Serv. 3, p. 16.

Wells, B.R., P.K. Bollich, W. Ebelhar, D.S. Mikkelsen, R.J. Norman, D.M. Brandon, R.S. Helms, F.T. Turner, and M.P. Westcott. 1989. Dicyandiamide (DCD) as a nitrification inhibitor for rice culture in the United States. Comm. Soil Sci. Plant Anal. 20:2023–2047.

Westerman, R.L., M.G. Edlund, and D.L. Minter. 1981. Nitrapyrin and etridiazole effects on nitrification and grain sorghum production. Agron. J. 73:697–702.

Wilson, R.J.R., Ed. 1988. Advances in Nitrogen Cycling in Agricultural Ecosystems, Proceedings of Symposium of 1987 in Australia, C.A.B. International, Wallingford, England, p. 451.

Winteringham, F.P.W. 1980. *Soil N as Fertilizer or Pollutant*, International Atomic Energy Agency, Vienna, pp. 307–344.

Zacheri, B. and A. Amberger. 1990. Effect of nitrification inhibitors dicyandiamide, nitrapyrin and thiourea on *Nitrosomonas europoea*. Fert. Res. 22:37–44.

PLATE 1. Nitrogen deficiency (see p. 162).

PLATE 2. Phosphorus deficiency (see p. 193).

PLATE 3. Calcium deficiency (see p. 250).

PLATE 4. Magnesium deficiency (see p. 251).

PLATE 6. Manganese deficiency (see p. 265).

PLATE 5. Iron deficiency (see p. 265).

PLATE 7. Iron toxicity (see p. 266).

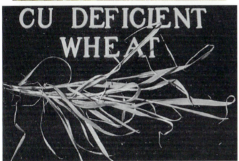

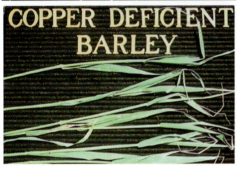

PLATE 8.   Copper deficiency (see p. 277).

PLATE 9.   Zinc deficiency (see p. 278).

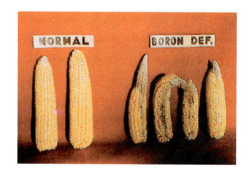

PLATE 10. Boron deficiency (see p. 289).

PLATE 11. Molybdenum deficiency (see p. 294).

9   PHOSPHORUS

In contrast to nitrogen, which constitutes 79% of the earth's atmosphere, phosphorus is present as mineral deposits, which are a nonrenewable natural resource. There is global concern about the energy and costs involved in mining the phosphate rock and its transport to manufacturing sites, as well as in the manufacture of different fertilizers and their transport to farm fields and application to the crops. This problem is very real and serious for a large number of countries having few or no phosphate rock deposits to meet their phosphate needs. Mining phosphate minerals and spreading phosphate fertilizers over the landscape is not sustainable because phosphate mineral deposits are limited. This problem will have to be addressed by future generations.

Another contrast between nitrogen and phosphorus results from the fact that, while nitrogen is easily lost from the soil by various mechanisms such as ammonia volatilization, leaching, and denitrification, the bulk of the phosphorus remains where it is applied due to its immobilization by reaction with Ca, Fe, and Al ions present in soil solution. Thus soluble phosphate fertilizer compounds once applied in the field soon revert to less soluble or insoluble forms. Only 15 to 20% of the applied phosphorus fertilizer becomes available to the crop, and a still smaller fraction to the succeeding crops (residual effect). Efficient phosphorus management therefore could be quite complicated, especially on high P–fixing, tropical ultisols and oxisols.

9.1. SOIL PHOSPHORUS

Total phosphorus content of soils is generally less than that of total N or K: about one-tenth to one-fourth that of nitrogen and one-twentieth that of K (Brady, 1990). Total P content in surface soil and subsoil may vary from a few mg kg^{-1} to over 1 g kg^{-1}. Also in contrast to soil N, which is concentrated in surface soil, P content in subsoil may be less than, equal to, or greater than that in surface soil (Figure 9.1). The data in Figure 9.1 also show that P is present in soils both in organic and inorganic forms. Both organic and inorganic P continuously undergo transformation as shown in Figure 9.2.

171

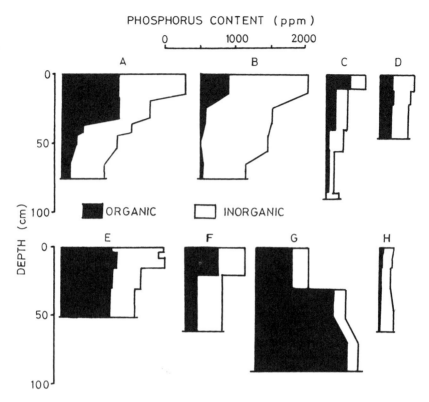

Figure 9.1. Distribution of phosphorus in various soil profiles. (From Anderson, G. 1986. *Phosphorus in Agriculture,* **Khasawneh, F.E., Sample, E.C., and Kamprath, J.E., Eds. pp. 411–432. With permission of the Soil Scientific Society of America.)**

9.1.1. Inorganic P

Inorganic P in soil is mostly present as compounds of Ca, Fe, and Al; Ca-phosphates dominate in neutral to alkaline soils, while Fe- and Al-phosphates dominate in acidic soils. At any specific time very small amounts of phosphate are present in soil solution in equilibrium with the solid inorganic phase. The concentration of P in soil solution is commonly approximately 0.05 mg l⁻¹ and seldom exceeds 0.3 mg l⁻¹ in unfertilized soils. When a water-soluble phosphate fertilizer such as super phosphate or ammonium phosphate is applied to soil, immediately after its dissolution the phosphate ions in solution react with Ca, Fe, or Al ions present in soil solution and are precipitated as insoluble compounds, or become adsorbed on the surface of clay particles. These processes are known as fixation or reversion of phosphate, and the compounds formed are known as "phosphate reaction products." A discussion on these is provided in a later section of this chapter.

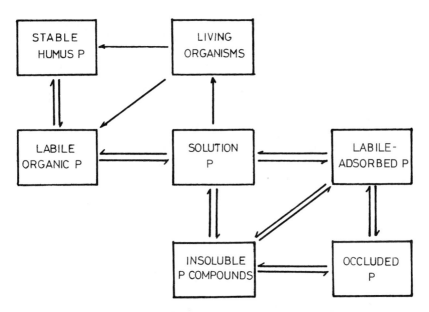

Figure 9.2. Phosphorus transformations in soil. (From Fixen and Grove, 1990. *Soil Testing and Plant Analysis*, Westerman, R.L., Ed. pp. 141–180. With permission of Soil Scientific Society of America.)

The ionic forms of inorganic P are pH dependent, as shown in Figure 9.3. Between pH values of 4.0 and 6.0, most of the P in soil solution is present as the $H_2PO_4^-$ ion, the form in which it can be readily absorbed by plant roots because this ionic form is soluble in water. Between pH 6.5 and 7.5, P in soil solution is present partly as the $H_2PO_4^-$ and partly as the HPO_4^{2-} ion. HPO_4^{2-} ions can also be taken up by the plant roots but not as readily as $H_2PO_4^-$ ions. Between pH 8.0 and 10.0 the HPO_4^{2-} ion is dominant. Under such conditions Na^+ ions dominate the soil cation exchange complex, and some phosphate is present as sodium phosphate, making $H_2PO_4^{2-}$ ions available on hydrolysis. Beyond pH 10.0 the dominant ionic form of P is PO_4^{3+}, and unless present as sodium phosphate, P is not available to crop plants. At the other extreme, that is, below pH 3.0, which is generally not found in cultivated soils, P would be present in H_3PO_4 (phosphoric acid) form, a very reactive form. That is why in highly acidic ultisols and oxisols, phosphate fixation or reversion is rapid and large amounts of phosphate fertilizers are required to obtain good crop growth.

9.1.2. Organic P

The amount of P present in organic form in soils varies from a few milligrams to about 0.5 g kg^{-1} soil (20 to 80% total P) (Figure 9.1). Organic P content depends upon a number of factors such as climate, vegetation, soil

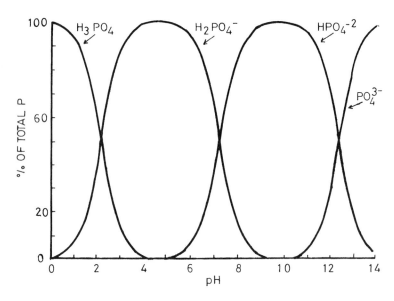

Figure 9.3. Distribution of orthophosphate species in solution as a function of pH.

texture, land use pattern, fertilizer practices, drainage and irrigation, etc.; a number of these factors are interdependent. In general, organic P in surface soils is a smaller fraction of total soil P in the warm regions than it is in the cool regions of the world; about 35.2% as organic P in equatorial regions (countries between 40° parallels) and 48.6% in colder regions (Harrison, 1987). Also more soil P is present as organic P in organic soils and peats (Histosols) than in mineral soils (Table 9.1). Again in mineral soils, high clay soils have a greater percentage of their P in organic form than do sandy soils. In general, organic P tends to accumulate in surface soil because organic P is a part of soil organic matter. However, there are soils such as uncultivated pine bogs where the subsoil is richer in P than the surface soil (Figure 9.1).

In the initial phase of pedogenesis most P in rocks is present as apatite (Figure 9.4), which is gradually brought into solution by chemical and biological processes. The P coming into soil solution is then reprecipitated as secondary inorganic phosphates or is utilized by microorganisms and plants, which after their death and decomposition make organic P available. Thus as pedogenesis proceeds, organic P accumulates. However, in the advanced stages of pedogenesis, when the bulk of the bases and silica has leached out (as in ultisols or oxisols) and soil mineral matter is dominated by Fe- and Al-hydroxides and oxyhydroxides, Fe- and Al-phosphates are formed and are precipitated. If Fe- or Al-hydroxides are precipitated on the surface of Fe- and Al-phosphates, the phosphate becomes occluded. Soils in the advanced stages of pedogenesis may contain sizable amounts of occluded secondary inorganic phosphate. Amount of organic P in such soils is considerably reduced. Ultisols

Table 9.1 Organic P Content of Surface Mineral/Organic Soils in Relation to Soil Texture

Soil texture	Number of samples	Organic P (mg kg⁻¹) Soil	% of total P
Mineral soils			
Sands	194	121	34.1
Loams	663	250	39.9
Clay loams and clays	309	332	41.4
Organic soils			
Organic loams	5	523	58.9
Peats	85	579	65.4

Adapted from Harrison (1987).

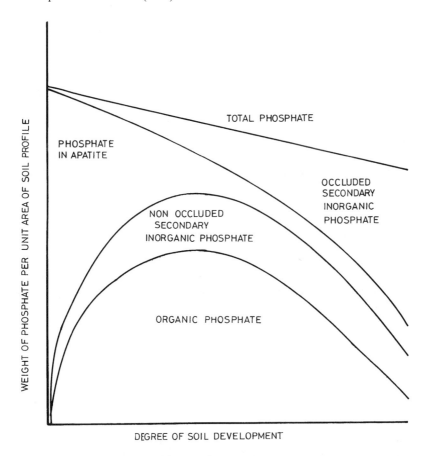

Figure 9.4. Changes in the magnitudes of phosphate fractions during pedogenesis. (From Williams and Walker, 1969. Soil Sci. 107:213–219. With permission of Williams & Wilkins.)

Table 9.2 Forms of Organic P in Soils

Form	Soil (mg kg^{-1})	% of organic P
Inositol phosphate	1.4–356	0.3–62
Nucleic acids	0.1–97	0.1–65
Phospholipids	0.4–17	0.03–5.4

Adapted from Harrison (1987).

and oxisols are found in warm regions of the world and in many cases organic matter content itself is very low, thereby reducing the organic P in soil.

Although often present in appreciable amounts, soil organic P has not been as intensively investigated as inorganic P. Three groups of organic-P compounds have so far been identified. These are inositol phosphates (phosphate esters of inositol $C_6H_6(OH)_6$), nucleic acids, and phospholipids. Together these forms may account for 50 to 70% of soil organic P (Table 9.2).

Inositol phosphate.   Inositol phosphate is thought to be of microbial origin and can exist in several stereoisomeric forms; phosphate esters of myo-, scyllo-, neo-, and chrio-inositol have been characterized in soil (Cosgrove, 1962). Myo-inositol hexophosphoric acid (phytic acid) is usually a major pool of organic P. It is fairly stable in an alkaline medium, but gradually hydrolyzes to a range of intermediate inositol phosphates and finally to inositol in acidic media, the optimum for hydrolysis being near pH 4.0. Enzymes phytases also hydrolyze myo-inositol phosphates. Inositol phosphates make up from less than 1 to 62% of total organic P in soil (Table 9.2). In a wide range of soils from the United States, Canada, and Australia, inositol phosphates account for less than 20% of the organic P.

Nucleic acids.   RNA (ribonucleic acid) and DNA (deoxyribonucleic acid) are found in all living beings. For a time it was believed that at least half of the organic P in soils was present in nucleic acids, but when specific methods of identification and measurements were applied, much lower values were found. For example, in the fulvic acid fraction of two Iowa soils not more than 1.2 to 6 mg P kg^{-1} was present as nucleic acids; this was equivalent to 0.2 to 1.8% of the organic P. Harrison (1987), after surveying the world literature, reported values of 0.1 to 97 mg P kg^{-1} as nucleic acid (0.1 to 65% of organic P); very high values should, however, be taken with caution.

Phospholipids.   Most reported values of soil phospholipids fall within the range of 0.2 to 14 mg P kg^{-1} soil (Kowalenko and McKercher, 1971) representing less than 5% of the organic P (also see Table 9.2). Phosphatidyl choline (lecithin) and phosphatidyl ethanolamine are the predominant phospholipids in soil.

Microorganisms are probably the major source of much of the remainder of the organic P (30 to 50%) in soils. Mention may be made of teichoic acids found in the cell walls of Gram-positive bacteria and of phosphorylated carboxylic acids other than uronic acid isolated from a number of Scottish soils.

Regarding the role of organic P in crop production, opinions differ; indeed some of the most P-deficient soils contain large amounts of organic P. Nevertheless, the importance of mineralization of organic P in the replenishment of P available to plants has frequently been well demonstrated. In soils with high organic matter content, only a small amount of organic P in soils needs to be mineralized to provide a substantial proportion of the P requirements of crops or natural vegetation.

9.2. PHOSPHATE RETENTION OR FIXATION IN SOIL

P is highly immobile in soils (Table 9.3) and generally is fixed near the point of application. To explain P fixation in soil, a wide variety of sorption and precipitation mechanisms has been suggested, with no real consensus as to the relative magnitudes of their contributions. Researchers using dilute P solutions, usually in the millimolar range, have developed several adsorption equations, while those working with concentrated P solution, usually in the molar range, have observed precipitates forming separately at the surfaces of soil constituents. There are four steps in the precipitation process that occurs when dissolved P reacts with soil: (1) formation of a surface-adsorbed P complex; (2) dissolution of clay minerals, which increases the P-reactive metal ion concentration in solution; (3) slow desorption of surface-adsorbed P compounds; and (4) slow nucleation, crystallization, and recrystallization of P compounds (Talibuddin, 1981). If the P retention processes are viewed throughout the entire zone influenced by fertilizer application and over an entire growing season or even a longer period, P retention in soil should be considered as a continuum embodying precipitation, chemisorption, and adsorption (Sample et al., 1986). Phosphorus is retained in soils by hydroxides and oxyhydroxides of Fe and Al, alumino-silicate minerals, soil carbonates, and soil organic matter.

9.2.1. P Retention by Hydroxides and Oxyhydroxides of Fe and Al

Retention of P by interaction with hydroxides and oxyhydroxides of Fe and Al in dilute solutions of P involves replacement of OH^- by PO_4^{3-}. One or both the hydroxyls may be involved. When only one hydroxyl is involved, the adsorption is said to be reversible (Figure 9.5), and when two hydroxyls are involved, the adsorption is said to be irreversible (Hingston et al., 1974). The other terms used for P adsorption involving two hydroxyls are bidentate and binucleate fixation of P. Surface complexes between Al^{3+} and phosphate are more labile than those involving Fe^{3+}. Furthermore, noncrystalline Fe-oxides, which have surface areas of 400 m^2 g^{-1}, may play a wider role than the

Table 9.3 Extent of Movement of Phosphate Solution[a,b]

Cumulative time (hr)	Thickness (mm) of the wetted soil shell		Thickness (mm) of drained time DCP shell in tablet	
	Moisture equivalent			
	0.5	1.0	0.5	1.0
2	1.0	1.0	0.5	0.5
18	2.2	2.0	1.0	0.9
44	3.7	3.1	2.0	1.1
69	4.5	3.3	2.3	1.3
164	5.4	3.5	4.0	3.0
216	5.4	3.5	4.0	3.0

[a] Recrystallized monocalicum phosphate monohydrate (MCP). Reagent grade was compressed to tablets of 3 mm in thickness and 8 mm in diameter and contained 0.1 g of P. Tablets were inserted to a depth of 2.54 cm in 100-g soil samples (wetted to 0.5 or 1.0 ME [moisture equivalent] and loaded into a glass cell), and the cell was sealed promptly. Microscopic observations were then made.

[b] Means of five soils.

Adapted from Lehr et al. (1959).

crystalline Fe-oxides, which have a surface area of about 100 m^2 g^{-1}. Different Fe-oxides can have widely different P sorption capacities. For example, synthetic hematite could adsorb only 0.2 µmol P m^{-2}, while synthetic goethites are reported to adsorb around 2.5 µmol P m^{-2}.

One of the consequences of the specific adsorption of phosphate or other anions by Fe- and Al-hydroxides and oxyhydroxides is an increase in the negative charge (CEC). For example, Nakaru and Uehara (1972) reported that P adsorption increased the CEC of ferruginous tropical soils up to 100 cmol kg^{-1}; the ratio of change being approximately 0.8 cmol/cmol of P added. Schalscha et al. (1974) demonstrated that CEC of some volcanic soils as determined with KH_2PO_4 was much greater than that with KCl or NaCl and that P adsorption increased the CEC.

9.2.2. P Retention by Clay Minerals

As with Fe- and Al-hydroxides, several researchers invoke a low-concentration sorption and high-concentration precipitation mechanism. With low P concentrations reactions with silicate minerals are much slower than with Fe- and Al-hydroxides. The edge faces of kaolinite (have OH$^-$) behave similar to the Al-hydroxides in the adsorption of P (Hingston et al., 1972). The P adsorption capacity by clay minerals depends, among other factors, on the proportion of surface area occupied by edge faces. Accordingly, kaolinite adsorbs more P per unit surface area than 2:1 clay minerals.

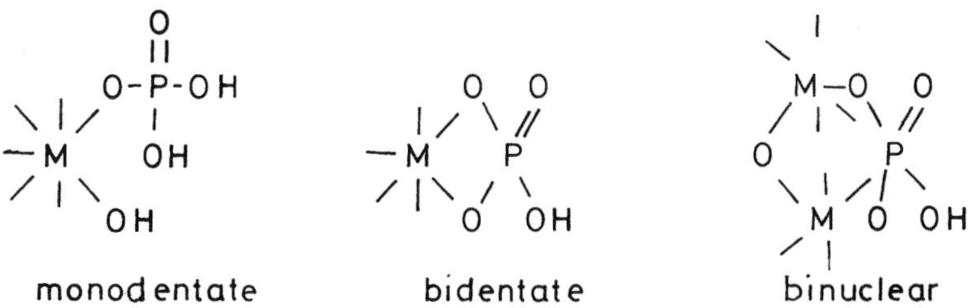

Figure 9.5. Reversible (monodentate) and irreversible (bidentate and binuclear) adsorption of P. (From Hingston et al., 1974; Fixen and Grove, 1990.)

With strong P solutions such as 1 M (ammonium, potassium, and sodium phosphate) illite, kaolinite, and monmorillonite behave the same way as Fe- and Al-hydroxides and oxyhydroxides and give precipitated phases consisting of Fe- and Al-phosphates containing the cations furnished by the P solution. Also concentrated P solutions decompose kaolinite (Kittrick and Jackson, 1956), resulting in precipitation of an Al-P compound. Lindsay and Stephenson (1959) reacted successive increments of Hartsells FSL (fine, sandy loam) (pH 4.6, from Tennessee) and Rosebud loam (pH 7.6, from Nebraska) with a triple-point solution TPS (the solution in equilibrium with $Ca(H_2PO_4)_2 H_2O$ and $CaHPO_4$; the most concentrated solution resulting from the dissolution of mono-calcium phosphate in water having pH 1.01 and 4.50 M P and 1.34 M Ca). As successive soil increments were contacted by the TPS, simulating its movement from a fertilizer granule, the dissolution processes of Fe, Al, and Ca continued and the pH of the solution gradually increased (Figure 9.6), though it still remained strongly acid (pH 2 to 2.4). Soluble Fe and Al concentrations in both the soils continuously increased. With 1 hour contact time the concentration of soluble Fe in Hartsells FSL was 150 mmol l^{-1} with ten increments of soil, the corresponding value for Rosebud loam was slightly more. Soluble Al concentration after 1 hour of contact with ten increments of soil was about 260 mmol l^{-1} in Hartsell FSL and about 140 mmol l^{-1} in Rosebud loam. Concentration of soluble Al increased with time of contact in both soils. However, for Fe, concentration decreased in Rosebud loam showing precipitation of Fe-phosphate. These data clearly show that fixation or retention of phosphate in soils involves dissolution and precipitation.

9.2.3. Retention by Soil Carbonates

The chemistry of calcareous soils is dominated by soil carbonates. Thus reactions of P with pure carbonates have been studied in some detail. P interactions with calcite involve two reactions: the first reaction at low P concentrations consists of adsorption of P by calcite surface, while the second process is a nucleation process of calcium phosphate crystals (Griffin and

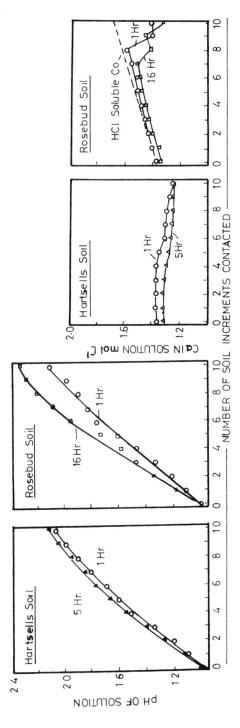

Figure 9.6. Soil pH, Ca, Fe, and Al concentration in solution brought about by bringing successive increments of soil in contact with a triple-point solution. (Adapted from Lindsay and Stephenson, 1959.)

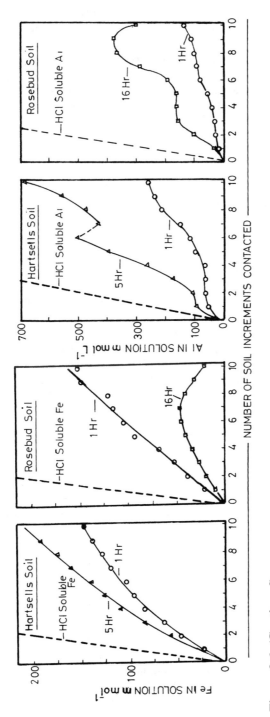

Figure 9.6. (Continued)

Jurinak, 1973, 1974). At high P concentrations, precipitation reactions are both copious and rapid. There is evidence to suggest that P is less strongly bound to $CaCO_3$ than to the hydrous oxides of Fe and Al and hence is more available to crop plants. The general consensus is that in soils where Al- and Fe-oxide are available, carbonates may play only a secondary role in P sorption. However, in soils where Fe-oxide and clay contents are very low, such as in some calcareous Vertisols and soils of arid regions, carbonates have a major effect on P solubility.

9.2.4. Retention by Soil Organic Matter

Soil organic matter is involved in retention and release of P. In association with cations such as Fe, Al, and Ca soil humus is able to retain significant amounts of P. On the other hand, researchers working with dilute P solutions have shown that organic matter complexes Ca, Fe, and Al and therefore makes more P available to plants.

When fertilizer P is applied to soils, organic matter coatings in the microsite are dissolved and move with the advancing front of the fertilizer. Giordano et al. (1971) showed that up to 10% of the soil organic matter was dissolved by mono-ammonium phosphate (MAP) and ammonium pyrophosphate (APP); APP dissolved twice as much as MAP. Removal of organic matter coatings on soil minerals may expose new surfaces for P precipitation at the microsite of fertilizer granule application. On the other hand, the dissolved organic matter may be reprecipitated in P-retention reactions. Soil organic matter can thus change the site of P retention.

9.3. FACTORS AFFECTING THE RETENTION OF PHOSPHOROUS BY SOIL

In addition to the soil components directly involved in the retention of phosphorus in soils, namely, Fe- and Al-oxides, clay minerals, carbonates, and organic matter, a number of other soil factors may also affect P retention by soil. These are briefly discussed.

9.3.1. pH

Soil pH has a profound influence on the availability of Ca, Al, Fe, and other ions in soil solution and thus has an overriding influence on the retention of phosphorus by soil. Phosphorus availability (solubility) in soils is greatest in the pH range 6 to 6.5 (Figure 9.7). At lower pH, increased solubility of Fe and Al results in retention of applied P as Fe- and Al-phosphates, while at higher pH (above 7.0), Ca plays a dominant role in phosphate retention. Liming of acid soils increases the availability of phosphorus by decreasing P fixation. For example, in a study on an oxisol in Brazil (Table 9.4), response to P applied

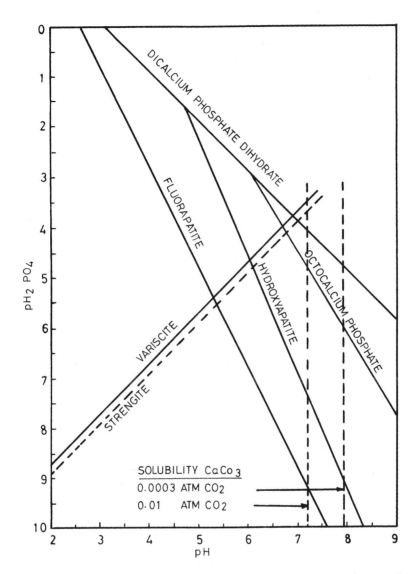

Figure 9.7. Solubility diagram for phosphate compounds in soils at 25°C and 0.005 M Ca concentrations (pCA = 2.50). (Adapted from Lindsay and Moreno, 1960.)

at lower rates was much greater in the presence of lime; but at higher levels of P application the effect of liming was not as apparent because the crop was able to meet its P requirements. In another study in Australia four times as much superphosphate was required to produce equal yields of alfalfa on a soil of pH 5.1 and 49% Al saturation as on a limed soil at pH 6.1 with no Al (Munns, 1965).

**Table 9.4 Effect of Lime on Seed Yield (kg ha^{-1})
of Common Bean (*Phaseolus vulgaris* L.) Grown at
Different P Rates in an Oxisol in Central Brazil**

P level (kg ha^{-1})	Without lime	With lime
0	37	110
26	777	900
52	953	1207
104	1427	1380
208	1323	1363

From Fageria et al. 1990. *Crops and Enhancers of Nutrient Use*, Baligar, V.C. and Duncan, R.R., Eds., p. 675. With permission of Academic Press.

9.3.2. Cations Present in Soil Solution

In addition to Fe, Al, and Ca, which are directly involved in phosphorus retention, presence of other cations can also influence the retention of phosphorus by soil. For example, Mg has been shown to prevent adsorption of phosphate onto calcite (Yadav et al., 1984). In sodic soils, where Na ions play a dominant role, phosphorus becomes associated with Na and is readily available to plants, resulting in a net reduction in the retention of soil phosphorus.

9.3.3. Anion Effects

Specifically sorbed ions such as hydroxyl, sulfate, and molybdate can compete with phosphate for the available reaction sites; nevertheless, phosphate is a stronger competitor.

9.3.4. Temperature

An increase in temperature generally decreases the availability of unorganic P. A study at the Northern Plains Field Station at Mandan, North Dakota, (Power et al., 1964) showed that both the water and NaHCO$_3$-extractable P extracted from soils to which phosphate fertilizer had been added decreased when the soil incubation temperatures were above 15°C. Beaton et al. (1965) reported that there was a 33% reduction in water-soluble phosphorus for each 15°C increase in temperature. This corollary explains why soils in warm regions of the world are generally much greater fixers of phosphorus than the soils of cooler regions. On the other hand, mineralization of phosphorus from soil organic matter or crop residues generally increases with an increase in soil temperature. For example, Campbell et al. (1984) from a 12-year study in Saskatchewan reported that the bicarbonate-soluble P increased between late fall and spring thaw.

9.4. PHOSPHATE FERTILIZER REACTION PRODUCTS IN SOIL

Even under conditions of low soil water, dissolution of phosphate granules containing water-soluble P is fairly rapid. For example, Lawton and Vomocil (1954) found that the water content of ordinary superphosphate granules in a soil at more than 1.5 MPa water tension increased to 16.2% within 24 hours. Thus ample water had diffused into the granules to initiate dissolution.

Once the saturated solution in and around the fertilizer granule or band is formed, an osmotic potential gradient is established between the concentrated fertilizer solution and the soil water. Water is drawn into this concentrated zone by vapor transport; the more the inward movement of water, the more is outward movement of solution. As long as the original salt remains, this process continues to produce a nearly saturated solution until the concentration is decreased by dilution or by reaction of fertilizer and soil constituents or until the osmotic gradient disappears. This phenomenon is also true for fertilizer materials of low-water solubility such as dicalcium phosphate and apatites, with the only difference being that the saturated solutions are more dilute (as dictated by the solubility products of the chemical compounds) and create a very small osmotic gradient between the fertilizer solution and soil solution. The chemical environments near the application of fluid fertilizers are quite similar to those surrounding highly water-soluble, solid fertilizers.

The solubility product (SP) principle has been widely used in identifying the phosphate reaction products, and a brief discussion follows. The SP principle applies to chemical compounds, which are sparingly soluble, for example $BaSO_4$. When a small quantity of $BaSO_4$ is placed in water, despite the sparingly soluble characteristic of the compound, very low activities of Ba (A_{Ba}) and SO_4 (A_{SO_4}) can be measured in the water. The product of the activities of Ba and SO_4 ($A_{Ba} \times A_{SO_4}$), symbolized as SP_{BaSO_4}, has been found to be constant, irrespective of the source of $BaSO_4$. The term pK SP is defined as $\log \left(\dfrac{1}{SP} \right)$ following the terminology for pH. pK_{sp} values for such sparingly soluble compounds have therefore been used to identify unknown substances by matching the values obtained with those of the known compounds. Since many phosphate reaction products are sparingly soluble, this technique has been used in the identification of some of these compounds (Lindsay et al., 1959; Wright and Peach, 1980). pK_{sp} values for some phosphate compounds and stable forms of Fe, Al, and Ca (the ions that react with phosphate in soil solution) are given in Table 9.5.

Phosphate in soil solution is in a dynamic state and is continuously reacting with Ca, Fe, and Al. The rate of reaction depends on soil pH and the activity of reacting cations. Phosphate reaction products are essentially different phases in an overall reactive and dynamic state of soil solution. The phosphate phase present at a particular point can be predicted by using phase

Table 9.5 Solubility Expressions and Solubility Products of Some Phosphate Compounds at 25°C

Compound or species	Chemical formula	Solubility expression when activity of $H_2O = 1$	pK sp
Gibbsite	$Al(OH)_3$	$pK_g = pAl + 3pOH$	33.8
Variscite	$Al(OH)H_2PO_4$ [a]	$pK_v = pAl + 2pOH + pH_2PO_4$	30.5
Goethite	$FeOOH$	$pK_{gt} = pFe + 3pOH$	unknown
Ferric hydroxide	$FeOOH$	$pK_{fh} = pFe + 3pOH$	38.1
Strengite	$Fe(OH)_2H_2PO_4$ [a]	$pK_{st} = pFe + 2pOH + pH_2PO_4$	33.6–35.0
Dicalcium phosphate dihydrate	$CaHPO_4 \cdot 2H_2O$	$pK_{dcpd} = pCa + pHPO_4$	6.56
Dicalcium phosphate anhydrous	$CaHPO_4$	$pK_{dcpa} = pCa + pHPO_4$	6.66
Octocalcium phosphate	$Ca_4H(PO_4)_3 \cdot 3H_2O$	$pK_{ocp} = 4pCa + pH + 3pPO_4$	46.91
Hydroxyapatite	$Ca_{10}(PO_4)_6(OH)_2$	$pK_{ha} = 10pCa + 6pPO4 + 2pOH$	113.7
Fluorapatite	$Ca_{10}(PO_4)_6F_2$	$pK_{fa} = 10pCa + 6pPO_4 + 2pF$	118.4
Fluorite	CaF_2	$pK_{ft} = pCa + 2pF$	9.84
Calcite	$CaCO_3$	$pK_{cc} = pH - 1/2\ pCa + 1/2 \log pCO_2$	4.93
Phosphoric acid	H_3PO_4	$pK_1 = pH + pH_2PO_4 - pH_3PO_4$	2.12
Dihydrogen phosphate ion	$H_2PO_4^{-}$	$pK_2 = pH + pHPO_4 - pH_2PO_4$	7.20
Monohydrogen phosphate ion	HPO_4^{-}	$pK_3 = pH + pPO_4 - pHPO_4$	12.32

[a] The chemical formulate for variscite and strengite may also be expressed as $AlPO_4 \cdot 2H_2O$ and $FePO_4 \cdot 2H_2O$, respectively.

Adapted from Lindsay and Moreno (1960).

diagrams such as the one shown in Figure 9.7. In Figure 9.7 lines represent the activity solubility isotherms for the respective phosphate compounds. Points falling above these isotherms represent supersaturation of the soil solution with respect to that compound, indicating that that compound is likely to be precipitated. Since in calcareous soils pH is governed predominantly by the solubility of $CaCO_3$ and the partial pressure of CO_2, pH values are indicated in Figure 9.7 at two partial pressures of CO_2 by broken vertical lines for the value pCa = 2.5.

The best documented example of P-reaction products is the dicalcium phosphate (DCP) residues remaining at application sites of fertilizers containing monocalcium phosphate (MCP). For example when MCP, OSP (ordinary superphosphate), or CSP (concentrated superphosphate) granules were placed in five soils at two water levels, 20 to 34% of the applied P remained as DCP after complete dissolution of the MCP component (Lehr et al., 1959). Dicalcium phosphate dihydrate (DCPD) is the dominant initial reaction product of these fertilizers, with monocalcium phosphate (MCP), formed in most alkaline and calcareous soils.

The reactions products and their amounts resulting from fertilizer mixtures are reported to be different than those from the pure phosphate fertilizers. For example, Bouldin et al. (1960) placed granules containing MCP + $(NH_4)_2 SO_4$ in moist soil and found that only about 2% P remained as dicalcium phosphate anhydrate (DCPA) residue after 3 weeks as compared with 21% with MCP alone. Bouldin et al. (1960) reported the following sequence of reaction products from granules of MCP + KCl.

Time (hours)	Reaction Products
2	MCP and KCl bulk phases; $Ca_2K H_7 (PO_4)_4 . 2H_2O$ minor phase
17.5	KCl and $Ca_2K H_7 (PO_4)_4 2H_2O$ major phase; MCP and DCPD (dicalcium phosphate dihydrate) minor phase
21	$Ca_2K H_7 (PO_4)_4 2H_2O$ major phase; DCPD and KH_2PO_4 minor phases
23.5	DCPD major phase; Abundant $Ca_2K H_7 (PO_4)_4 . 2H_2O$ and a trace of KH_2PO_4
25.5	DCPD major phase; $Ca_2K H_7 (PO_4)_4. 2H_2O$ minor phase

With alkaline phosphate fertilizer DAP, MAP, and MCP, the products $Ca_8 H_2 (PO_4)_6 5 H_2O$, $Ca HPO_4 . 2H_2O$ and $Mg NH_4 PO_4 .6H_2O$ were identified in slightly acidic and calcareous soils (Bell and Black, 1970). With monoammonium (MAP) and monopotassium phosphate (MKP), ammonium taranakite and traces of variscite and potassium taranakite were the dominant products in an acidic soil from Tripura, while brushite, struvite, and newberryite with MAP and brushite and monetite with MKP were the main products on a calcareous alluvial soil from Bihar (Das and Datta, 1969). With monocalcium

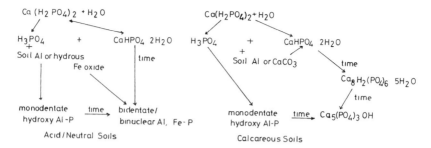

Figure 9.8. Transformations of monocalcium phosphate in soils. (Adapted from Thomas and Peaslee, 1973.)

phosphate there was considerable dissolution of Fe and Al, which precipitated as amorphous Fe- and Al-phosphates, in addition to brushite and monetite in both the soils (Das and Datta, 1969).

Lindsay et al. (1962) added soil and soil constituents to saturated solutions of the common fertilizers and identified 30 crystalline P compounds, in addition to colloidal precipitates of variable composition. In acidic soils the initial reaction products such as the taranakites and amorphous Fe- and Al phosphates, as well as calcium phosphates, change with time to variscite-like and strengite-like crystalline compounds.

With triammonium pyrophosphate (TPP), a major constituent of ammonium polyphosphate, in most Indian soils under study (alfisol, oxisol, entisol, mollisol, and vertisol) the most abundant products were $Ca(NH_4)_2 P_2O_7 . H_2O$ and $Mg (NH_4)_2 P_2O_7 . 4H_2O$ (Yadav and Mistry, 1984). Some other reaction products identified were $Ca(NH_4)_4 H_2 (P_2O_7)_2$; $Ca_3 (NH_4)_4 H_6 P_2O_7 . 3H_2O$; $Fe NH_4 P_2O_7$; $Ca (NH_4)_2 P_2O_7 . H_2O$; $Mg (NH_4)_2 P_2O_7 . 4H_2O$ and $NH_4 Al_{0.33} Fe_{0.67} P_2O_7$.

In nature the ultimate products are thought to be hydroxy and fluoroapatites in alkaline and calcareous soils and variscite and strengite in acidic to neutral soils (Hsu and Jackson, 1960). For ordinary or triple superphosphates containing monocalcium as the P compound, the transformations as given by Thomas and Peeslee (1973) are shown in Figure 9.8. As soil weathers with a decrease in pH, Ca-phosphates change to amorphous and crystalline Al-phosphates, which in turn change to Fe-phosphates (Chang and Jackson, 1958; Hsu and Jackson, 1960).

A summary of compounds formed from the reaction of phosphate fertilizers with soils and soil constituents is given in Table 9.6.

9.5. INTENSITY (I) AND QUANTITY (Q) FACTORS IN PHOSPHORUS AVAILABILITY

The quantity of P present in soil solution at a given time is a measure of its intensity. P concentration may be measured as molar concentration of total

Table 9.6 Summary of Compounds Formed from the Reaction of Phosphate Fertilizers with Soils or Soil Constituents

Compound	Mineral name	Compound	Mineral name
$AlPO_4 \cdot 2H_2O$	Variscite	$FePO_4 \cdot 2H_2O$	Metastrengite
$AlPO_4 \cdot 2H_2O$	Metavariscite	$Fe_3(PO_4)_2 \cdot 8H_2O$	Vivianite
$Al(NH_4)_2H(PO_4)_2 \cdot 4H_2O$	—	$FeNH_4(HPO_4)_2$	—
$Al_2(NH_4)_2H_4(PO_4)_4 \cdot H_2O$	—	$Fe_3NH_4H_6(PO_4)_6 \cdot 6H_2O$	—
$Al_5(NH_4)_3H_6(PO_4)_8 \cdot 18H_2O$	NH_4-taranakite	$Fe_3KH_6(PO_4)_6 \cdot 6H_2O$	—
$AlNH_4PO_4OH \cdot 2H_2O$		$Fe_2K(PO_4)_2OH \cdot 2H_2O$	K-leucophosphite
$AlNH_4PO_4OH \cdot 3H_2O$	—	$MgHPO_4 \cdot 3H_2O$	Newberryite
$Al_2NH_4(PO_4)_2OH \cdot 2H_2O$	—	$Mg_3(PO_4)_2 \cdot 4H_2O$	—
$Al_2NH_4(PO_4)_2OH \cdot 8H_2O$	—	$Mg_3(PO_4)_2 \cdot 22H_2O$	—
$AlKH_2(PO_4)_2 \cdot H_2O$	—	$MgNH_4PO_4 \cdot 6H_2O$	Struvite
$Al_5K_3H_6(PO_4)_8 \cdot 18H_2O$	K-taranakite	$Mg(NH_4)_2(HPO_4)_2 \cdot 4H_2O$	Schertelite
$Al_2K(PO_4)_2OH \cdot 2H_2O$	Leucophosphite	$Mg_3(NH_4)_2(HPO_4)_4 \cdot 8H_2O$	Hannayite
$AlKPO_4OH \cdot 0.5H_2O$	—	$MgKPO_4 \cdot 6H_2O$	—
$AlKPO_4OH \cdot 1.5H_2O$	—	$Mg_2KH(PO_4)_2 \cdot 15H_2O$	—
$Al_2K(PO_4)_2(F,OH) \cdot 3H_2O$	Minyulite	$Al(NH_4)_2P_2O_7OH \cdot 2H_2O$	—
$CaHPO_4$	Monetite	$Ca_2P_2O_7 \cdot 2H_2O$	—
$CaHPO_4 \cdot 2H_2O$	Brushite	$Ca_2P_2O_7 \cdot 4H_2O$	—
$Ca_8H_2(PO_4)_6 \cdot 5H_2O$	Octocalcium phosphate	$Ca_3H_2(P_2O_7)_2 \cdot 4H_2O$	—

Table 9.6 Summary of Compounds Formed from the Reaction of Phosphate Fertilizers with Soils or Soil Constituents (Continued)

Compound	Mineral name	Compound	Mineral name
$Ca_{10}(PO_4)_6(OH)_2$	Hydroxyapatite	$Ca(NH_4)_2P_2O_7 \cdot H_2O$	—
$Ca_{10}(PO_4)_6F_2$	Fluorapatite	$Ca_3(NH_4)_2(P_2O_7)_2 \cdot 6H_2O$	—
$CaAlH(PO_4)_2 \cdot 6H_2O$	—	$Ca_5(NH_4)_2(P_2O_7)_3 \cdot 6H_2O$	—
$CaAl_6H_4(PO_4)_3 \cdot 20H_2O$	—	$CaNH_4HP_2O_7$	—
$CaNH_4PO_4 \cdot H_2O$	—	$Ca_2NH_4H_3(P_2O_7)_2 \cdot 3H_2O$	—
$Ca(NH_4)_2(HPO_4)_2 \cdot H_2O$	—	$CaK_2P_2O_7$	—
$Ca_2NH_4H_7(PO_4)_4 \cdot 2H_2O$	NH_4-Flatt's salt	$Ca_3K_2(P_2O_7)_2 \cdot 2H_2O$	—
$Ca_2(NH_4)_2(HPO_4)_3 \cdot 2H_2O$	—	$Ca_6K_2(P_2O_7)_3 \cdot 6H_2O$	—
$CaKPO_4 \cdot H_2O$	—	$Ca_2KH_3(P_2O_7)_2 \cdot 3H_2O$	—
$CaK_3H(PO_4)_2$	—	$CaNa_2P_2O_7 \cdot 4H_2O$	—
$Ca_2KH_7(PO_4)_4 \cdot 2H_2O$	K-Flatt's salt	$Fe(NH_4)_2P_2O_7 \cdot 2H_2O$	—
$CaFe_2H_4(PO_4)_4 \cdot 5H_2O$	—	$Mg(NH_4)_2P_2O_7 \cdot 4H_2O$	—
$CaFe_2H_4(PO_4)_4 \cdot 8H_2O$	—	$Mg(NH_4)_6(P_2O_7)_2 \cdot 6H_2O$	—
$Ca_3Mg_3(PO_4)_4$	—	$Mg(NH_4)_2H_4(P_2O_7)_2 \cdot 2H_2O$	—
$FePO_4 \cdot 2H_2O$	Strengite	$Ca(NH_4)_3P_3O_{10} \cdot 2H_2O$	—

From Sample et al., 1986. *The Role of Phosphorus in Agriculture*, Khasawneth, F.E., Sample, E.C., and Kamprath, E.J., Eds., p. 284. With permission of American Society of Agronomy.

soluble P, including all species of orthophosphate ions, as well as ion pairs in association with other cations. It may also be expressed in ppm or mg l^{-1}. The other terms used for intensity measurements are chemical potential of H_2PO_4, and monocalcium phosphate potential (Schofield's phosphate potential).

When P intensity is diminished by withdrawal of P from solution, solid-phase P goes into solution to replenish the loss. The quantity of solid-phase P that acts as a reserve is called the "quantity factor"; some authors refer to it as the "capacity factor." However, the recent trend has been toward adopting the word "quantity" to describe these reserves and to assign "capacity" to gradients that relate quantity to intensity. The quantity factor or reserves of soil P refer to both the labile and non-labile forms of soil P (see Figure 9.2); the labile P is directly linked with solution P (intensity factor) with a high dissociation rate, while the nonlabile P is indirectly linked with solution P with a low dissociation rate. (Note the similarity between nonexchangeable and exchangeable components of the quantity factor of K and solution K — Chapter 10.)

Basically, labile P is the fraction of soil P that is isotopically exchangeable with ^{32}P within a specified time (Olsen and Khasawneh, 1980). Isotopically exchangeable P, utilized by a growing plant over the span of a growing season, is called the L value. The E value is the laboratory equivalent of the L value, measured over a shorter but specified period of equilibration. These definitions clearly indicate that labile P does not present a precisely defined and a clearly distinct phase of solid-phase P, but one that has arbitrary boundaries of time and other procedural factors. For a general relationship between quantity (Q) and intensity (I) factors, the student is advised to read Chapter 10 on potassium.

Soil test extractants discussed in the next section tend to estimate the quantity factor. For a group of closely related soils, there would exist a certain degree of correlation between labile P and P extractable by different extractants, and it is this relationship that provides support for measuring the P quantity factor via chemical extractants.

9.6. SOIL TESTING FOR PHOSPHORUS

Soil testing for P is done using a chemical extractant. A large number of extractants have been suggested by various workers such as ammonium bicarbonate-DPTA, Brays 1 & 2, citric acid, bicarbonate-ammonium fluoride - EDTA, Mehlich I, II, and III, Morgan's acetic acid - sodium acetate, Olsen's 0.5 M sodium bicarbonate, and Truog's sulfuric acid - ammonium sulfate (Fixen and Grove, 1990). Surveys of procedures in use in the United States have indicated that Bray's P_1, Mehlich III, and Olsen's extract are most often used. Bray's P_1 extract is 0.03 M NH_4F + 0.025 M HCl, while Mehlich III is 0.015 M NH_4F + 0.2 M CH_3COOH + 0.25 M NH_4NO_3 + 0.013 M HNO_3. Both these extractants are suited for acidic soils. Olsen's extracting solution is 0.5 M $NaHCO_3$ of pH 8.5 and is best suited for neutral, calcareous, and alkaline soils.

There are four general reactions contributing to P release from soil by different chemical extractants. These are (1) acid dissolution; (2) anion exchange; (3) cation complexation; and (4) cation hydrolysis. Chemical extractants containing acids such as Bray's P_1 or P_2 or Mehlich I, II, and III will dissolve soil P by acid dissolution. The solubility of soil phosphate in decreasing order is Ca-P > Al-P > Fe-P, Ca-P being the most soluble and Fe-P being the least soluble. Extractants containing bicarbonate, citrate, lactate, and sulfate can replace adsorbed soil P by anion exchange. Fluoride and certain organic anions (citrate and lactate) can complex Al, and extractants containing these anions can release P from Al-P compounds. Similarly, bicarbonate precipitates soluble Ca as $CaCO_3$, causing release of Ca-P. It is argued (Thomas and Peaslee 1973) that bicarbonate and F remove similar soil P forms, though F is more competitive/aggressive. Cation hydrolysis occurs at high pH values where the OH^- ions dissolve a portion of Al-P and Fe-P by hydrolysis of Al and Fe. Extractants containing HCO_3 can extract some soil P by this process. The exact mechanism of release of P from soil organic compounds is not well understood, but the most successful extractant contains both fluoride and dilute acid. A knowledge of extractant-soil interaction chemistry is most useful when understanding divergent extractant behavior because of (1) differences in available P sources on soils of otherwise similar properties (Admont et al., 1986) or (2) differences among soils fertilized with a common P source (Varvel et al., 1981).

Another approach to available P assessment in soil is to use models. Models may or may not use extractable soil P. Barrow (1980) outlined a descriptive model that simulates the initially rapid, but ultimately slow, gradual decline in available P after fertilization while taking into account P removal by the harvested crop. A rather more mechanistic model on the same line was proposed by Wolf et al. (1987). Other proposed models such as those of Cox et al. (1981) and Jones et al. (1984) rely on soil test information. If models are to become more important in agronomic/environmental management (reducing reliance on conventional soil testing), more validation work needs to be done.

9.7. PHOSPHORUS-DEFICIENCY SYMPTOMS IN PLANTS

Phosphorus is readily mobilized in plants and when a deficiency occurs, the P contained in older tissues is transferred to the active meristematic region. Small leaves, darker green colored than normal with reddish purple cast and dying tips indicate P deficiency (Figure 9.9). Purpling in young plants is more distinct. Other symptoms in small grain crops such as wheat include stunted growth, poor tillering, and delayed maturity.

9.8. PHOSPHATE FERTILIZERS

All phosphate fertilizers are produced from phosphate rock; the only exceptions are polyphosphates produced from elemental P. Global phosphate

Figure 9.9. A young plant (above) and a stunted more mature plant (below) showing phosphorus deficiency. The leaves develop a purplish color. (From *Corn Field Manual*, J.R. Simplot Company, Minerals & Chemical Division, Pocatello, ID, ©1984. With permission.) See Plate 2 following p. 170.

rock reserves are estimated at 41,000 million Mg. The countries having major reserves are Morocco (20,000 million Mg), former USSR (8000 million Mg), United States (5700 million Mg), Republic of South Africa (1800 million Mg), China (1000 million Mg), Western Sahara (850 million Mg), and Australia (800 million Mg) (Stowasser, 1979). Other countries important from the viewpoint of reserves of phosphate rock are Tunisia, Senegal, Algeria, Egypt, and Togo in Africa; Mexico; Brazil and Peru in South America; and Jordan and Syria in Asia.

Phosphate rock basically consists of fluor-, hydroxy-, chlor-, and carbonate apatites—fluroapatite being the dominant mineral. Apatites have an insular structure, and therefore the phosphates they contain are not available to growing crop plants, unless the insular structure created by apatite bondage is broken down. This can be achieved by two ways, namely, reacting the phosphate rock with a strong acid such as sulfuric, nitric, hydrochloric, or phosphoric acid or by employing heat (very high temperatures, generally above 1400°C) and high pressure. When an acid is employed it disintegrates the mineral and its anionic component such as SO_4^{2-}, PO_4^{3-}, NO_3^-, or Cl^- reacts with calcium in the rock. When heat is employed, the fluoride, chloride, hydroxy, or carbonate ions present in the mineral are released as F, Cl, or CO_2 gases or water vapors, and the crystal structure of mineral is broken down. When finely ground rock phosphate is applied to acid soils, soil acids disintegrate the apatite minerals in the same way as strong acids, but the slower rate of reaction depends upon soil pH and potential acidity. Phosphate rock contains 11 to 16% P (25 to 37% P_2O) and 33 to 36% Ca. The most commonly manufactured and used phosphate fertilizers, along with their composition, are listed in Table 9.7, and a brief discussion of them follows.

9.8.1. Terminology

Phosphate in fertilizers may be present in one or more forms of orthophosphate ion ($H_2PO_4^-$, HPO_4^{2-}, PO_4^{3-}, or condensed polyphosphates such as pryo- and tripolyphosphate). Because these forms differ in their solubility, specific terms are used to define them.

9.8.1.1. Water-Soluble Phosphorus.
A small sample of fertilizer is extracted with distilled water for a prescribed period (AOAC, 1960) and the slurry is then filtered. The amount of P in the filtrate is determined and expressed as a percentage by weight of the fertilizer. This fraction is called water-soluble P. Monocalcium phosphate [$Ca(H_2PO_4)_2$], (the main P component of ordinary or concentrated phosphate), monoammonium phosphate [$NH_4H_2PO_4$], diammonium phosphate [$(NH_4)_2 HPO_4$], and potassium phosphate [KH_2PO_4] are water soluble.

9.8.1.2. Citrate-Soluble Phosphorus.
The fertilizer residue left after extracting water-soluble P is extracted with a neutral (pH 7.0) solution of 1 N ammonium citrate for a prescribed period (AOAC, 1960), and the slurry is then filtered. The phosphorus content in the filtrate is then determined and expressed as a percentage by weight of the fertilizer. This fraction is called citrate-soluble P. Some European countries use alkaline ammonium citrate solution for extracting phosphorus from fertilizers (Terman et al., 1964) in place of neutral ammonium citrate solution; alkaline extracting solution will extract lesser amounts of P (Figure 9.10) and is considered by some workers (Terman et al., 1964) to be better correlated with plant growth and P uptake. However, most countries in the world use neutral ammonium citrate solution for extracting citrate-soluble P.

Table 9.7 General Composition of Phosphate Fertilizers

Material	Total nitrogen (%)	Total potassium (%)	Total sulfur (%)	Total calcium (%)	Total magnesium (%)	Phosphorus Total[a] (%)	Phosphorus Available[b] (% of total)
Ordinary superphosphate (OSP)	—	—	11–12	18–21	—	7–9.5	97–100
Conc. (triple) superphosphate (CSP)	—	—	0–1	12–14	—	19–23	96–99
Enriched superphosphate	—	—	7–9	16–18	—	11–13	96–99
Ammoniated OSP	2–5	—	10–72	17–21	—	6.1–8.7	96–98
Ammoniated CSP	4–6	—	0–1	12–14	—	19–21	96–99
Dicalcium phosphate	—	—	—	29	—	23	98
Ammonium phosphates[c]							
21-53-0	21	—	—	—	—	23	100
21-61-0	21	—	—	—	—	27	100
11-48-0	11	—	0–2	—	—	21	100
16-48-0	16	—	0–2	—	—	21	100
18-46-0	18	—	0–2	—	—	20	100
16-20-0	16	—	14	—	—	8.7	100

Table 9.7 General Composition of Phosphate Fertilizers (Continued)

Material	Total nitrogen (%)	Total potassium (%)	Total sulfur (%)	Total calcium (%)	Total magnesium (%)	Phosphorus Total[a] (%)	Phosphorus Available[b] (% of total)
Ammonium phosphate nitrate	30	—	—	—	—	4	100
Ammonium polyphosphate	15	—	—	—	—	25	—
Potassium phosphate		29–45	—	—	—	18–22	100
Magnesium ammonium phosphate	8		—	—	—	14	17
Raw rock phosphate			—	33–36	—	11–17	14–65
Basic slag			0.2	32	3	3.5–8	62–94
Defluorinated phosphate rock			—	20	—	9	85
Phosphate rock-magnesium silicate			—	20	8.4	10	85
Rhenania phosphate		—	—	30	0.3	12	97
Potassium metaphosphate		29–32	—	—	—	24–25	—

[a] Values given as elemental P. Fertilizers are generally marketed giving $\%P_2O_5$, which can be obtained by multiplying P value with 2.27. For example, see the grades of ammonium phosphates in this table.

[b] By neutral 1.0 N ammonium citrate procedure.

[c] Ammonium phosphate grades expressed as $\%N$, $\%P_2O_5$, $\%K_2O$.

From Tisdale et al., 1993. *Soil Fertility and Fertilizers*, 5th ed., p. 208. With permission of Prentice-Hall, Inc., Upper Saddle River, NJ.

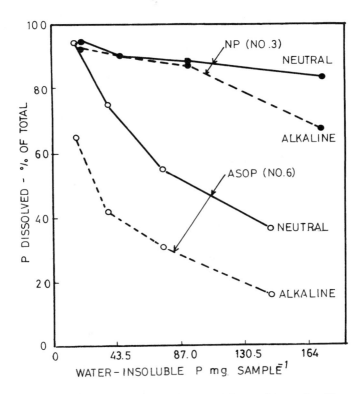

Figure 9.10. Water and citrate-soluble phosphorus in two fertilizer materials as affected by extraction with either neutral or alkaline ammonium citrate (From Terman et al., 1964. Adv. Agron. 16:59–100. With permission of Academic Press, Inc.)

9.8.1.3. Available Phosphorus. The sum of the water-soluble and citrate-soluble P represents an estimate of the fraction available to plants and is termed as available phosphorus.

9.8.1.4. Total Phosphorus. The sum of water-soluble, citrate-soluble, and citrate-insoluble P (P in residue left after extracting citrate-soluble P) represent the total P in fertilizer. Total P can be and is generally determined directly in a single step.

9.8.2. Phosphate Fertilizers

9.8.2.1. Ordinary (Single) Superphosphate

It is made by reacting phosphate rock with sulfuric acid and contains 7 to 9.5% P (16 to 22% P_2O_5) as monocalcium phosphate (MCP); about 90%

is water soluble. In addition, it contains 11 to 12% S as calcium sulfate. This material is ideally suited in areas where S is deficient.

9.8.2.2. Concentrated (Triple) Superphosphate

It is made by reacting phosphate rock with phosphoric acid and contains 19 to 23% P (44 to 52% P_2O_5) as MCP. All P present is water soluble.

9.8.2.3. Enriched Superphosphates

These are made by reacting phosphate rock with a mixture of sulfuric and phosphoric acids, and they contain 11 to 13% P (25 to 30% P_2O_5) of which 90 to 95% is water soluble.

9.8.2.4. Ammoniated Superphosphates

These are produced by reacting ordinary or triple superphosphate with anhydrous or aqua ammonia, and they contain 2 to 6% N and 6 to 21% P (14 to 49% P_2O_5). Ammoniation of superphosphate offers an inexpensive way of adding nitrogen to fertilizer, but reduces the water-soluble P content—less than 20% in ordinary superphosphate but nearly 50% in concentrated (triple) superphosphate (Terman et al., 1964). For crops responding to a high degree of water-soluble P, a high degree of ammoniation of ordinary superphsophate will have a depressing effect on the plant availability of P in fertilizer (Terman et al., 1964).

9.8.2.5. Ammonium Phosphates

These are made by reacting ammonia with phosphoric acid. Both monoam-monium phosphate (MAP) and diammonium phosphate (DAP) are widely used. The common MAP grades vary from 11-48-0 (21% P) to 11-55-0 (26% P), while DAP grades are 16-48-0 (21% P), 18-46-0 (20% P), and 21-53-0 (23% P).

When sulfuric acid is mixed with phosphoric acid and ammoniation is done, the product is ammonium phosphate sulfate (16-20-0) (8.6% P). Similarly, urea can be added to DAP to obtain urea ammonium phosphate (28-28-0) (12.2% P).

9.8.2.6. Nitric or Nitrophosphates

These are made by reacting phosphate rock with nitric acid. Since one of the reaction products is calcium nitrate, which is highly hygroscopic, steps are taken to remove it or modify it to some other form. Calcium nitrate can be removed by refrigeration and centrifugation, or it can be converted to calcium carbonate by injecting carbon dioxide. Some processes employ the use of sulfuric or phosphoric acid along with nitric acid to convert part of the calcium nitrate to calcium sulfate or calcium phosphate. The acidified slurry is then ammoniated. The final product contains a complex assortment of salts such as ammonium phosphate, dicalcium phosphate, ammonium nitrate, calcium sulfate, and others. The water solubility of nitric phosphates may vary from 0 to 80%, depending upon the process used. When used on crops responding to water-soluble P, nitric phosphates containing 30% or less water-soluble P may be inferior to P fertilizers containing a higher degree of water-soluble P (Prasad and Dixit, 1976). Nitric phosphates give best results on acidic soils and with crops with a relatively long growing season such as turf and sod crops (Tisdale et al., 1985) or sugarcane. Also when used in a cropping system and the study is extended to more than one crop, nitric phosphates may be as effective as highly water-soluble, P-containing fertilizers (Venugopalan and Prasad, 1989a).

9.8.2.7. Ammonium Polyphosphate

This product was developed by the Tennessee Valley Authority (TVA) in the United States by ammoniation of mixtures of electric-furnace superphosphoric acid and up to 30% of the P as wet-process orthophosphoric acid. This process results in a granular product with a grade of 15-62-0 (27% P). Both liquid and solid ammonium polyphosphate are available.

Before being taken up by plants, polyphosphates must undergo hydrolysis to change into orthophosphate; the hydrolysis is brought about by enzymes such as pryophosphatase (Dick and Tabatabai, 1986), which are generally present in soil. Venugopalen and Prasad (1989b) reported that the half-life values for the polyforms of P in liquid ammonium polyphosphate ranged between 1.6 to 2.0 days under anaerobic conditions and 5.2 to 8.7 days under aerobic conditions; the corresponding values for solid ammonium polyphosphate were 3.9 to 9.2 days under anaerobic and 12.5 to 27.0 under aerobic conditions. In this study the hydrolysis rate was fastest for a laterite, intermediate for a sodic, and slowest for an alluvial soil. Soil factors that may affect hydrolysis of ammonium polyphosphate are pH (greater hydrolysis on acidic than on alkaline soils), temperature (increases with a rise in temperature), texture, and water content (more hydrolysis on flooding).

In addition to high nutrient content, ammonium polyphosphates have the advantage of chelating micronutrients and reduced P-fixation due to the time required for their conversion to orthophosphates.

9.8.2.8. Thermal Phosphates

These fertilizers are made by the heat treatment of phosphate rock. There are two processes, namely, "calcination" and "fusion" by which thermal phosphates are made. In calcination the heating is done below the melting point of the mixture, while in fusion it is done above the melting point, so that the material fuses together. The calcinated material is sintered and porous in appearance, while the fused material is a glossy product. Some important fertilizer materials in this group are as follows:

1. *Defluorinated phosphate rock or Coronet phosphate* is made by combining finely ground phosphate rock, high silica tailings from the phosphate rock mining process, and enough water to form a slurry. The slurry is passed into an oil-fired kiln having temperatures between 1480 to 1590°C and heated for 30 minutes. The product is then quenched and ground to allow 60% of it to pass a 200-mesh screen. It contains 9% total P (21% P_2O_5) out of which 8% P (18% P_2O_5) is citrate soluble.

2. *Rhenenia phosphate* is prepared by calcining a mixture of phosphate rock, soda ash, and silica at 1100 to 1200°C. It is then quenched and ground. It contains 12% total P (28% P_2O_5); 11.8% P (27.5% P_2O_5) is citrate soluble.

3. *Phosphate rock-magnesium silicate glass* is formed by fusing phosphate rock and olivine or serpentine in a furnace at 1550°C. It contains 10% total P (22.5% P_2O_5) - 8% P (19% P_2O_5) is citrate soluble.

4. *Basic or Thomas slag* is a by-product of the steel industry. Basic slag in the United States contains about 3% P_2O_5, while that in Europe contains about 14 to 18% P_2O_5. Indian basic slags contain 1.5 to 3% P_2O_5.

The major disadvantages of thermal phosphates are (1) no water-soluble P; (2) high energy requirement for production, making it more expensive to produce; and (3) no value to the fertilizer industry because products cannot be ammoniated and thus manufacture of NPK fertilizers is not possible.

9.8.2.9. Partially Acidulated Rock Phosphate (PARP)

In countries where native S is not available and thus manufacture of sulfuric acid or phosphoric acid involves costs and foreign exchange, partial acidulation with sulfuric acid or other acids can be utilized. Generally, about half the quantity of acid is used. On a highly acidic and high P-fixing, dark-red latosol (oxisol), PARP was as good as triple superphosphate for rice and beans (Fageria et al., 1991). Sometimes PARPs may even perform better than ordinary or concentrated superphosphate (Marwaha, 1989). It is possible that the drop in pH in the soil surrounding a dissolving PARP granule is less than that around the fully acidulated fertilizer. In acidic soils this would result in less dissolution of Fe and Al and consequently less fixation of P; residues in the PARP granule and the reaction products in the soil may thus sustain a higher solution-P concentration than for ordinary or concentrated superphosphate (Sharpley et al., 1992).

9.8.2.10. Rock Phosphate

Ground phosphate rock contains 11 to 16% total P (25 to 37% P_2O_5). As already discussed, most of the P is in apatite form and thus there is no water-soluble P. Citrate solubility can vary from 5 to 17% of the total P, depending upon the chemical nature of the rock and the degree to which it is ground. Direct application of finely ground phosphate is recommended for acidic soils (pH 6 or below). On highly acidic (pH 4.9) and high P-fixing, dark-red latosol (oxisol), ground Brazilian rock phosphates gave as good yields of rice and beans as triple superphosphate from the second year onward; in the first year there was very little response to ground-rock phosphate (this would be expected because soil acids solubilize rock-phosphate P rather slowly) (Fageria et al., 1991). In another study relative agronomic effectiveness (taking a value of 100% for concentrated superphosphate) ranged from 36 to 100% for ground (<0.075 mm) North Carolina phosphate rock (Chien and Friesen, 1992).

Thus phosphate fertilizers differ not only in their content of total P, but also in the form in which P is present. Choice of the fertilizer will depend upon the soil, the crop to be grown, and the price per unit of P.

9.9. EFFICIENT PHOSPHATE MANAGEMENT

Efficient phosphate management has three major components: (1) strategies for efficient utilization of native soil P; (2) strategies for efficient utilization of fertilizer P; and (3) strategies for the direct use of rock phosphate.

9.9.1. Strategies for Efficient Utilization of Native Soil P

9.9.1.1. Use of P-Efficient Crops and Their Cultivars

Crops and cultivars within crop species differ in their abilities to survive, grow, and produce at low levels of available P in soils. According to Loneragan (1978), the differences among plants in their ability to absorb P from soils may be due to at least three distinct root attributes: (1) the physiological ability to absorb P from dilute solutions; (2) metabolic activity resulting in solubilization of sorbed P; and (3) the ability of the root system to explore the soil mass. Plant species having deeper root systems can feed better on native soil P. Even within a crop, cultivars differ in their ability to feed and grow on native soil P. From a practical viewpoint, cultivars that produce well under a low level of P and respond well to added P are the most desirable. Fageria et al. (1988) screened 25 rice cultivars on an oxisol in Brazil and found that seven cultivars utilized P more efficiently than others.

9.9.1.2. *Use of Vesicular Arbuscular Mycorrhizae (VAM)*

The effect of mycorrhizae on increasing root extraction of P is well known for many species (Tinker, 1978; Kucey et al., 1989). Three primary mechanisms by which VAM enhance soil P utilization by plants are (1) the increased physical exploration of the soil; (2) chemical modification of the rhizosphere; and (3) physiological difference between VAM and plant roots (Sharpley et al., 1992; Gianinazzi-Pearson, 1986).

Extensive hyphal growth of VAM reduces the distance for diffusion of P in soil and thereby increases P uptake by plants. Furthermore, the smaller diameter of VAM hyphae (2 to 4 μm) compared with root hair (7 to 10 μm) affords a greater absorptive surface area for hyphae and enables the entry of hyphae into soil pores and organic matter that cannot be entered by root hairs.

VAM may chemically modify the rhizosphere through the exudation of chelating compounds or phosphatases, which help in solubilizing slowly soluble soil P.

Generally, the majority of work on the use of VAM has been done under controlled conditions. The use of these organisms to increase the availability of native soil P will much depend on the selection of strains that have the ability to increase P uptake and to compete with native soil microflora under natural conditions of crop production.

9.9.1.3. *Use of Phosphobacterium*

Bacterium *Bacillus megatherium* var. *phosphobacterium*, called a biofertilizer phosphobacterium (Cooper, 1959), was first used in the USSR for increasing the availability of native soil P. The bacteria was credited for mineralizing organic P in soils. In about 50% of the crops inoculated with this bacteria, yield increases of the order of 1 to 20% were obtained in Soviet soils. The best results were obtained on neutral to alkaline soils with high organic-matter content. A positive response to phosphobacterium has also been reported from India. However, no response was obtained in field tests with wheat, oats, and sorghum in Alaska, Minnesota, Montana, North Dakota, and Texas. There is a need to identify the soil and climatic conditions under which phosphobacterium can function.

9.9.2. Strategies for Efficient Utilization of Fertilizer Phosphorus

Due to the fast reactivity of soluble P in fertilizers with the cations in soil solution and cations and anions on the surface of clay and organic matter particles, P does not move far from the point of placement. Therefore the key to efficient P utilization is deep placement near the growing young roots of the crop (Figure 9.11). Although a few recommendations are available for foliar application of soluble phosphate fertilizers, most fertilizer phosphate is applied to soil before or at seeding of a crop. Deep placement in a standing

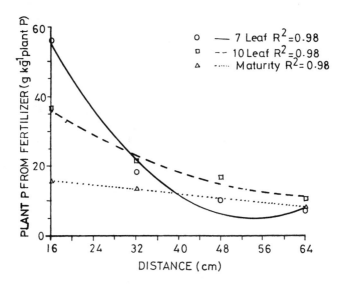

Figure 9.11. Effect of P distance from plants on P uptake from fertilizer at three sampling times over 2 years, 1984 and 1985. (From Eghball and Sander. 1989. Soil Sci. Soc. Am. J. 53:282–287. With permission of SSSA.)

crop without damaging the root system is neither possible nor desirable due to the immobile nature of P. However, in recent years some recommendations have emerged for split application of P in rice; top dressing in such cases has to be quite early in the crop's life span and can provide a solution in situations where phosphate fertilizers are not applied before transplanting.

A knowledge of the early rooting habits of various crop plants is helpful in determining the most satisfactory method of placing phosphate fertilizer. If a vigorous taproot is produced early, such as for cotton, tobacco, and most grain legumes, application may best be made directly under the seed. If many lateral roots are formed early, such as in cereals, side placement may be best. Increased root growth in the P-fertilized volume of soil compared with unfertilized soil is well known.

A rule of thumb is that, for row-sown crops, phosphate must be placed; the wider the interrow spacing, the greater the advantage of P placement. Only for those crops that are sown broadcast, incorporation in soil may be as good as deep placement. By banding, fertilizer P comes into less direct contact with soil, reducing the opportunity for fixation of soluble P. Likewise, using larger fertilizer P particles, which are subject to less fixation, is sometimes more effective than using finely ground P fertilizer.

In countries where rice is the principal crop, phosphate fertilizer is imported and expensive, and soils are usually high P-fixers. Dipping the rice seedlings before transplanting in a slurry of soil and phosphate fertilizer can increase the efficiency of fertilizer P and economize on phosphate use.

Table 9.8 Results of 3 Years of Experiments on Responses to
P_2O_5 Application of Kharif[a] and Rabi[b] in
Some Cropping Systems Involving Wheat

		Total rotational response to 60 kg P_2O_5/ha	
Cropping sequence	Research centre	P_2O_5 applied to kharif crop (q/ha)	P_2O_5 applied to wheat crop (% change)
Rice-wheat	Raipur	13.4	+4
	Jabalpur	19.9	−6
	Kathulia	24.7	−26
	Bichpuri	13.0	−19
	Varanasi	6.7	−6
	Mashodha	11.9	+46
	Kharagpur	15.1	−53
Jowar-wheat	Indore	18.9	+1
	Siruguppa	25.5	+13
Maize-wheat	Palampur[c]	36.2	+17
	Ludhiana	14.2	+40
Bajra-wheat	Hissar	12.7	+6

[a] Rainy season, July–October.

[b] Fall–spring season, November–April.

[c] At 120 kg P_2O_5/ha.

From Goswami and Kamath. 1984. Fert. News 29(2):22–26.

In multiple cropping systems the recommendation is to give priority to those crops that are more responsive to P and are poor users of residual P (wheat, potato) as compared with crops that can benefit from residual P (rice, corn) (Goswami and Kamath, 1984) (Table 9.8).

9.9.3. Strategies for the Direct Use of Rock Phosphate

In general, rock phosphate can be directly used in acidic soils having pH below 5.5 to 6.0, but for this purpose the phosphate rock must be finely ground to assure an adequate rate of P release (Tisdale et al., 1985). Also, the dosages used with rock phosphate are generally several times those recommended for soluble P fertilizers (Kucey and Bole, 1984).

Plant species differ in the capacity to utilize P from rock phosphate. This may be related to cation exchange capacity (CEC) of roots; the higher the CEC, the greater the capacity to extract P from rock phosphate (Drake and Steckel, 1955). Also, a relationship has been observed between the rhizosphere pH and plant P uptake. For example, buckwheat (*Fayopyron esculentum* L.) can acidify its rhizosphere and cause dissolution of rock phosphate, but corn cannot.

Some of the strategies for increasing the efficiency of ground-rock phosphate are discussed as follows:

1. *Mixing rock phosphate with soluble P fertilizers.* It is possible to use finely ground rock phosphate at least in part mixed with soluble P fertilizers such as superphosphate. The proportion of the two materials will, of course, depend upon the soil conditions and the crop. Fageria et al. (1990) have also suggested the use of rock phosphate in combination with a soluble P source as a strategy for improving P status of acidic soils; they recommend broadcast application of rock phosphate and band application of soluble P sources.

2. *Mixing rock phosphate with elemental S or S-producing compounds.* Mixing of rock phosphate with elemental S was suggested in the early part of the century (Lipman et al., 1916) for increasing the availability of P in the rock. The principle involved is that the S in the mixture is oxidized to sulfuric acid by chemautotrophic bacteria (*Thiobacillus thioxidans* and *Thiobacillus thioparus*), which solubilize the P in rock phosphate. This has been demonstrated on a number of soils. The availability of rock-phosphate-sulfur mixtures appears to be greatly influenced by a number of factors such as soil pH, soil temperature, soil water, and particle size of the mixture.

   Swaby (1975) inoculated rock-phosphate-sulfur mixtures with *T. thiooxidans* and *T. thioparus* and called them biosupers; these mixtures were found to be superior to uninoculated mixtures for pasture production in tropical soils.

   Iron pyrites as a source of sulfur have also been suggested for mixing with rock phosphate. This strategy has been used for utilizing low-analysis phosphate rock (18 to 20% P_2O_5), which cannot be used for making soluble phosphate fertilizers (Tiwari, 1979).

3. *Use of phosphate-solubilizing (PS) microorganisms.* The major site of occurrence of PS microorganisms has been on soil or seed surfaces. Phosphate-solubilizing organisms have been found in almost all soils tested, although the numbers vary with soil, climate, and cropping history (Chhonkar and Subba-Rao, 1967; Kucey, 1983). However, after subculturing PS isolates, many of the bacterial isolates may lose their PS activity. In several greenhouse studies, P uptake by inoculated plants was equal to or greater than that from superphosphate.

   Several studies of PS microorganisms have also included the effect of VA mycorrhizae. The individual effects of VAM and PS (*Pseudomonas* and *Agrobacterium* spp.) appear to be additive for solubilizing P from rock phosphate (Azcon et al., 1976). Under natural conditions effects of PS organisms will depend on how they interact with the general microbial population and how they survive under the prevailing soil environment.

4. *Other techniques.* A number of other techniques have been suggested for increasing the dissolution of rock phosphate when used directly as a source of phosphorus. These include the following:

- Combining N fertilizers with rock phosphate by banding, granulating, or compacting.
- Composting with animal manure or organic residues (Mishra et al., 1984).
- Including fine-rooted legumes in rotation to generate a low-pH rhizosphere with low Ca concentrations (Sharpley et al., 1992).

REFERENCES

Admont, P.H., R. Boniface, J.C. Fardeau, M. Jahiel, and C. Morel. 1986. Currently used methods for measuring available soil P: their use in assessing the fertilizer value of rock phosphates. Fert. Agric. 92:39–50.

Anderson, G. 1986. Assessing organic phosphorus in soils, in *The Role of Phosphorus in Agriculture* (2nd Print), F.E. Khasawneh, E.C. Sample, and J.E. Kamprath, Eds., Am. Soc. Agron., Crop Sci. Soc. Am., and Soil Sci. Soc. Am., Madison, WI, pp. 411–432.

AOAC. 1960. *Official Methods of Analysis.* 9th ed. Association of Official Agricultural Chemists, Washington, D.C.

Azcon, R., J.M. Barea, and D.S. Hayman. 1976. Utilization of rock phosphate in alkaline soils by plants inoculated with mycorrhizal fungi and phosphate solbulizing bacteria. Soil Biol. Biochem. 8:135–138.

Barrow, N.J. 1980. Evaluation and utilization of residual phosphorus in soils, in *The Role of Phosphorus in Agriculture* (2nd Print), F.E. Khasawneh, E.C. Sample and J.E. Kamprath, Eds., Am. Soc. Agron., Crop Sci. Soc. Am., and Soil Sci. Soc. Am., Madison, WI, pp. 333–360.

Beaton, J.D., R.C. Speer, and G. Brown. 1965. Effect of soil temperature and length of reaction period on water solubility of phosphorus in soil fertilizer reaction zone. Soil Sci. Soc. Am. J. 29:194.

Bell, L.C. and C.A. Black. 1970. Comparison of methods for identifying crystalline products by interactions of orthophosphate fertilizers with soils. Soil Sci. Soc. Am. Proc. 34:579–582.

Bouldin, D.R., J.R. Lehr, and E.C. Sample. 1960. The effect of associated salts on transformation of monocalcium phosphate monohydrate at the site of application. Soil Sci. Soc. Am. Proc. 24:464–468.

Brady, N.C. 1990. *The Nature and Properties of Soils*, 10th ed., John Wiley & Sons, New York.

Campbell, C.A., D.W.L. Read, G.E. Winkelman, and D.W. Andrews. 1984. First 12 years of a long-term crop rotation study in southwestern Saskatchewan — bicarbonate P distribution in soil and P uptake by the plant. Can. J. Soil Sci. 64:125–137.

Chang, S.C. and M.L. Jackson. 1958. Soil phosphate fractions in some representative soils. 9:109–119.

Chhonkar, P.K. and N.S. Subbarao. 1967. Phosphate solubilization by fungi associated with legume root nodules. Can. J. Microbiol. 13:749–753.

Chien, S.H. and D.K. Friesen. 1992. Phosphate rock for direct application, in *Future Directions for Agricultural Phosphorus Research*, F.J. Sikora, Ed., National Fert. and Environ. Res. Ctr., TVA, Muscle Shoals, AL, p. 18.

Cooper, R. 1959. Bacterial fertilizers in the Soviet Union. Soils & Fert. 22:327–333.

Cosgrove, D.J. 1962. Forms of inositol hexaphosphate in soils. Nature (London) 194:1265–1266.

Cox, F.R., E.J. Kamprath, and R.E. McCollum. 1981. A description model of soil test nutrient levels following fertilization. Soil Sci. Soc. Am. J. 45:529–532.

Das, D.K. and N.P. Datta. 1969. Products of interaction of fertilizer phosphorus in acid soil of Tripura and alluvial calcareous soil of Bihar. J. Indian Soc. Soil Sci. 17:119–124.

Dick, R.P. and M.A. Tabatbai. 1986. Hydrolysis of polyphosphates in Soils. Soil Sci. 142:132–140.

Drake, M. and J.E. Steckel. 1955. Solubilization of soil and rock phosphate as related to cation exchange capacity. Soil Sci. Soc. Am. Proc. 19:449–450.

Eghball, B. and D.H. Sander. 1989. Distance and distribution effects of phosphorus fertilizer on corn. Soil Sci. Soc. Am. J. 53:282–287.

Fageria, N.K., B.C. Baligar, and R.J. Wright. 1991. Influence of phosphate rock sources and rates on rice and common bean production in an oxisol. Plant Soil 134:137–144.

Fageria, N.K., J.R. Wright, and V.C. Baligar. 1988. Rice cultivar evaluation for phosphorus efficiency. Plant Soil 111:105–109.

Fageria, N.K., V.C. Baligar, and D.G. Edwards. 1990. Soil-plant nutrient relationships at low pH stress, in *Crops and Enhancers of Nutrient Use*, V.C. Baligar and R.R. Duncan, Eds., Academic Press, New York, pp. 475–508.

Fixen, P.E. and J.H. Grove. 1990. Testing soils for phosphorus, in *Soil Testing and Plant Analysis*, R.L. Westeman, Ed., 3rd ed., Soil Sci. Soc. Am. Book Ser. No. 3, Madison, WI, pp. 141–180.

Gianinazzi-Pearson, V. 1986. Mycorrhiza: a potential for better use of phosphate fertilizer. Fert. Agric. 92:3–12.

Giordano, P.M., E.C. Saple, and J.J. Mortvedt. 1971. Effect of ortho- and pyrophosphate on Zn and P in soil solution. Soil Sci. 111:101–106.

Goswami, N.N. and Kamath, M.B. 1984. Fertilizer use research on P in relation to its utilization by crops and cropping systems. Fert. News 29(2):22–26.

Griffin, R.A. and J.J. Jurinak. 1973. The interaction of phosphate with calicte. Soil Sci. Soc. Am. Proc. 37:847–850.

Griffin, R.A. and J.J. Jurinak. 1974. Kinetics of phosphate interaction with calcite. Soil Sci. Soc. Am. Proc. 38:75–79.

Harrison, A.F. 1987. *Soil Organic Phosphorus — A Review of World Literature*. CAB Intl, Wallingford, U.K.

Hingston, F.J., A.M. Posner, and J.P. Quirk. 1972. Anion adsorption by goethite and gibbsite. I. The role of the proton in determining adsorption envelopes. J. Soil Sci. 23:177–192.

Hingston, F.J., A.M. Posner, and J.P. Quirk. 1974. Anion adsorption by goethite and gibbsite. II. Desorption of anions from hydrous oxide surfaces. J. Soil Sci. 25:16–26.

Hsu, P.H. and M.L. Jackson. 1960. Inorganic phosphate transformations by chemical weathering in soils as influenced by pH. Soil Sci. 90:16–24.

Jones, C.A., C.V. Cole, A.N. Sharpley, and J.R. Williams. 1984. A simplified soil and plant phosphorus model. I. Documentation. Soil Sci. Soc. Am. J. 48:800–805.

Kitrick, J.A. and M.L. Jackson. 1956. Electron microscope observations of the reactions of phosphate with minerals, leading to a unified theory of phosphate fixation in soils. J. Soil Sci. 7:81–88.

Kowalenko, C.G. and R.B. McKercher. 1971. Phospholipid P content of Saskatchewan soils. Soil Biol. Biochem. 3:243–247.

Kucey, R.M.N. and J.B. Bole. 1984. Availability of phosphorus from 17 rock phosphates in moderately and weakly acidic soils as determined by 32 P dilution, A value, and total P uptake methods. Soil Sci. 138:180–188.

Kucey, R.M.N., H.H. Janzen, and M.E. Legget. 1989. Microbially mediated increases in plant available phosphorus. Adv. Agron. 42:199–228.

Kucy, R.M.N. 1983. Phosphate-solubilizing bacteria and fungi in various cultivated and virgin Alberta soils. Can. J. Soil Sci. 63:674–678.

Lawton, K. and J.A. Vomocil. 1954. The dissolution and migration of phosphorus from granular superphosphate in some Michigan soils. Soil Sci. Soc. Am. Proc. 18:26–32.

Lehr, J.R., W.E. Brown, and E.H. Brown. 1959. Chemical behavior of monocalcium phosphate monohydrate in soils. Soil Sci. Soc. Am. Proc. 23:3–7.

Lindsay, W.L. and E.C. Moreno. 1960. Phosphate phase equilibria in soils. Soil Sci. Soc. Am. Proc. 24:177-182.

Lindsay, W.L. and H.F. Stephenson. 1959. Nature of the reactions of monocalcium phosphate monohydrate in soils. II. Dissolution and precipitation reactions involving iron, aluminum, manganese and calcium. Soil Sci. Soc. Am. Proc. 23:18–22.

Lindsay, W.L., A.W. Frazierand, and H.F. Stephenson. 1962. Identification of reaction products from phosphate fertilizers in soils. Soil Sci. Soc. Am. Proc. 26:446–452.

Lindsay, W.L., M. Peech and J.S. Clark. 1959. Solubility criteria for the existence of variscite in soils. Soil Sci. Soc. Am. Proc. 23:357–360.

Lipman, J.G., H.C. McLean, and H.C. Lont. 1916. Sulfur oxidation in soils and its effect on availability of mineral phosphates. Soil Sci. 2:499–538.

Loneragan, J.F. 1978. The physiology of plant tolerance to low phosphorus availability, in *Crop Tolerance to Suboptimal Land Conditions*, G.A. Jung, Ed., American Society of Agronomists, Madison, WI, pp. 329–342.

Marwaha, B.C. 1989. Rock phosphate holds the key to productivity in acid soils. Fert. News 34:23–29.

Mishra, M.M., A.L. Khurrana, S.S. Dudeja, and K.K. Kapoor. 1984. Effect of phosphocompost on the yield and uptake of red gram. Trop. Agric. 61:174.

Munns, D.N. 1965. Soil acidity and growth of a legume. I. Interactions of lime with nitrogen and phosphate on growth of *Medicao sativa* L. and *Trifolium subterraneum* L. Aust. J. Agri. Res. 16:733–741.

Nakaru, T. and G. Uehara. 1972. Anion adsorption in ferruginous tropical soils. Soil Sci. Soc. Am. Proc. 36:296–300.

Olsen, S.R. and F.E. Khasawneh. 1980. Use and limitations of physical-chemical criteria for assessing the status of phosphorus in soils, in *The Role of Phosphorus in Agriculture*, F.E. Khasawneh, E.C. Sample, and E.J. Kamprath, Eds., Am. Soc. Agron., Madison, WI, pp. 361–410.

Power, J.F., D.L. Grunes, G.A. Reichman, and W.O. Willis. 1964. Soil temperature effects on phosphorus availability. Agron. J. 56:545.

Prasad, R. and L.A. Dixit. 1976. Fertilizers containing partially water soluble or no water soluble phosphate. Indian Council of Agric. Res. Bull., p. 34.

Sample, E.C., Soper, R.J., and G.J. Racz. 1986. Reactions of phosphate fertilizers in soils, in *The Role of Phosphorus in Agriculture* (2nd Print), F.E. Khasawneh, E.C. Sample, and E.J. Kamprath, Eds., Am. Soc. Agron., Crop Sci. Soc. Am., and Soil Sci. Soc. Am., Madison, WI, pp. 263–310.

Schalscha, E.B., P.F. Pratt, and D. Soto. 1974. Effect of phosphate adsorption on the cation exchange capacity of volcanic ash soils. Soil Sci. Soc. Am. Proc. 38:539–540.

Sharpley, A.N., J.J. Meisinger, J.F. Power, and D.L. Suarez. 1992. Root extraction of nutrients associated with long-term soil management. Adv. Soil Sci. 19:151-217.

Stowasser, W.F. 1979. Mineral commodity profiles. U.S. Department of Interior, Bureau of Mines, Pittsburgh, PA.

Swaby, R.J. 1975. Biosuper-Biological superphosphate, in *Sulfur in Australian Agriculture*, K.D. McLacklan, Ed., Sydney University Press, Sydney, pp. 213–220.

Talibuddin, O. 1981. Precipitation, in *The Chemistry of Soil Processes*, D.J. Greenland and M.H.B. Hayes, Eds., John Wiley & Sons, Chichester, U.K., pp. 81–116.

Terman, G.L., W.M. Huffman, and B.C. Wright. 1964. Crop response to fertilizers in relation to content of available phosphorus. Adv. Agron. 16:59–100.

Thomas, G.W. and D.E. Peaslee. 1973. Testing soils for phosphorus, in *Soil Testing and Plant Analysis*, Revised ed., L.M. Wash and J.D. Beaton, Ed., Soil Sci. Soc. Am., Madison, WI, pp. 115–132.

Tinker, P.B. 1978. Effects of vesicular-arbuscular mycorrhizae on plant growth. Physiol. Veg. 16:793–851.

Tisdale, S.L., W.L. Nelson, and J.D. Beaton. 1985. *Soil Fertility and Fertilizers*, 4th ed., Macmillan, New York, p. 754.

Tiwari, K.N. 1979. Efficiency of Mussorie rock phosphate, pyrites, and their mixtures in some soils of Uttar Pradesh. Indian Soc. Soil Sci. Bull 12:519–526.

Varvel, G.E., F.N. Anderson, and G.A. Peterson. 1981. Soil test correlation problems with two phosphorus methods on similar soils. Agron. J. 73:516–520.

Venugopalan, M.V. and R. Prasad. 1989a. Relative efficiency of ammonium polyphosphate and orthophosphate for wheat and their residual effects on succeeding cowpea fodder. Fert. Res. 20:109–114.

Venugopalan, M.V. and R. Prasad. 1989b. Hydrolysis of ammonium polyphosphate in soils under aerobic and anaerobic conditions. Biol. Fert. Soils 8:325–327.

Williams, J.D.H. and T.W. Walker. 1969. Fractionation of phosphate in a maturity sequence of New Zealand basaltic soil profiles. Soil Sci. 107:213–219.

Wolf, J., C.T. deWit., B.H. Janssen, and D.J. Lathwell. 1987. Modelling long-term crop response to fertilizer phosphorus. I. The Model. Agron. J. 79:445–451.

Wright, B.C. and M. Peech. 1960. Characterization of phosphate reaction products in acid soils by the application of solubility criteria. Soil Sci. 90:32–43.

Yadav, B.R., K.V. Paliwal, and N.M. Nimgade. 1984. Effect of magnesium rich waters on phosphate sorption by calcite. Soil Sci. 138:153–157.

Yadav, V.V. and K.B. Mistry. 1984. Reaction products of tri-ammonium pyrophosphate in different Indian soils. Fert. Res. 5:423–434.

10   POTASSIUM

In most soils potassium is present in much larger amounts than nitrogen and phosphorus; the earth's crust contains about 1.9% K and 0.11% P. The potassium content in surface soil may vary from a few hundred kilograms per hectare in light, sandy soils to about 50,000 kilograms per hectare in heavy, clayey soils rich in micas and 2:1 layer silicates. In addition to larger amounts, potassium offers two more contrasts to nitrogen and phosphorus in soils. These are (1) almost all potassium is present in inorganic form, and (2) potassium is fairly well distributed throughout the profile, and in some cases subsoils may even have more potassium than surface soils (Figure 10.1).

10.1. FORMS OF SOIL POTASSIUM

Potassium is present in soils in four different forms, namely, primary-mineral, fixed, exchangeable, and solution K. These four forms of K are interrelated as shown in Figure 10.2. Normally, there is essentially no organic form of K in soils. Potassium contained in manures and crop residues returned to soils is rapidly leached out of the organic material and is dissolved in the soil solution, where it can react with the clay minerals.

10.1.1. Primary-Mineral K

The potassium bearing minerals in soils are the feldspars, orthoclase $[(K,Na)AlSi_3O_8]$ and microline $[Na,K)AlSiO_4]$, and the micas, muscovite $[KAl_3Si_3O_{10}(OH)_2]$, biotite $[K(Mg,Fe)_3AlSi_3O_{10}(OH)_2]$, and phlogopite $[KMg_3 Al Si_3 O_{10} (OH)_2]$. While feldspars are generally present in the coarser fraction of soil, micas are predominantly in the clay fraction (Table 10.1). Potassium in mineral form in soils may vary from 5000 to 25,000 mg K kg^{-1} soil (0.5 to 2.5%) (Tisdale et al., 1985).

10.1.2. Nonexchangeable or Fixed K

Nonexchangeable K is distinct from mineral K in that it is not bonded covalently within the crystal structure of soil mineral particles; instead, it is

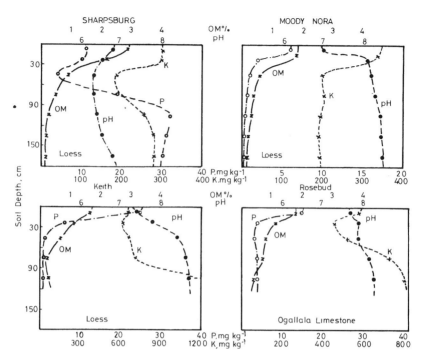

Figure 10.1. Distribution of organic matter, available P, available K, and pH in some Nebraska soils. (From McCallister et al., 1987. Soil Sci. Soc. Am. J. 51:1646. With permission of SSSA.)

held between adjacent tetrahedral layers of micas, vermiculites, and intergrade minerals (Sparks, 1987). Micas (muscovite, biotite, and phlogopite) have K fixed in the interlayer spaces (see Chapter 4). Bonding of K is stronger in dioctahedral mica (muscovite) than in tri-octahedral micas (biotite and phlogopite). Weathering of micas releases K in soils. Due to the variation in binding strength, the rate of release of K from different micas differs. The rate of potassium liberation from biotite is 13 to 16 and 75 to 105 times faster than from phlogopite and muscovite, respectively (Huang et al., 1968). The sequence of K release from K-bearing minerals by oxalic or citric acid has been found to be as follows: biotite > microline ≈ orthoclase > muscovite (Table 10.2).

Because micas have their entire negative charge satisfied by K (Chapter 4), the release of K results in the formation of secondary clay minerals such as illite (hydrous micas) and vermiculite, with accompanying gain of water or OH_3^+ and swelling of the lattice (Figure 10.3). Once this happens, K changes from a non-exchangeable form to an exchangeable form. Depletion of K from soil solution by plants or by leaching lowers the K concentration in solution and induces the liberation of interlayer, fixed K. In one of our unpublished

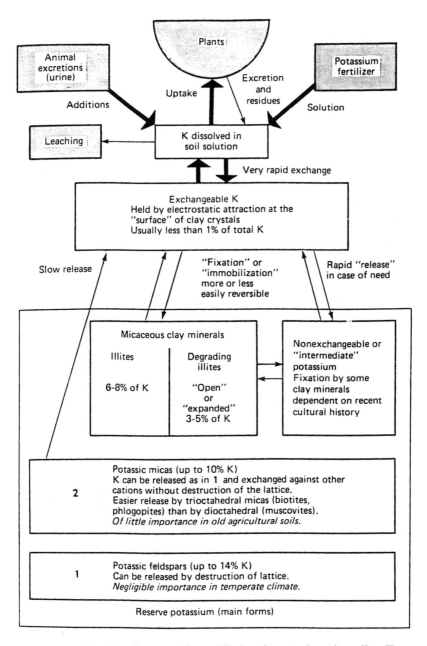

Figure 10.2. **Possible forms and equilibria of potassium in soils. (From Tisdale et al., 1993.** *Soil Fertility and Fertilizers,* **5th ed., p. 232. With permission of Prentice-Hall, Inc., Upper Saddle River, NJ.)**

Table 10.1 Distribution of Feldspar and Mica K in Different Size Fractions of a Haverhill Ap Soil in Saskatchewan, Canada

Particle size (μm)	Total K	Feldspar K	Mica K	% Fraction of K Feldspar	Mica
		g kg⁻¹ soil			
<0.2	21.1	—	21.1	0	100
0.2–2	26.1	0.8	5.3	3	97
2–5	20.1	6.1	14.0	30	70
5–20	17.3	8.7	8.6	50	50
20–50	15.4	9.7	5.7	63	37
50–500	12.7	7.6	5.1	60	40
500–2000	17.3	17.3	—	100	—
<2000+	16.9	6.8	10.1	40	60

From Somasiri et al. 1971. Soil Sci. Soc. Am. Proc. 35:500. With permission of SSSA.

Table 10.2 Potassium Released From K-Bearing Minerals by 0.01 mol L⁻¹ Solution of Oxalic and Citric Acids at the End of a 10-Day Reaction Period

	K released (g kg⁻¹ of structural K)	
Mineral	Oxalic acid	Citric acid
Biotite	44	11
Muscovite	1.1	0.9
Microcline	2.3	1.4
Orthoclase	2.2	1.2

From Song and Huang. 1983. Agron. Abst. 222. With permission.

studies it was found that, compared with grass sod, 120 years of cropping had greatly reduced the nonexchangeable K content, and it showed that illite was changing to vermiculite.

Nonexchangeable K in soils is generally measured by extracting the soil with 1 N boiling nitric acid and is reported as HNO_3 extractable K. Nonexchangeable, fixed K in soils varies from 50 to 705 mg K kg⁻¹ soil (Tisdale et al., 1985), depending upon the nature and amount of clay minerals present.

10.1.3. Exchangeable K

This is the potassium held in the exchange complex of 2:1 layer silicates. Thus soils containing smectite have more exchangeable K than those containing illite, which in turn have more than the soils containing kaolinite (Table 10.3). The amount of exchangeable K in soils may vary from 40 to 600 mg K kg⁻¹ soil (Tisdale et al., 1985).

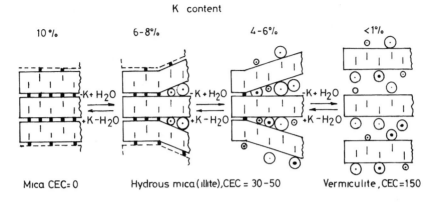

Figure 10.3. Schematic weathering of micas and their transformation into clay minerals: a matter of potassium release and fixation. (From McLean, 1979. *Potassium in Soil and Crops*, G.S. Sekhon, Ed., pp. 1–13. With permission of Potash Research Institute of India.)

Table 10.3 Relationship Between Clay Content, Kinds of Minerals, CEC, and Exchangeable Potassium

Soil	Depth cm	Clay %	Minerals[a]	CEC	Ex K
				cmol kg^{-1}	
Houston clay (Vertisol) (Clark County, AR)	0–10	66	MKI[b]	32.9	0.59
Vista coarse, sandy loam (Inceptisol) (San Diego County, CA)	0–8	10.2	IK	17.2	0.20
Cecil fine, sandy loam (Ultisol) (Iredell County, NC)	1–10	10.4	KIV	7.8	0.08

[a] Minerals in the order of dominance.

[b] *M*, Montmorillonite; *K*, kaolinite, *I*, illite, *V*, vermicullite.

From Buol et al. 1980. *Soil Genesis and Classification*, 2nd ed., p. 367. With permission of Iowa State University Press.

Exchangeable K in soils is generally determined by extracting soil with neutral 1 N ammonium acetate and therefore includes water soluble K; the entire value is known as available K in soils (this is the conventional method in soil testing). For the precise determination of exchangeable K, water-soluble K should be determined by extracting soil with distilled water separately and the value obtained should be subtracted from the 1 N ammonium acetate extractable K value to obtain exchangeable K. Normally, water-soluble K is usually a fraction of exchangeable K, except possibly in sandy soils.

10.1.4. Soil-Solution K

This is the potassium present in soil solution and is measured by extracting the soil with distilled water. Amounts of water-soluble K are generally comparable to those determined by electroultrifiltration (EUF_{10}-K desorbed during 10 minutes), a technique proposed by Nemeth (1979) (Sekhon et al., 1992). Water-soluble K in soil may vary from 1 to 10 mg K kg^{-1} soil. Solution K concentration is important for successful crop production. Singh and Jones (1975) at the University of Idaho suggested a critical concentration of 8.7 mg L^{-1} for crops that need lesser amounts of K (tomatoes and beans) and 14.5 mg L^{-1} for crops such as celery and potatoes that have a high K requirement. Levels of water-soluble K below 8 mg L^{-1} may suggest K deficiency.

There is a continuous transfer of mineral K to exchangeable K and fixed K and from there to solution K; the process may reverse to the fixed-K stage under some soil conditions or when heavy K dressings are made. The entire equilibrium may be represented as below:

$$\text{Nonexchangeable K}^{\text{slow}} \rightleftharpoons \text{exchangeable K}^{\text{rapid}} \rightleftharpoons \text{soil solution K}$$

Figures 10.4 and 10.5 show the above relationships for a series of Indian soils. As can be seen from these figures, the release of exchangeable to solution K, as well as from nonexchangeable to exchangeable K, varies from soil to soil and depends much upon the dominant clay minerals present. In general illite- and kaolinite-dominant soils have a larger proportion of water-soluble to exchangeable K than smectite-dominant soils (Figure 10.4). On the other hand, the proportion of exchangeable to nonexchangeable K is greater in smectite-dominant soils than illite- and kaolinite-dominant soils; the proportion for the soils with mixed mineralogical composition is in between (Figure 10.5).

Data on different forms of K, expressed as percent of total K for some Indian soils, are presented in Table 10.4. These data illustrate that the mineral forms of K comprise 92 to 97%, nonexchangeable (fixed K) 2 to 7.5%, and available K (exchangeable and water-soluble) 0.1 to 2% of total K. This general relationship would apply to most soils.

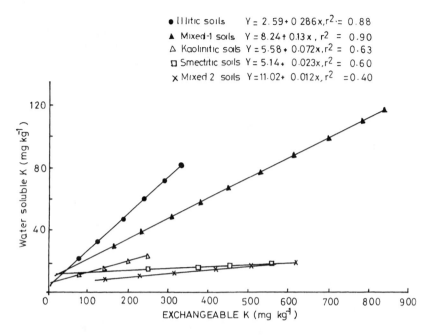

Figure 10.4. Relationship between exchangeable K and water-soluble K content for different soil groups. (From Sekhon et al., 1992. Special Pub. 3, p. 25. With permission of Potash Research Institute of India.)

10.2. QUANTITY/INTENSITY RELATIONSHIPS

Cations other than K present in soil solution largely influence the effectiveness of solution K for plant nutrition. The activity ratio (AR^K) in a solution in equilibrium with soil therefore provides a better and more satisfactory measure of the availability of K to plants.

$$AR^K = \frac{\text{activity of K}}{\sqrt{\text{activity of Ca + Mg}}} \quad \text{or} \quad \frac{a_K}{\sqrt{a_{Ca+Mg}}} = (ak)/a(Ca + Mg)^{1\,2}$$

where K, Ca, and Mg are expressed in moles L^{-1}.

AR^K is an intensity measure of labile K and indicates the amount of K that is immediately available to plants.

However, as pointed out earlier, solution K tends to be in equilibrium with exchangeable K, which is the total quantity (Q) of K available in soil. The Q/I concept in potassium was developed by Beckett (1964ab) and is in line with the ratio law enunciated by Schofield (1947). This relationship is explained in Figure 10.6, which shows that the different available quantity (Q) in two different soils can have the same intensity of K (AR^K) depending upon the

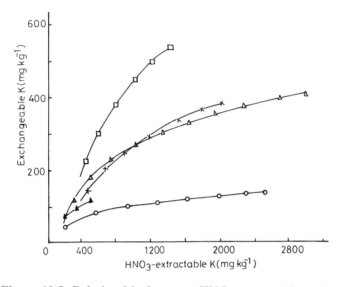

□ Smectitic soils	Y =-1530.6 + 285.5 lnX	r² = 0.64
✗ Mixed 2 soils	Y = - 932.0 + 173.21 lnX	r² = 0.78
△ Mixed 1 soils	Y = - 653 4 + 133.1 lnX	r² = 0.95
▲ Kaolinitic soils	Y = - 201.5 + 50.3 lnX	r² = 0 84
○ Illitic soils	Y= - 160.1 + 37.6 lnX	r² = 0 19

Figure 10.5. Relationship between HNO₃-extractable and exchangeable K content for different soil groups. (From Sekhon et al., 1992. Special Pub. 3, p. 27. With permission of Potash Research Institute of India.)

soil's mineralogical makeup, which largely determines its CEC and K saturation of the exchange complex. When a certain quantity of K (ΔQ) is removed from the soil solution by plants (or leaching), the change in intensity (ΔI) is much greater in soil B (sandy soil with low CEC) than in soil A (clayey soil with higher CEC).

The ratio (slope) $\Delta Q / \Delta I$ is referred to as the potassium buffering capacity (PBC^K). A large PBC^K signifies a good K-supplying power of a soil, while a small value suggests the need for frequent fertilization.

There are two more points to be noted in Figure 10.6:

1. The value of AR^K when ΔK is zero represents the current status of soil K.
2. The variation from the linearity of the curve as it approaches an activity ratio of zero (dotted extension of the curves below the x-axis) is labile K; it is a rough measure of the potential for a K source to exchange available K from currently unavailable sources. It may be noted that soil B has a much smaller potential than soil A.

**Table 10.4 Forms of Potassium Expressed as Proportion of
Total Potassium in Different Soils**

Soil	Form of K (% of total K)		
	Water-soluble	Exchangeable	Nonexchangeable
Alluvial			
Calcareous illitic (8)[a]	0.12	0.28	7.52
Acidic kaolintic (2)	0.32	0.78	2.34
Vertisols and vertic type			
Smectitic (7)	0.25	2.12	5.56
Red and Laterite			
Kaolintic (5)	0.19	0.49	3.14

[a] Numbers in parentheses refer to the number of soil series.

From Sekhon et al. 1992. Special Pub. 3, p. 23. With permission of Potash Research Institute of India.

10.3. POTASSIUM FIXATION

Potassium fixation is defined as the conversion of soil solution or exchangeable K into nonexchangeable forms and was once considered a negative soil property causing a drastic reduction of plant available K. However, this view is no longer held and K-fixation is considered beneficial because it reduces K losses by leaching and luxury consumption, yet it maintains a potentially available K pool (Pearson, 1952). In most well-fertilized soils, the nonexchangeable K–exchangeable K equilibrium is dynamic and fertilizer K that is fixed on addition is also available for more immediate release (Bertsch and Thomas, 1985). This is evidenced by the observation that the release of nonexchangeable K by intensive cropping results in increased fixation of subsequently applied K (Munn and McLean, 1975).

A number of factors affect potassium fixation in soils. These are briefly discussed.

10.3.1. Clay Minerals

The amount of K fixed by a soil depends much upon its clay content. Both quantity and quality are important, but in general, the greater the clay content, the greater the K fixation. Regarding the kind of clay minerals, illite,

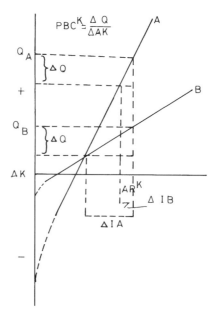

Figure 10.6. Q/I relationship in soil potassium. Soil A represents a fine-textured soil rich in illite and smectite, while Soil B represents a coarse-textured soil having kaolinite as the dominant mineral. (Adapted from Bertsch and Thomas, 1985.)

weathered mica, vermiculite, smectite, and interstratified minerals fix K, while kaolinite fixes very little (Figure 10.7).

10.3.2. Soil pH

In acid soils the presence of Al^{3+} and aluminum hydroxide cations and their polymers occupy the K-selective binding sites on clay minerals. Also, the presence of hydroxyl aluminum-iron interlayer groups under acid conditions prevents the collapse of silica layers in expanded clay minerals and therefore prevents the entrapment of K. Thus when pH of acid soil is raised, it affects K-fixation in two complementary ways: (1) precipitation of Al^{3+} due to higher OH^- ion concentration and (2) progressive hydroxylation of monomeric and polymeric forms of hydroxyl-Al ions, whose positive charge is gradually neutralized. Liming will thus increase K-fixation, especially in temperate-region soils where 2:1 layer silicates having Al^{3+} dominate. Liming may have little effect on K-fixation in tropical soils where kaolinite dominates (Ritchey, 1979; Malavolta, 1985).

As already discussed in Chapter 4, raising soil pH above neutrality increases the negative charge on oxides and hydroxyoxides of iron and aluminum, which results in increased adsorption of K ions and consequent reduc-

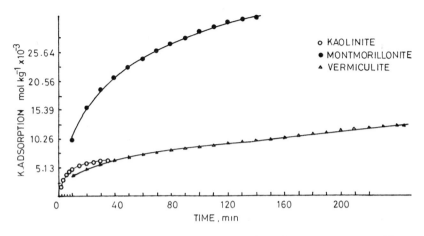

Figure 10.7. Potassium adsorption versus time in pure systems. (From Sparks and Jardine, 1984. Soil Sci. 138(2):115–122. With permission of Williams & Wilkins.)

tion in soil solution K (Figure 10.8). Thus the overall effect of an increase in soil pH is increased K-fixation (Figure 10.9).

Thus liming in general lowers the concentration of K in soil solution. In terms of the Q/I relationship discussed earlier, the AR^K value is reduced (Mielniczuk, 1977), as is shown in Figure 10.10. However, in both soils 1 and 2 in this study, the K-buffering capacity (PBC^K), that is, $\Delta Q/\Delta I$ or the slope of the linear curve, increased from 0.08 to 0.24 in soil 1 and from 0.84 to 1.35 in soil 2.

The overall effect of liming is beneficial because it reduces the leaching of K and it prevents the luxury consumption of K. Even on very acid tropical soils the effects of temporarily lowering K in soil solution (AR^K) are over-compensated by the removal of the restrictive effect of Al on root growth and vigor (Tisdale et al., 1985). This is supported by data on K uptake studies by crops. For example, Fageria et al. (1989) showed that the K concentration in alfalfa shoots grown on an acid inceptisol from Tennessee increased with liming.

10.3.3. Wetting and Drying

Data on K-fixation after wetting or drying are presented in Table 10.5. These data show that despite great differences between temperate and tropical soils, K fixation was 2 to 3 times greater after drying than after wetting (Malovolta, 1985). Thus air drying of soils high in exchangeable K will result in fixation and a decline in exchangeable K. This needs to be considered when interpreting soil test data for available K, which is the sum of exchangeable and water-soluble K and is generally determined by extraction with 1 N ammonium acetate after air drying the soil samples.

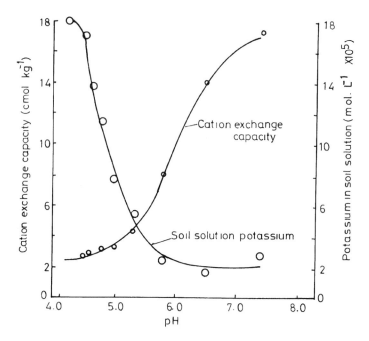

Figure 10.8. The influence of increased pH resulting from lime additions on the pH-dependent, cation-exchange capacity of a soil and the level of potassium in the soil solution. As the cation-exchange capacity increases, some of the soil solution potassium is attracted to the adsorbing colloids. (From Brady, N.C. 1990. *The Nature and Properties of Soils,* **p. 376. With permission of Prentice-Hall, Inc., Upper Saddle River, NJ.)**

However, drying of field-moist soils, particularly subsoils, with low to medium levels of K is reported to increase exchangeable K. This has been attributed to the exfoliation of edge-weathered micas and exposure of inter-layer K.

10.3.4. Potassium Fertilization

Adding large amounts of fertilizer K generally results in increased K fixation because solution K concentration is greatly increased, pushing the equilibrium between soluble and fixed-K toward the fixed K pool.

10.3.5. Freezing and Thawing

Alternate freezing and thawing may result in increased exchangeable K in some soils; however, the reverse may also happen in illitic soils having high exchangeable K (Tisdale et al., 1985).

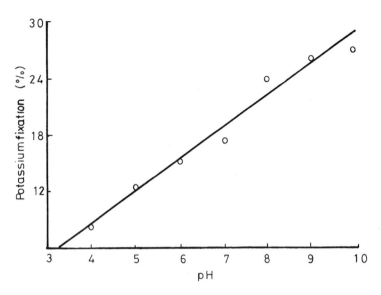

Figure 10.9. The effect of pH on the fixation of potassium soils in India.
(From Grewal and Kanwar, 1976. *Potassium and Ammonium Fixation in Indian Soils — Review,* Indian Council of Agricultural Research, New Delhi.

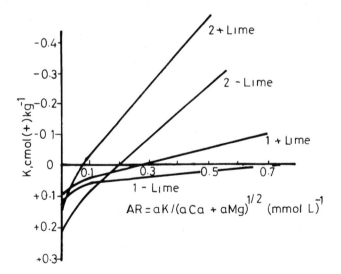

Figure 10.10. Effects of lime on Q/I values of Brazilian soils. (Mielniczuk, 1977. Rev. Bras. ci Solo 1:55–56.)

**Table 10.5 Potassium Fixation After the
Wet and Dry Methods[a]**

	K fixation %	
Region	Wet	Dry
Temperate		
Range	Trace–63	15–95
Average	25	68
Tropical and subtropical		
Range	2–40	15–67
Average	17	24

[a] Shaking with K^+ solution is the wet method; drying of the
soil with K^+ solution at 80°C is the dry method.

From Malovolta 1985. *Potassium in Agriculture*, R.D. Munson, Ed., p. 163. With permission of ASA.

10.4. LEACHING OF POTASSIUM

Potassium leaching from a soil fluctuates in accordance with the quantity, timing, and intensity of rainfall. For example, in a study on a sandy soil, K leaching was 19 kg K ha^{-1} yr^{-1} in dry years but increased to 57 kg K ha^{-1} yr^{-1} in wet years (Jurgens-Gschwind and Jung, 1979). Despite low K content in tropical soils, considerable K may be lost by leaching due to heavy rains. For example, McColl (1970) reported a loss of 52 kg K ha^{-1} yr^{-1} from forests of Costa Rica receiving an annual rainfall of 380 mm yr^{-1}. Godefroy et al. (1975) in the Ivory coast (oxisols on Schists) estimated that 50 to 60% of the K fertilizer (1100 kg K or 1325 kg K_2O yr^{-1}) applied to bananas (*Musa paradisiaca* L.) was lost by leaching. Ritchey (1979) showed that 1 month after burning the vegetation for clearing the land in Peru, the exchangeable K content to a depth of 30 to 50 cm increased three times the preclearing value (Figure 10.11). However, after 6 to 10 months, the K level returned to the original value due to excessive leaching of K.

Liming generally reduces K leaching (due to increased K fixation as discussed earlier). For example, Shaw and Robinson (1960) reported K leaching losses in a loam soil of 55 kg ha^{-1} on the unlimed soil (pH 4.5) and 21.6 kg K ha^{-1} with an application of 12.5 Mg dolomite limestone ha^{-1}.

10.5. POTASSIUM FERTILIZERS

There are two major potash fertilizers, namely, potassium chloride or muriate of potash (KCl) (50 to 52% K or 60 to 63% K_2O) and potassium sulphate (K_2SO_4) (40 to 44% K or 48 to 53% K_2O). Both of these fertilizers

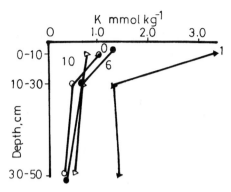

Figure 10.11. Extractable K measured before burning (0) and 1, 6, and 10 months after jungle vegetation in Peru. (From Ritchey, 1979. Cornell Int. Agric. Bull. 37, Cornell University Press.)

are mined as minerals in Canada, the United States, and Germany. Production of potassium chloride is about 20 times that of potassium sulphate. Potassium chloride therefore makes up the bulk of fertilizer K consumed in the world. Potassium sulfate also contains about 17% S, and this is an advantage for areas having sulfur deficiency.

Small amounts of potassium are also marketed as double salts of potassium and magnesium. Kainite (KCl Mg SO$_4$ 3H$_2$O) contains 15.99% K (19.2% K$_2$O), 9.94% Mg, and 13% S. Langbeinite, which is marketed as K-Mag or Sul-Po-Mag, has a theoretical composition of 18.85% K (22.7% K$_2$O), 11.71% Mg, and 23.18% S. These materials have the advantage of supplying Mg and S in addition to K.

Very small amounts of potassium are applied as potassium nitrate (36.7% K or 44% K$_2$O) and potassium phosphate (13 to 26% P or 30 to 60% P$_2$O$_5$ and 25 to 41.7% K or 30 to 50% K$_2$O). In addition, in the United States, some potassium is applied as potassium polyphosphate (17.5 to 26% P or 40 to 60% P$_2$O$_5$ and 18.3 to 40% K or 22 to 48% K$_2$O).

10.5.1. Choice of K Fertilizers

Potassium is applied to a majority of crops as potassium chloride or muriate of potash. The recognition that the Cl$^-$ ion, quite apart from its role in plant nutrition, suppresses some disease organisms, such as those causing take-all of wheat and stalk-rot of corn, and may influence the selection of muriate of potash as a source.

Potassium sulphate is commonly applied to tobacco because excessive chloride uptake may impair the burning quality of the cured leaf. Superiority of potassium sulfate for potatoes is suggested for higher specific-gravity and starch yields. Potassium sulfate is also preferred on soils having S deficiency, for example, for alfalfa in Wisconsin and Nebraska.

10.6. EFFICIENT USE OF POTASSIUM FERTILIZERS

For efficient use of potassium fertilizers, one has to consider the following: (1) soil factors; (2) weather factors, especially precipitation; and (3) crop factors. Soils having 2:1 layer silicates and interstratified minerals can fix appreciable K and therefore these soils do not pose a serious leaching problem. On the other hand, sandy soils and soils having kaolinite as the dominant mineral, as in tropical and subtropical soils (which are also in high-rainfall regions), present serious leaching problems (already discussed under leaching in this chapter). Method and time of potassium fertilizer application therefore assume considerable importance. Fageria (1982) found that in upland rice in Brazil, application of moderate amounts of K (30 to 45 kg ha^{-1}) in the planting furrow gave the same yield as twice as much K broadcast and incorporated in the soil. Application of K in the planting furrow is advantageous when soils are poor in K and the levels applied are low to moderate; application of high rates in the planting furrow may result in a localized salinity problem, causing seedling injury. On soils susceptible to K leaching, broadcast application followed by incorporation to the 20-cm depth is considered superior to band application (Ritchey, 1979). Application of 100 kg K ha^{-1} by incorporation raised exchangeable K by 1.2 mmol kg^{-1} soil, while banding 80 kg K ha^{-1} in a band 8 cm wide every 80 cm increased exchangeable K to 10 mmol K kg^{-1} soil in the bands, which the soil was not able to retain. This resulted in excessive K-leaching. On the other hand, with high-K-fixing soils and low to moderate levels of K application, banding is advantageous for row crops such as corn.

Split application of K (part at planting and part side-dressed later) is recommended to avoid salinity effects and leaching losses both in annual and perennial crops. The beneficial effects of split application of K have been reported in a number of crops and countries: lowland rice in India, Japan, Bangladesh, and Indonesia; bananas and pineapple (*Ananes cosmosus* [L], Merr) in West Africa; and coffee, sugarcane, cotton, and cassava (*Manihot essulenta*, Crantz) in Brazil.

Genotypic differences in crop species with respect to K nutrition are well known. For example, Glass and Perley (1980) detected fairly large differences in K$^+$ influx in barley varieties as early as 6 days after germination. There is now a growing interest in identifying plant genotypes that can tolerate stresses to different plant nutrients (including potassium), and a symposium on this topic was organized at the University of Nebraska in Lincoln in 1993 under the auspices of the INTSORMIL, a sorghum/millet collaborative research support program of the U.S. Agency for International Development (USAID).

REFERENCES

Beckett, P.H.T. 1964a. Studies in soil potassium. I. Confirmation of the ratio law: measurement of potassium potential. J. Soil Sci. 15:1–8.

Beckett, P.H.T. 1964b. Studies on soil potassium. II. The immediate Q/I relations of labile potassium in the soil. J. Soil Sci. 15:9–23.

Bertsch, P.M. and G.W. Thomas. 1985. Potassium status of temperate region soils, in *Potassium in Agriculture*, R.D. Munson, Ed., Am. Soc. Agron., Madison, WI, pp. 131–163.

Brady, N.C. 1990. *The Nature and Properties of Soils*, Prentice-Hall, Upper Saddle River, NJ.

Buol, S.W., F.D. Hole, and R.J. McCracken. 1980. *Soil Genesis and Classification*, Iowa State University Press, Ames, p. 406.

Fageria, N.K. 1982. Nutricao e adubacao potassica do aroz no Brasil, in *Potassio na Agriculture Brasiliera*, T. Yameda, Ed., Institute de Potassa e Fosfato (EUA) and Instituto Internacional de Potassa (Suica), Piracicaba, Brasil, pp. 421–436.

Fageria, N.K., V.C. Baligar, and R.J. Wright. 1989. Growth and nutrient concentrations of alfalfa and common bean as influenced by soil acidity. Plant Soil 119:331–333.

Glass, A.D.M. and J.E. Perley. 1980. Varietal differences in potassium uptake by barley. Plant Physiol. 65:160–164.

Godefroy, J., E.J. Roose, and M. Muller. 1975. Estimation des pertes par les eaux de ruissellment et de drainage des elements fertilisant dans un sol de bananaire du sud du Cote d'Ivoire. Fruits 30:223–235.

Grewal, J.S. and J.S. Kanwar. 1976. *Potassium and Ammonium Fixation in Indian Soils — Review*. Indian Council of Agricultural Research, New Delhi, India.

Huang, P.M., L.S. Crossan, and D.A. Rennie. 1968. Chemical dynamics of K release from potassium minerals common in soils. Trans. 9th Int. Congr. Soil Sci. 2:612–705.

Jurgens-Gschwind, S. and J. Jung. 1979. Results of lysimeter trials at the Limburgerhoff facility, 1927–1977: the most important finding from 50 years of experiments. Soil Sci. 127:146–160.

Magdoff, F.R. and R.J. Bartlett. 1980. Effect of liming acid soils on potassium availability. Soil Sci. 129:12–14.

Malovolta, E. 1985. Potassium status of tropical and subtropical region soils, in *Potassium in Agriculture*, R.D. Munson, Ed., American Society of Agronomy, Madison, WI, pp. 163–200.

McCallister, D., C.A. Shapiro, W.R. Rawn, F.N. Anderson, G.W. Rehm, O.P. Engelstad, M.P. Russelle, and R.A. Olson. 1987. Rate of phosphorus and potassium buildup/decline with fertilization for corn and wheat on Nebraska mollisols. Soil Sci. Soc. Am. J. 51:1646–1652.

McColl, J.G. 1970. Properties of some natural waters in tropical wet forest of Costa Rica. BioScience 20:1096–1100.

McLean, E.O. 1979. Influence of clay content and clay composition on potassium availability, in *Potassium in Soil and Crops*, G.S. Sekhon, Ed., Potash Research Institute of India, Guragaon, pp. 1–13.

Mielniczuk, J. 1977. Formas de potassio em solos do Brasil. Rev. Bras. ci Solo 1:55–56.

Munn, D.A. and E.O. McLean. 1975. Soil potassium relationships as indicated by solution equilibrations and plant uptake. Soil Sci. Soc. Am. Proc. 39:1072–1076.

Nemeth, K. 1979. The availability of nutrients in the soil as determined by electro-ultrafiltration (EUF). Adv. Agron. 31:155–188.

Pearson, R.W. 1952. Potassium supplying power of eight Alabama soils. Soil Sci. 74:301–309.

Ritchey, K.D. 1979. Potassium fertility in oxisols and ultisols of humid tropics. Cornell Int. Agric. Bull. 37, Cornell University, Ithaca, NY.

Schofield, R.K. 1947. A ratio law governing the equilibrium of cations in the soil solution. Proc. Int. Congr. Pure Appl. Chem. 3:257–261.

Sekhon, G.S., M.S. Brar, and A. Subba Rao. 1992. Potassium in some benchmark soils of India. Potash Research Institute of India, Gurgaon, Sp. Pub. 3, p. 82.

Shaw, W.M. and B. Robinson. 1960. Reaction efficiency of liming material as indicated by lysimeter leachate composition. Soil Sci. 89:209–218.

Singh, B.B. and J.P. Jones. 1975. Use of sorption-isotherms for evaluating potassium requirements of some Iowa soils. Soil Sci. Soc. Am. J. 39:881–885.

Somasiri, S., S.Y. Lee, and P.M. Huang. 1971. Influence of certain pedogenic factors on potassium reserves of selected Canadian prairie soils. Soil Sci. Soc. Am. Proc. 35:500–505.

Song, S.K. and P.M. Huang. 1983. Dynamics of potassium release from potassium-bearing minerals as influenced by oxalic and citric acids. Agron. Abstr. 222.

Sparks, D.L. 1987. Potassium dynamics in soils. Adv. Soil Sci. 6:1–63.

Sparks, D.L. and P.M. Jardine. 1984. Comparison of kinetic equations to describe K-Ca exchange in pure and mixed systems. Soil Sci. 138:115–122.

Tisdale, S.L., W.L. Nelson, and J.D. Beaton. 1985. *Soil Fertility and Fertilizers*, 4th ed., Macmillan, New York, p. 754.

11   SULFUR

Sulfur along with calcium and magnesium is listed as a secondary plant nutrient. Because the criterion for grouping plant nutrients as primary, secondary, or micro is based on their amounts removed by the crop plant, it may not be long before this classification of plant nutrients has to be reexamined because sulfur and magnesium are taken up by the crop plants in about the same amounts as phosphorus. In some instances calcium may be taken up in even larger amounts than phosphorus by some species. Crop uptake of S in relation to P, as well as in relation to N, is shown in Figure 11.1. Sulfur uptake by most crops is 10 to 15% of N uptake. S uptake by crops in relation to P varies considerably. While in rapeseed-mustard S uptake is about 175% of P uptake, in most other oilseed crops in general uptake is about the same for S and P. In cereals S uptake is about 60 to 75% that of P.

Over the years there have been increasing reports from all parts of the world, especially the tropics and subtropics (Pasricha and Fox, 1993), of sulfur deficiency in crop plants and responses of crops to sulfur. Australia (Anderson, 1952), the United States (Mitchell and Mullins, 1990; Rechcigl, 1992), Central America (Raun and Barreto, 1992), India (Tandon, 1992, Kumar et al., 1992), and Pakistan (Rashid et al., 1992) are some of these countries. This is primarily due to the following reasons:

1. Increased use of high-analysis fertilizers, which contain no or only traces of sulfur; for example, the use of diammonium phosphate (DAP), urea ammonium phosphate (UAP), and ammonium poly-phosphates (APP) in place of ordinary superphosphate and the use of urea or ammonium nitrate in place of ammonium sulfate.
2. Increased crop yields due to the introduction and cultivation of nitrogen-responsive, high-yielding hybrids, composites, and varieties of cereals, resulting in rapid depletion of soil sulfur.
3. Reduction in the emission of sulfur dioxide (SO_2) due to environmental pollution control regulations. In the United States SO_2 emission from the combustion of fossil fuels, petroleum refining, and other industries declined from 32 million Mg (18 million Mg sulfur)

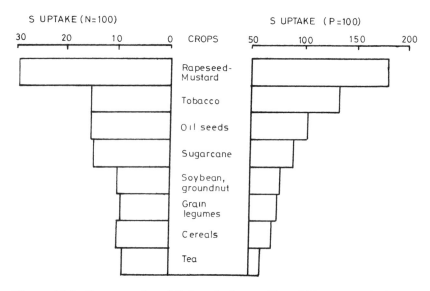

Figure 11.1. Crop uptake of S in relation to N and P uptake. (Adapted from Tandon, 1991.)

in 1972 to 26 million Mg (13 million Mg sulfur) in 1980 (Tisdale et al., 1985). A contour map of sulfate concentration in rain in different parts of the United States for the period 1979 to 1982 is shown in Figure 11.2. Sulfate concentration in rain was highest in the Michigan and southern Ontario region of the United States and Canada followed by that in the upper Chesapeake Bay region of the East Coast. It was much less in the northwest, central, and southern United States. Since most of the emitted SO_2 returns to the soil in rainfall (acid rain), this source of replenishment of soil sulfur is declining. Estimates range from 1 kg S ha^{-1} yr^{-1} in rural areas to about 100 kg S ha^{-1} yr^{-1} near industrial townships. This is just a small component of the entire S cycle (Figure 11.3).

4. Decreased use of farm manure even in developing countries.

11.1. SULFUR IN SOILS

Total S in soils may vary from a few to 1000 mg S kg^{-1} soil (0.1%); higher values can be encountered in problem soils such as saline and acid-sulfate soils (Takkar, 1988; Ganeshmurthy et al., 1989). Sulfur in soils is present both in organic and inorganic forms. While inorganic forms are important because most of the S is taken up by plants as SO_4^{2-} (sulfate), organic forms are important because they often make up the bulk of soil S. Because S is an integral part of soil organic matter, total S is generally greater in fine-textured than in coarse-textured soils. In general, soils containing greater amounts of organic matter contain a larger fraction of their S in organic form.

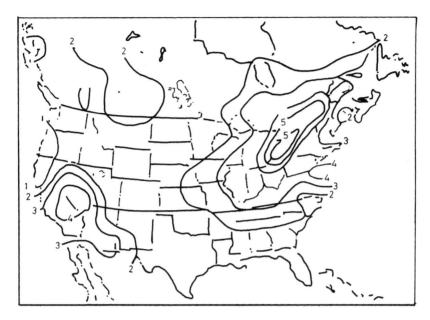

Figure 11.2. Precipitation weighted sulfate concentration contour map of North America for the period 1979–1982. (From Mohnen and Wilson, 1985. *Acid Rain in North America: Concepts and Strategies*, **p. 441. With permission of Plenum Press).**

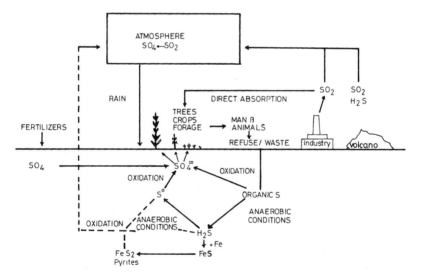

Figure 11.3. Sulfur cycle.

11.1.1. Organic S

Since the bulk of S in plant tissue is present as the S-containing amino acids cystine, cystein, and methionine, organic S is generally associated with proteins in plants and in their residue when added to the soil. S is therefore as much an integral part of soil organic matter as N, and the N:S ratio in most non-saline soils falls within the narrow range 6 to 8:1 (Tisdale et al., 1985). Most of the biological processes involved in mineralization of organic N in soils therefore are also important in organic S mineralization.

Well over 90% of the S in surface layers of well-drained, nonsaline soils is present in organic form. Organic S is divided into two major groups, namely, C-bonded S (amino acids) and non-C-bonded S (ester sulfates such as phenolic sulfates and sulfated polysaccharides). Non-C-bonded S is reduced to H_2S by hydriodic acid (HI) and is estimated this way. Carbon-bonded S can be estimated by subtracting non-C-bonded S from total organic S (Stevenson, 1987). In a study involving 24 Australian soils, organic S was 93% of total S; of this, 52% was non-C-bonded and 41% was C-bonded (Freney, 1967).

11.1.2. Inorganic S

Inorganic S in most soils is present as SO_4^{2-} ions associated with monovalent (Na, K) and divalent cations (mostly with Ca and Mg; traces with Cu, Mn, Zn, and Fe). Sulfate salts present in the soil solution (soluble S) may be adsorbed onto soil colloids, or the salts may be present as insoluble compounds.

Soluble S.   The sulfate content of the soil solution (soluble S) is variable and depends upon a number of factors.

1. Weather conditions, particularly temperature that determines the rate of mineralization of soil organic matter.
2. Precipitation — heavy rains can lead to excessive leaching.
3. Associated cations — leaching losses are greatest when monovalent cations such as Na and K are mostly associated with sulfate.
4. Soil water content — soil water affects sulfate content in soil solution in two ways. First of all, the sulfate concentration in soil solution generally decreases as the water content increases (dilution effect). Secondly, as a soil dries out due to high evapotranspiration rates, sulfate from lower soil layers moves upward by capillary action with water and increases the concentration of sulfate in the surface-layer soil solution. With sufficient drying, sulfates may then precipitate on the soil surface, particularly calcium sulfate.
5. Application of sulfur containing fertilizers — as expected, this will increase soil solution sulfur content.

A soluble-S content of 5 mg kg^{-1} soil is generally adequate for the growth of most crops; *Brassicas* require somewhat more S.

Adsorbed S. Sulfur as SO_4^{2-} can be adsorbed on clay minerals by salt adsorption; on hydroxides and oxyhydroxides of iron and aluminum, which acquire a positive charge under low-pH conditions; and on soil organic matter, which develops positive charges under certain conditions.

Among the clay minerals 1:1 layer silicates such as kaolinite adsorb more sulfate than 2:1 layer silicates. Due to sulfate retention tendencies of hydroxides and oxyhydroxides of aluminum and iron (more in aluminum), adsorbed sulfate is an important fraction of the total S contributing to crop plants on oxisols and many ultisols. Such soils are generally low in soil organic matter and also release of organic S is likely to be slow. Adsorbed S is also an important component in the subsoils of these soils, which are generally very low in organic matter content. Also sulfate leached from surface soils tends to accumulate in subsoil, either by precipitation or adsorption.

Insoluble S. Calcium sulfate can coprecipitate with calcium carbonate and this coprecipitated sulfate forms an important fraction of total S in calcareous soils. Unless brought into solution, which is difficult and very slow, this form of S is relatively unavailable to plants. A number of factors, particularly the size of the calcium carbonate particles, common ion effects, and soil water content determine the rate of dissolution of insoluble S. Grinding of soil samples while preparing it for analysis may reduce the particle size of calcium carbonate and may lead to increased dissolution of insoluble S. This results in higher available S values as determined in a soil test.

Soils in arid and semiarid regions have large deposits of gypsum ($CaSO_4$ $2H_2O$). When mined and used for reclaiming sodic soils, gypsum also adds large amounts of S to soils. As a matter of fact gypsum is a an extremely cheap source of S and is being used in India for this purpose. Often sodic soils may have a layer of gypsum deposited in the subsoil which, if mixed with the surface (deep plowing) can solubilize the exchangeable Na^+ and help reclaim the sodic soil.

Under submerged soil conditions SO_4^{2-} present in soil is reduced to H_2S (sulfide). Also, the anaerobic decomposition of soil organic matter leads to the production of H_2S (this is the reason for the foul smell coming from swamps). Because reduced soil conditions also lead to the transformation of Fe^{3+} to Fe^{2+}, the presence of H_2S leads to the precipitation of S as iron sulfide, which later undergoes conversion to pyrite (FeS_2). Pyrite is the principal insoluble-S form in flooded rice paddies. Availability of this form of S requires the oxidation of pyrites.

11.2. ELEMENTAL SULFUR AND ITS OXIDATION

As discussed earlier, H_2S and FeS_2 are produced under anaerobic conditions. Under the same reduced conditions in some instances elemental $S°$ may also be deposited. In well-drained soils elemental S or H_2S is oxidized to SO_4^{2-}.

This oxidation is brought about by *Thiobacilli* (*T. thioxidans, T. thioparus, T. ferrooxidans*, etc.) (Kuenen, 1975; Starkey, 1966), which are a group of autotrophic bacteria that derive their energy from the oxidation of S to SO_4^{2-} for the fixation of CO_2 into organic matter. The general equation, as given by Tisdale et al., (1985), is as follows:

$$CO_2 + S + 1/2 O_2 + 2 H_2O \rightarrow \left[CH_2O\right] + SO_4^{2-} + 2 H^+$$

Oxidation of elemental S to SO_4^{2-} by *Thiobacilli* is controlled by a number of factors, including soil pH, temperature, and aeration. *Thiobacillus thiooxidans* grows best in the pH range 1.0 to 4.0, while *T. thioparus* has an optimum pH range of 4.5 to 7.5 (Tisdale et al., 1985). Thus S oxidation is most rapid under low soil pH conditions (this is in contrast to nitrifying bacteria, which perform best in neutral or somewhat higher soil pH). Regarding temperature, the oxidation of S proceeds most rapidly between 20 and 40°C. S-oxidizing bacteria need oxygen, and their activity is reduced when anaerobic conditions prevail. Normally, well-drained soils suitable for most crops are also suitable for S-oxidizing bacteria.

Under anaerobic conditions the oxidation of S may be carried out by photosynthetic bacteria *Chlorobium*, the green S bacteria, and *Chromatium, Thiocystis*, and *Thiocapsa*, the purple S bacteria (Fenchel and Blackburn, 1979). These bacteria can also oxidize H_2S with the intermediate production of $S°$, but are inhibited by high concentrations of H_2S; the tolerance limits are 4 to 8 mM L^{-1} for green bacteria and 0.4-2 mM L^{-1} for the purple bacteria (Pfening, 1967). Photosynthetic $S°$ oxidation is dependent on light and a source of reduced S compounds; such a situation is present on marine muds and sands, where sulfide diffuses from below and oxygen penetration is low. Such a habitat is known as "sulfuretum" (Fenchel and Blackburn, 1979). The general equation as given by Tisdale et al. (1985) is as follows:

$$CO_2 + 2 H_2S \rightarrow \left[CH_2O\right] + H_2O + S°$$

11.3. OXIDATION OF PYRITES

The pyrite oxidation cycle as given by Fenchel and Blackburn (1979) is shown in Figure 11. 4, The organism responsible is *Thiobacillus ferroxidans*, which can also oxidize elemental S and reduce S compounds such as H_2S. *T. ferroxdians* is found where pyrite is exposed to atmospheric O. The initial process of pyrite oxidation as given by Fenchel and Blackburn (1979) is as follows:

$$2 FeS_2 + 2 H_2O + 7 O_2 \rightarrow 2 Fe^{2+} + 4 SO_4^{2-} + 4 H^+ \qquad (1)$$

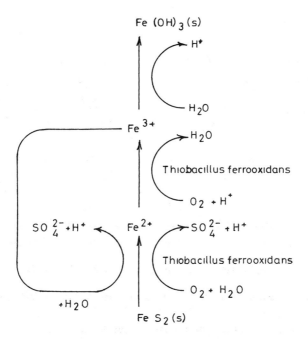

Figure 11.4. The pyrite oxidation cycle. (From Fenchel and Blackburn, 1979. *Bacteria and Mineral Cycling*, p. 225. With permission of Academic Press.)

Ferrous is further oxidized to Fe^{3+}

$$4\ Fe^{2+} + O_2 + 4\ H^+ \rightarrow 4\ Fe^{3+} + H_2O \tag{2}$$

The Fe^{3+} formed is itself an oxidant and oxidizes pyrite abiologically as below (at low pH values):

$$FeS_2 + 14\ Fe^{3+} + 8\ H_2O \rightarrow 15\ Fe^{2+} + 2\ SO_4^{2-} + 16\ H^+ \tag{3}$$

Fe^{2+} produced works as a substrate for *T. ferrooxidans*. As the pH of the effluent water increases, the Fe^{3+} is hydrated to $Fe(OH)_3$, which precipitates.

$$Fe^{3+} + 3\ H_2O \rightarrow Fe(OH)_3 + 3\ H^+$$

Pyrite oxidation can also take place abiologically, but the rate is greatly increased (by a factor of 10^6) when *Thiobacilli* are involved. Oxidation of pyrites is important when used as a soil amendment for reclaiming sodic soils.

11.4. ASSESSING S NEEDS OF SOILS

A number of methods for estimating available S in soils have been pro-posed by several workers; the most common among these are as follows:

1. Calcium chloride extractable
2. Heat-soluble S
3. Monopotassium or monocalcium phosphate (500 mg P L^{-1}) solution extractable
4. Morgan's extract (sodium acetate–acetic acid buffer of pH 4.0)
5. 0.5 N ammonium acetate + 0.25 N acetic acid extractable
6. Olsen's 0.5 M sodium bicarbonate extractable S

Critical levels for some crops in India for different extracting solutions are given in Table 11.1. The critical levels are usually in the range of 11 to 13 mg S kg^{-1} soil with calcium chloride, monopotassium phosphate, heat-soluble, and Morgan's extracting-solution methods.

One needs to remember that critical levels of S will vary with the method of analysis, soil type, and crop, and therefore critical levels need to be deter-mined for different crops grown under different soil-climate conditions.

Sulfur recommendations in the United States vary from state to state, and within a state these differ for soils and crops. In Maryland the recommendation for sandy or sandy loam soils for corn, soybean, and small grains are 44 kg S ha^{-1} when calcium phosphate extracted SO_4^{2-}-S is less than 6 mg kg^{-1} soil; lesser amounts are recommended for soils testing higher. The recommenda-tions in southern states (South Carolina, Alabama, Arkansas, Louisiana) in general are for lesser amounts (11 kg S ha^{-1}) for crops other than forages; for forages the recommendations in Louisiana and Arkansas are from 28 to 56 kg S ha^{-1} (Messick, 1992).

A soil test for available S generally determines the inorganic SO_4^{2-}-S in soil at the time of sampling, but does not estimate the amount of S that may be available to a crop during its growth period from the oxidation of soil organic matter, from the subsoil reserves, and that which may be added by rain. Researchers in Oklahoma, South Dakota, and Wisconsin have therefore developed models to account for various aspects of the S cycle (Messick, 1992). However, a large database is necessary for making the models predict-able and more useful than the routine soil testing.

11.5. SULFUR DEFICIENCY SYMPTOMS IN PLANTS

The behavior of S in plants is similar to N, a corollary to the similarity of S and N in regard to function in the plant and distribution in the soil. Sulfur deficiency in plants resembles N deficiency, and S deficient plants also turn pale yellow. However, S is less mobile than N in plants, and the younger leaves

Table 11.1 Critical Levels of Available Soil-S (ppm) for Different Crops According to Common Methods of Extraction

Crop	CaCl$_2$	KH$_2$PO$_4$	CaH$_2$PO$_4$	Morgan	AmAc	Heat-S	Olsen	Mean
Rice	10	10	10	13	11	16	20	12.8
Wheat	13				25	13		17.0
Maize	13	10	14	25	30			15.5
Groundnut	10					11		10.5
Rapeseed-mustard	10	13	13	10	9			10.4
Sunflower	18	10			30			18.8
Soybean	14		14	8	8		18	12.4
Clusterbean				10		10		10.0
Alfalfa		9				20		14.5
Mean	12.6	10.4	12.7	13.2	18.8	14.0	19.0	

Adapted from Tandon 1992. Sulfur Agric. 16:20–23. With permission of The Sulfur Institute, Washington, D.C.

Table 11.2 Deficiency and Sufficiency Levels of S in Plants

Crop	Part of plant sampled	Deficient	Sufficient
		(g kg^{-1} dry matter)	
Alfalfa	Whole tops at bud stage (10% bloom)	1.5–2.3	2.3
Barley	Boot-stage tissue	1.2	1.4
Wheat	Boot-stage tissue	1.2	1.4
Soybeans	Upper, fully developed, trifoliate leaves prior to pod set	1.4	2.2–2.8
Tobacco	Leaves	1.1–1.8	1.5–2.6
Red clover	Whole tops at bud stage (10% bloom)	2.0 (critical)	
Cauliflower	Whole tops at the curd stage	1.8	1.9
Rapeseed-mustard		2.1 (critical)[a]	
Pigeon pea	Shoots before flowering	1.34 (critical)[b]	
Groundnut	Plant	2.0 (critical)[c]	

[a] Pasricha et al. (1988)

[b] Singh (1991)

[c] Cheema and Arora (1984)

Adapted from the Authority of the Atlantic Provinces Agricultural Services Coordinating Committee, Canada (1988).

may generally turn pale yellow, while the older leaves may remain green. Note that the reverse is the case in N deficiency. There are frequent exceptions to these symptoms, however. The general deficiency symptoms are pale yellow, stoopy plants with short and slender stalks; these symptoms will not disappear with N application. Plant species differ considerably in expressing S-deficiency symptoms. Deficiency and sufficiency levels of S in the different plant parts for some crops are given in Table 11.2.

11.6. SULFUR NEEDS OF CROPS

Crops and cultivars within crops vary considerably in their S requirements. Spencer (1975) has divided crops into three broad groups (Table 11.3). Group I includes Crucifers and *Brassicas* which have high S requirement (20 to 80 kg S ha^{-1}). Group II includes plantation crops, which have moderate S requirements (10 to 50 kg S ha^{-1}). Group III includes cereals, forages, and other field crops and has low S requirement (5 to 25 kg S ha^{-1}). As a rule of thumb, Tandon (1991) gives the following S requirements (kg S Mg^{-1} grain): 3 to 4 kg for cereals, 8 kg for grain legumes (beans), and 12 kg for oilseeds (rapeseed mustard, sunflower, groundnut, soybeans, etc.).

**Table 11.3 A Tentative Classification of Crops According to Their
S Fertilizer Requirement**

Crop	Fertilizer requirement in deficient areas[a] kg S/ha
Group I (high)	
Cruciferous forages	40–80
Alfalfa	30–70
Rapeseed	20–60
Group II (moderate)	
Coconuts	50
Sugar cane	20–40
Clovers and grasses	10–40
Coffee	20–40
Cotton	10–30
Group III (low)	
Sugar beet	15–25
Cereal forages	10–20
Cereal grains	5–20
Peanuts	5–10

[a] Figures cited for the high end of the range apply where the potential yield is
high, accessions in rainfall are low, the soil is low in available S, and there
is considerable loss in effectiveness of applied S. Figures cited for the low
end refer to the opposite situation. For perennials, a further consideration is
whether the requirement refers to a corrective fertilizer dressing or to an annual
maintenance dressing. The amount for the former is typically about four times
that of the latter.

From Spencer. 1975. *Sulfur in Australian Agriculture*, K.D. McLachlan, Ed.,
p. 103.

11.7. SULFUR FERTILIZATION

Sulfur has been applied in the past along with ammonium sulfate, ordinary
superphosphate, and potassium sulfate. However, with the advent of high-
analysis fertilizers such as anhydrous ammonia, urea, diammonium phosphate
(DAP), and ammonium polyphosphate (APP), application of S has been gradually reduced. Thus, where necessary, additional S may be applied as elemental
S, gypsum, or pyrite, depending upon the availability of the material and the
needs of crops and soils. The most common S and S-containing fertilizer
materials are listed in Table 11.4.

Table 11.4 Sulfur and Sulfur Containing Fertilizer Materials

Fertilizer	Chemical composition	S content (%[a])
S Fertilizers		
Elemental S	S	99.6
Agric-S	S	90.0
Gypsum	$CaSO_4 \cdot 2H_2O$	18.6
Commercial gypsum	$CaSO_4 \cdot 2H_2O$ + impurities	13–14
Pyrites	FeS_2	53.5
N Fertilizers		
Ammonium sulfate	$(NH_4)_2SO_4$	23.7
Urea sulfur		10
Ammonium phosphate sulfate	$(NH_4)_2SO_4 + NH_4H_2PO_4 +$ $(NH_4)_2HPO_4$	15.5
P Fertilizers		
Ordinary superphosphate (OSP)	$Ca(H_2PO_4)_2 + CaSO_4 \cdot 2H_2O$	13.9
Concentrated superphosphate (CSP)	$Ca(H_2PO_4)_2$	1.5
Ammoniated OSP		12
Ammoniated CSP		1.4
K Fertilizers		
Potassium sulfate	K_2SO_4	17.6
Potassium magnesium sulfate	$K_2SO_4 \cdot 2\ MgSO_4$	22.0
Others		
Copper sulfate	$CuSO_4 \cdot 5H_2O$	12.8
Zinc sulfate	$ZnSO_4 \cdot H_2O$	17.8
Manganese sulfate	$MnSO_4 \cdot 4H_2O$	14.5
Magnesium sulfate (Epsom salt)	$MgSO_4 \cdot 7H_2O$	13.0
Ammonium thiosulfate	$(NH_4)_2S_2O_3$	43.3

[a] Commercial grades generally contain somewhat lesser values.

REFERENCES

Anderson, A.J. 1952. The significance of sulfur deficiency in Australian soils. J. Australian Inst. Agric. Sci. 18:135–139.

Authority of the Atlantic Provinces Agricultural Services Coordinating Committee, Canada. 1988. Sulfur in soils and crops in Atlantic Canada. Publ No. 538-88, p. 7.

Cheema, H.S. and C.L. Arora. 1984. Sulfur status of soils, tube-well waters and plants in some areas of Ludhiana under ground-nut–wheat cropping system. Fert. News 29(3):28–31.

Fenchel, T. and T.H. Blackburn. 1979. *Bacteria and Mineral Cycling*, Academic Press, New York, p. 225.

Freney, J.R. 1967. Sulfur containing organics, in *Soil Biochemistry*, A.D. McLaren and G.H. Peterson, Eds., Marcel Dekker, New York, pp. 229–259.

Geneshmurthy, A.N., A.D. Mongia, and N.T. Singh. 1989. Forms of S in soil profiles of Andaman and Nicobar Islands. J. Indian Soc. Soil Sci. 37:825–829.

Keunen, J.G. 1975. Colorless sulfur bacteria and their role in the sulfur cycle. Plant Soil 43:49–76.

Kumar, V., G.C. Shrotriya, and S.V. Kaore. 1992. *Crop Response to Sulfur Application*, Indian Farmers Fertilizer Cooperative Ltd., New Delhi, p. 50.

Messick, D.L. 1992. Soil test interpretation for sulfur in the United States — an overview. Sulphur in Agric. 16:24–25.

Mitchell, C.C. and G.L. Mullins. 1990. Sources, rates and time of sulfur application to wheat. Sulphur in Agric. 14:20–24.

Mohnen, V.A. and J.W. Wilson. 1985. Acid rain in North America: concepts and strategies, in *Acid Deposition — Environmental, Economic and Policy Issues*, D.D. Adams and W.P. Page, Eds., Plenum Press, New York, pp. 439–452.

Pasricha, N.S. and R.L. Fox. 1993. Plant nutrient sulfur in tropics and subtropics. Adv. Agron. 50:209-269.

Pasricha, N.S., M.S. Aulakh, G.S. Bahl, and H.S. Baddesha. Fertilizer use research in oilseed crops. 1988. Fert. News 33(9):15–22.

Pfening, N. 1967. Phtotosynthetic bacteria. Ann. Rev. Microbiol. 21:285–324.

Rashid, M., M.I. Bajwa, R. Hussain, M. Naeem-ud-Din, and F. Rahman. 1992. Rice response to sulfur in Pakistan. Sulphur in Agric. 16:3–5.

Raun, W.R. and H.J. Barreto. 1992. Maize grain yield response to sulfur fertilization in Central America. Sulphur in Agric. 16:26–29.

Rechcigl, J.E. 1992. Sulfur fertilization of Bahiagrass forage. Sulphur in Agric. 16:39–42.

Singh, M.V. 1991. Results of practical utility. Proc. 18th Workshop, All India Coordinated Scheme of Micro and Secondary Nutrients and Pollutant Elements in Soils and Crops, Indian Institute of Soil Sciences, Bhopal, India.

Spencer, K. 1975. Sulfur requirements of crops, in *Sulphur in Australian Agriculture*, K.D. McLachlan, Ed., Sydney University Press, Sydney, pp. 98–108.

Starkey, R.L. 1966. Oxidation and reduction of sulfur compounds in soil. Soil Sci. 101:297–306.

Stevenson, F.J. 1987. *Cycles of Soil — Carbon, Nitrogen, Phosphorus, Sulfur, Micronutrients*, John Wiley & Sons, New York, p. 380.

Takkar, P.N. 1988. Sulfur status of Indian soils. Proc. The Sulphur Institute — Fertilizer Association of India Symp. Sulfur in Indian Agriculture, New Delhi, 5/1/2/1-31.

Tandon, H.L.S. 1991. *Sulphur — Research and Agricultural Production in India*, 3rd ed., The Sulphur Institute, Washington, D.C., p. 140.

Tandon, H.L.S. 1992. Sulfur in Indian agriculture, update 1992. Sulphur in Agric. 16:20–23.

Tisdale, S.L., W.L. Nelson, and J.D. Beaton. 1985. *Soil Fertility and Fertilizers*, 4th ed., Macmillan, New York, p. 754.

12   CALCIUM AND MAGNESIUM

Calcium and magnesium, two secondary plant nutrients, have several common characteristics. Some of these are as follows:

1. Both have only one active valence 2^+.
2. Both are taken up by plants as cations.
3. Both are basic or base-forming elements.
4. They occur together in nature as dolomitic limestone.

However, they differ in regard to their presence and functions in plants. Calcium is present in the cell wall, is involved in cell division, and is therefore an important component of plant structure; it is generally considered an immobile element in plants. Calcium in soils serves as an excluder or detoxifier of heavy metals such as Ni, as well as other elements that might otherwise be toxic. In addition, Ca provides protection against drought, salinity, and mechanical stress (Foy, 1992). Magnesium, on the other hand, is the core cation in the structure of the chlorophyll molecule and is thus vital to photosynthesis. Magnesium also serves as a structural component in ribosome and thus plays an important role in protein synthesis. It is fairly mobile in plants.

12.1. CALCIUM AND MAGNESIUM IN SOIL

Calcium is present in the earth's crust in much larger amounts (3.64%) than magnesium (1.93%). Highly weathered, coarse, sandy soils in humid regions may contain 0.1% total Mg and 0.1 to 0.3% total Ca, while fine-textured, clay soils rich in 2:1 layer silicates may contain 0.7 to 3% total Ca and often equal quantities of Mg. Values greater than 3% total Ca in soils indicate the presence of calcium carbonate. Such soils are known as calcareous soils, which may contain less than 1 to more than 25% total Ca. These soils may have a hard pan of calcium carbonate and calcium silicate at a rather shallow depth in soil profile. In soils developed from calcareous glacial tills, Ca is usually leached to the depth of rooting where it precipitates as calcium carbonate. Also, some temperate soils subjected to temporary flooding may contain small, calcareous, mollusk shells on or near the soil surface.

Calcium in soils is derived from the minerals anorthrite, pyroxenes (augite), amphiboles (hornblende), and albite. Calcite ($CaCO_3$) is the most important source of Ca in calcareous soils; when present, dolomite ($CaMg [CO_3]_2$) also contributes to soil Ca. In arid and semiarid regions gypsum ($CaSO_4 2H_2O$) could be an important source of Ca.

Magnesium in soils is derived from minerals biotite, phologpite, hornblende, olivine, and serpentine. In calcareous soils dolomite, when present, is an important source of soil Mg. In arid and semiarid regions substantial amounts of mineral epsomite ($MgSO_4 7H_2O$) may be present and may contribute to soil Mg.

Both Ca and Mg are present in nonexchangeable and exchangeable forms and in soil solution; the latter two forms remain in a dynamic equilibrium. The degree of calcium saturation of exchange complex required for optimum plant growth varies from crop to crop, and often wide variations exist between different cultivars of a species. Thus the optimum range of Ca saturation of the exchange complex (CEC) is wide (12 to 75%); for temperate regions and for a number of crops the suggested optimum is 65% (Eckert and McLean, 1981). On the other hand, the range for critical Mg saturation levels is narrow, often 5 to 10% of CEC.

Exchangeable Ca in soils can range from <25 mg kg^{-1} to more than 5000 mg kg^{-1}. Ca in soil solution may range from 68 to 778 mg kg^{-1}. Exchangeable Mg generally constitutes 4 to 20% of the CEC, and the concentration in soil solution may range from 50 to 120 mg L^{-1}.

When solution Ca or Mg is depleted by leaching or plant uptake, more Ca or Mg is released from the solid phase.

12.2. FACTORS AFFECTING THE AVAILABILITY OF CALCIUM AND MAGNESIUM IN SOILS

Release of Ca^{2+} or Mg^{2+} from the exchange complex and their availability to crop plants depends upon the following factors; a number of these are interdependent:

1. Total Ca or Mg supply
2. Type of clay minerals present
3. Cation exchange capacity (CEC) of soil
4. Percentage saturation of CEC with Ca^{2+} or Mg^{2+}
5. Soil pH
6. Ratio of Ca^{2+} or Mg^{2+} to other cations in soil solution

Soils having 2:1 layer silicates have higher CEC and can thus retain larger amounts of Ca or Mg. In such acid soils, however, Ca level may be too low to be available to plants. Smectites require nearly 75% saturation of exchange complex before appreciable Ca^{2+} is released for plant uptake. On the other hand, soils having 1:1 layer silicates (kaolinites) will release exchangeable Ca into the soil solution at only 20 to 40% Ca saturation of the exchange complex.

Soil pH is inversely related to exchangeable Ca. In acid soils, which can have high exchangeable Al, Ca concentrations become low. Liming such soils increases soil pH by increasing saturation of exchange complex with Ca^{2+}, which replaces exchangeable H^+. Absolute Ca deficiency and Ca deficiency symptoms are rarely seen in the field. Even in cases where the deficiency symptoms are observed, they are more likely due to Al-Ca antagonism than to low Ca supply *per se* (Foy, 1992). For example, in a study with barley cultivars, Al-sensitive Kearney barley developed Ca-deficiency symptoms (rolling and eventual collapse of youngest leaves) (Figure 12.1), but Al-tolerant Dayton did not show Ca-deficiency symptoms (Long and Foy, 1970). Al also antagonizes Mg uptake, and there are reports indicating Al-toxicity is a factor in controlling the severity of Mg deficiency (Grimme, 1984). Nevertheless, critical experimentation using $CaSO_4$ as a source of Ca, $MgCO_3$, or MgO for correcting exchangeable Al and $Ca(OH)_2$ for correcting both exchangeable Al and Ca deficiency has illustrated that Ca deficiency could be a limiting factor for plant growth on some acid ultisols (Adam and Moore, 1983; Njohn et al., 1987). Also, in acidic, sandy soils (primarily quartz sands) Ca deficiencies may be expressed.

Availability of Ca^{2+} and Mg^{2+} and their uptake by plants is largely influenced by the ratio of these cations with other cations. A Ca:(Ca+Mg+K) ratio in solution of 0.1 to 0.2 is generally considered desirable for adequate Ca uptake. Excess Ca may adversely affect Mg uptake, and a Ca:Mg ratio in solution greater than 7:1 is not considered desirable (Tisdale et al., 1985). This would explain why continuous liming of coarse-textured soils may lead to Mg deficiency (due to an increased Ca:Mg ratio). On the other hand, Ca:Mg ratios less than about 2:1 can result in high exchangeable Mg restricting adequate Ca uptake, causing Ca-deficiency symptoms in soils developed from some marine shales. Similarly, K^+ antagonizes Mg uptake (Figure 12.2), and this could be a concern in low-Mg soils. The recommended K:Mg ratios are <5:1 for field crops; 3:1 for vegetables and sugar beets, and 2:1 for fruits and greenhouse crops (Tisdale et al., 1985).

12.3. LEACHING OF CALCIUM AND MAGNESIUM

Calcium is often the dominant cation in drainage waters, springs, streams, ponds, and lakes. The annual leaching loss of Ca and Mg from a soil will depend on the total amounts present in a soil, its CEC, and the frequency and intensity of precipitation and in irrigated areas on the amount of irrigation water applied and its Ca and Mg content. The provision of a plant canopy such as grass cover can greatly reduce the leaching of Ca and Mg from a soil (Table 12.1). Excessive leaching of Ca is one factor responsible for the development of acidity in oxisols and ultisols. Excessive leaching of Ca and Mg in humid regions is also the reason why Ca- and Mg-responsive soils are mostly present in humid regions of the Appalachian range and in the south coastal

Figure 12.1. Al-induced Ca deficiency in Al-sensitive Kearney barley. No deficiency symptoms were developed in Al-tolerant Dayton barley. (From Long and Foy, 1970. Agron. J. 62:679–681. With permission of ASA.)

United States. In less humid regions, during soil development, some Ca and Mg was leached from the surface soil and precipitated mainly as carbonates and sulfates in the subsoil. This is common in the mollisols of the midwestern United States.

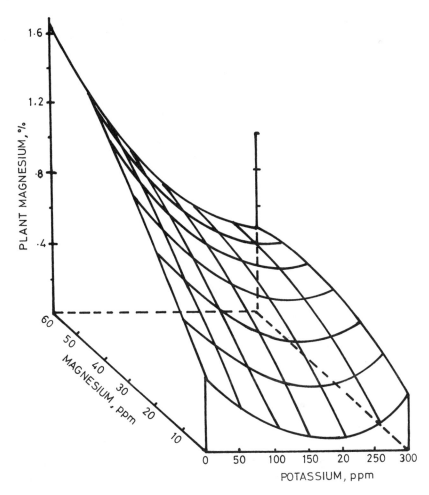

Figure 12.2. Predicted response surface of magnesium concentration in top tissue of sorghum to increasing levels of potassium and magnesium concentrations in nutrient solution at 75 days after seedling emergence. (From Ologunde and Sorensen, 1982. Agron. J. 74:44. With permission of ASA.)

12.4. DETERMINING AVAILABLE CALCIUM AND MAGNESIUM

The most common extractant for determining exchangeable (available) Ca and Mg is molar ammonium acetate at pH 7.0 (Lanyon and Heald, 1982). Exchangeable Ca levels, at which crops are no longer expected to respond to Ca application, vary considerably and range from 250 mg kg^{-1} in sandy soils to 500 mg kg^{-1} in silty clay soils. Several researchers prefer the Ca^{2+} saturation

Table 12.1 Estimated Annual Ca and Mg Drainage Losses from a Sandy, Loam Soil in Orchard Lysimeters at Summerland, B.C.

Vegetation	1978	1979	1980
Calcium (kg ha^{-1} yr^{-1})			
Grass[a] cover	259	117	367
No grass	1045	275	578
Magnesium (kg ha^{-1} yr^{-1})			
Grass cover	82	35	136
No grass	293	71	194
Precipitation and irrigation (mm)			
Precipitation	330	240	320
Irrigation[b]	670	690	860

[a]   Kentucky bluegrass (*Poa pratensis*).

[b]   Irrigation water generally contained 28×10^{-6} kg Ca L^{-1}, and 9×10^{-6} kg Mg L^{-1}.

Adapted from Nielsen and Stevenson (1983).

percentage of CEC as the criterion for determining Ca availability in soils (this has already been discussed). For determining Ca needs of peanut or groundnut (*Araclis hypogaea* L.) the soil is generally extracted with Mehlich I (0.05 mol L^{-1} HCl + 0.0125 mol L^{-1} H$_2$SO$_4$) (Mehlich, 1953). A value of 290 kg Ca ha^{-1} is considered the upper limit at which a response to Ca application can be expected. A Cate-Nelson (1971) graph relating peanut yield and log Mehlich Ca is shown in Figure 12.3.

Similarly, the exchangeable Mg level recommended for agronomic crop production ranges from 25 to 180 mg kg^{-1} soil. Again, several researchers prefer the Mg^{2+} saturation percentage of CEC as the criterion for determining available Mg. For example, response of cotton to Mg application on heavy, clay soils was found to be excellent at 3% Mg^{2+} saturation of soil CEC, and no response was obtained at 6.4% saturation (Lancaster, 1958). The general consensus is that 5% Mg^{2+} saturation of CEC of soil is needed for most crops other than alfalfa, corn silage, and cool season forage crops, which need higher Mg^{2+} saturation (10% of CEC) to avoid grass tetany in ruminant animals. Grass tetany is a disorder of ruminants caused by Mg concentrations less than 2 g kg^{-1} forage (dry-matter basis). The factors responsible for this include K concentration >30 g kg^{-1}, K/(Ca+Mg) equivalent ratio >2.2 and N concentration >40 g kg^{-1} in the plant (dry matter basis) (Grunes and Mayland, 1975).

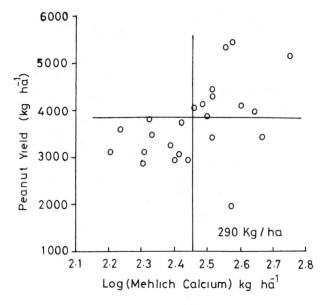

Figure 12.3. Relationship between pod yield and Log of Mehlich I Ca in Lakeland sand 90 days after planting. The vertical and horizontal coordinates show the partitioning of the data points according to the Cate-Nelson technique. The point of intersection of the vertical coordinate with the x-axis represents Ca level beyond which yield response is very minimal. (From Alva et al., 1989. Commun. Soil Sci. Plant Anal. 20:1742. With permission of Marcel Dekker, Inc.)

12.5. CALCIUM AND MAGNESIUM DEFICIENCY SYMPTOMS

Since Ca is generally immobile in plants, there is very little translocation of Ca in phloem. This leads to a poor supply, and consequent deficiency symptoms often result in storage organs and fruits. Bitter pit development in apples is a good example of this. Blossom-end rot in tomatoes and moldy, diseased, and low quality soybeans are some other examples. Terminal buds and apical tips of roots in plants fail to develop. Acute Ca deficiency in corn prevents the emergence and unfolding of young leaves, the tips of which may be covered with a sticky gelatinous material (Figure 12.4); the leaves tend to stick together giving a ladder-like appearance.

As a contrast to Ca, Mg is fairly mobile in plants and its deficiency symptoms appear first on the lower leaves. In corn, Mg deficiency results in interveinal chlorosis of the lower leaves (Figure 12.5); only the veins remain green. In cotton the lower leaves develop a reddish purple cast that may gradually turn brown and necrotic.

Figure 12.4. Calcium deficiency. The leaf tips stick to the next lower leaf, giving a ladderlike appearance (left). The young leaves of new plants are affected first. They are often distorted and small, the margins are irregular in form, and the leaves frequently show spotted necrotic areas. There may be dieback of the growing tip (right). Root growth is markedly impaired. (From *Corn Field Manual*, J.R. Simplot Company Minerals & Chemical Division, Pocatello, ID, ©1984. With permission.) See Plate 3 following p. 170.

12.6. CALCIUM AND MAGNESIUM AMENDMENTS

Most Ca and Mg amendments are applied as liming materials on acidic soils, and these have already been discussed in Chapter 5. In addition to liming materials, gypsum ($CaSO_4$), because it has higher water solubility than limestone, has received special attention for supplying Ca to peanut or groundnut. Calcium absorbed by the roots is not translocated to the developing pods, which therefore absorb directly from the soil solution the Ca needed for their development during initial pegging and seed development. For this purpose gypsum is generally band applied or broadcast during the first-bloom stage; the term lime plastering is sometimes used for this practice. The release of Ca^{2+} from gypsum is influenced by the particle size, so finely powdered material is generally used. Data from a field experiment conducted on Lakeland sand in Georgia (Alva et al., 1989) are presented in Table 12.2. Gypsum application increased the pod yield, as well as the percentage of sound, matured kernels (SMK); crystalline and fine powder (wet) were better than other materials.

Dolomite is the most widely used Mg amendment. The other commonly used material is Epsom salt ($MgSO_4$), which can be also used in foliar sprays when a Mg deficiency is detected at a later crop growth stage. Mg deficiency in citrus orchards in California is frequently corrected by foliar spray of $Mg(NO_3)_2$. Mg-chelates are also marketed and used.

Figure 12.5. Magnesium deficiency. Interveinal chlorosis starts on the lower leaves (top). Yellow streaking sometimes changes to dead, round spots, which gives the impression of beaded streaking (below). The older leaves may become reddish purple on the tips, edges, and underside and die in extreme cases. (From *Corn Field Manual*, J.R. Simplot Company Minerals & Chemical Division, Pocatello, ID, ©1984. With permission.) See Plate 4 following p. 170.

**Table 12.2 Effect of Various Gypsum
Amendments on Pod Yields and Grades
in Peanut on Lakeland Soil in Georgia**

Gypsum materials	Pod yield[a] (Mg ha[-1])	SMK[b] (%)
Crystalline	4.48 a[c]	70 a
Fine powder (wet)	4.38 ab	69 a
Coarse powder	4.11 abc	68 a
Fine powder (dry)	3.86 abc	68 a
Granular 1	3.46 abc	70 a
Granular 2	3.30 bc	68 a
Pelleted	3.10 c	67 a
Control (no gypsum)	3.03 c	61 b

[a] Adjusted to 7% moisture.

[b] Sound, mature kernels.

[c] The means followed by same letter are not significantly
different at $p = 0.10$.

Adapted from Alva et al. (1989).

REFERENCES

Adams, F. and B.L. Moore. 1983. Chemical factors affecting root growth in subsoil
horizons of coastal plain soils. Soil Sci. Soc. Am. J. 47:99–102.

Alva, A.K., G.J. Gascho, and Y. Guang. 1989. Gypsum material effects on peanut and
soil calcium. Commun. Soil Sci. Plant Anal. 20:1727–1744.

Cate, R.B. Jr. and L.A. Nelson. 1971. A simple statistical procedure for partitioning
soil test correlation data into two classes. Soil Sci. Soc. Am. Proc. 35:658–660.

Eckert, D.L. and E.O. McLean. 1981. Basic cation saturation ratios as a basis for
fertilizing and liming agronomic crops. I. Growth chamber studies. Agron. J.
75:795–799.

Foy, C.D. 1992. Soil chemical factors limiting plant growth. Adv. Soil Sci. 19:97–149.

Grimme, H. 1984. Aluminum tolerance of soybean plants as related to magnesium
nutrition. Proc. 6th Int. Colloq. for Optimization of Plant Nutrition, Montpelier
Cedex, France 1:243–249.

Grunes, D.L. and H.F. Mayland. 1975. Controlling grass tetany. USDA Leaflet 561.
U.S. Government Printing Office, Washington, D.C.

Lancaster, J.D. 1958. Magnesium status of Blackland soils of northeast Mississippi for
cotton production. Miss. State Univ. Agric. Exp. Stn. Bull. 560.

Lanyon, L.E. and W.R. Heald. 1982. Magnesium, calcium, strontium and barium, in
Methods of Soil Analysis, Part 2, 2nd ed., A.L. Page, Ed., Agron. Mongr. 9, Am.
Soc. Agron. and Soil Sci. Soc. Am., Madison WI, pp. 247–262.

Long, F.L. and C.D. Foy. 1970. Plant varieties as indicators of aluminum toxicity in
the A_2 horizon of a Norfolk soil. Agron. J. 62:679–681.

Mehlich, A. 1953. Determination of P, Ca, Mg, K, Na and NH4. N.C. Soil Test Mimeo.

Nielsen, G.H. and D.S. Stevenson. 1983. Leaching of soil calcium, magnesium and potassium in irrigated orchard lysimeters. Soil Sci. Soc. Am. J. 47:692–696.

Njohn, B.A., W.O. Enwezor, and B.I. Onzenakwe. 1987. Calcium deficiency identified as an important factor limiting maize growth in acid ultisols of eastern Nigeria. Fert. Res. 14:113–124.

Ologunde, O.O. and R.C. Sorensen. 1982. Influence of concentrations of K and Mg in nutrient solutions on sorghum. Agron. J. 74:41–46.

Tisdale, S.L., W.L. Nelson, and J.D. Beaton. 1985. *Soil Fertility and Fertilizers*, 4th ed., Macmillan, New York, p. 754.

13   IRON AND MANGANESE

Manganese is the key element that activates the enzyme system responsible for splitting the water molecule in the photosynthesis process, and therefore it is ultimately responsible for the presence of oxygen in the atmosphere. In biological systems, especially in animals, Fe is supreme and plays a dominant role, while Mn is secondary and displays some toxicity tendencies. In the oxidation-reduction reactions of soils, however, Fe is second to Mn in importance. Manganese and iron together provide the key to the establishment of the organic mantle, the humified soil top layer that covers the surface of the earth and serves as the nurturing home for the roots of all plants and carbon-recycling microorganisms (Bartlett and James, 1993). Manganese provides O_2 to the poorly ventilated interior soil pores, where Fe^{2+} and complex organic molecules would otherwise tend to remain unoxidized.

There are a number of similarities in the behavior of Fe and Mn in soils and plants. They are as follows:

1. Both Fe and Mn are transition elements and have more than one valency.
2. Both Fe and Mn are soluble in the +2 valency form (Fe^{2+} and Mn^{2+}).
3. Both Fe and Mn are present in soils as oxides, hydroxides, and oxyhydroxides.
4. Compared with the quantities present in soil, both Fe and Mn are taken up by plants in small amounts.
5. Under anaerobic conditions, especially in acidic soils, both Fe and Mn can reach toxicity levels.
6. In acidic soils both Fe and Mn are involved in phosphate retention.
7. While in soils Fe^{2+} (ferrous) and Mn^{2+} (manganous) forms are fairly mobile; both Fe and Mn are immobile in plants.
8. Both Fe and Mn are involved in the process of photosynthesis in plants.
9. Both nutrients can partly substitute for one other plant nutrient. Iron can partially substitute for Mo as the metal cofactor necessary for the functioning of nitrate reductase. Manganese can substitute to some extent for Mg in phosphorylation and group transfer reactions.

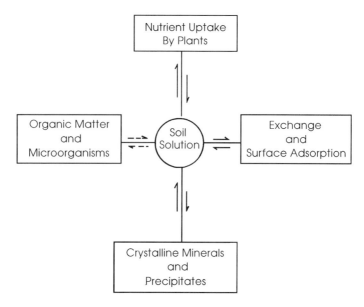

Figure 13.1. Dynamic equilibria for micronutrients occurring in soils. (From Lindsay, 1991. *Micronutrients in Agriculture,* **J.J. Mortvedt, F.R. Cox, L.M. Shuman, and R.M. Welch, Eds., pp. 89–112. With permission of SSSA.)**

10. Both Fe and Mn are involved in imparting specific color(s) to the soils, providing the base for such names as terra rosa, krasnozems, red earths, yellow earths, etc.

13.1. AMOUNTS AND FORMS OF IRON AND MANGANESE IN SOIL

Iron comprises about 50 g kg^{-1} (5%), while Mn is present at about 1 g kg^{-1} (0.1%) of the earth's crust. The amounts of Fe and Mn in a soil depend upon the kind and degree of weathering the parent material has undergone during soil formation. The amounts of these elements are greatest in highly weathered oxisols and ultisols.

As already indicated, Fe and Mn are mostly present as oxides, hydroxides, and oxyhydroxides, which have been discussed in Chapter 4. Small amounts of Fe and Mn may be present as chlorides, sulfates, and carbonates.

The relationship existing between different micronutrient pools (this applies to Fe as well) is shown in Figure 13.1.

13.2. SOIL SOLUTION IRON AND MANGANESE

Both Fe and Mn must exist in their +2 valence form to be soluble in soil solution and taken up by plants. Diffusion is the main process by which Fe

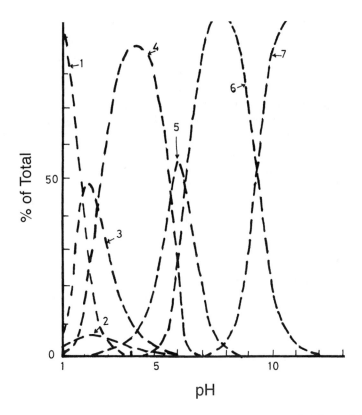

**Figure 13.2. Relative concentrations of the principal series in a 0.1 *M*
Fe(III) solution. 1 = Fe³⁺, 2 = FeOH²⁺; 3 = Fe₂(OH)₂⁺⁴;
4 = Fe₃(OH)₄⁵⁺; 5 = Fe(OH)₂⁺; 6 = Fe(OH)₃; 7 = Fe(OH)₄²⁻.
(From Baes and Mesmer, 1976. *The Hydrolysis of Cations.*
With permission of John Wiley & Sons.)**

and Mn ions are made available for uptake by plants. The other processes
involved in uptake of Fe^{2+} and Mn^{2+} by plants are mass flow and root interception. Chelation by fulvic and other low-molecular-weight humic acids
produced on the decomposition of soil organic matter plays an important role
in solubilizing Fe and keeping it in soil solution.

In aqueous solution Fe^{3+} is surrounded by six molecules of water, that is as
$Fe(H_2O)_6^{3+}$. Increasing pH removes H^+ from the coordinated water and gives rise
to various hydrolysis products as shown in Figure 13.2. The hydrolysis species
are important in soils because they increase total soluble Fe, which determines
transport from soil to plant roots, a rate-limiting step in Fe nutrition of plants.

pH has a significant influence on the solubility of Fe, which is minimum
in the pH range 7.4 to 8.5, characteristic of calcareous soils. This would explain
why Fe deficiency is most prevalent in calcareous alkaline soils (Loeppert and
Hallmark, 1985). A soil solution at pH 8.0 in equilibrium with soil Fe contains
only $10^{-21.3}$ *M* Fe^{3+}, whereas the total soluble Fe components also consists of

$10^{-10.4}$ M Fe $(OH)^0$, $10^{-10.9}$ M Fe $(OH)^-$, and $10^{-11.0}$ M Fe $(OH)_2^+$, giving a total soluble Fe of $10^{-10.20}$ M (Lindsay, 1988). This level of Fe is much below that needed to supply adequate Fe to most crops. Most plants need in excess of 10^{-8} M of soluble Fe in the soil solution to meet their nutritional needs (Schwab and Lindsay, 1989). Inorganic Fe^{+3} can maintain this level of soluble Fe only in soils with pH below 5.5 to 6. At higher soil pH levels, chelation by natural (organic) or synthetic chelates or reduction to Fe^{+2} is necessary to raise the solubility of Fe in soils to acceptable levels (Lindsay, 1991).

Because both Fe and Mn are transition metals and change valence while undergoing oxidation-reduction reactions, recently the redox parameter (pe + pH)* has been used to quantify the ionic species present in a solution (Figure 13.3). A value of zero for pe + pH corresponds to the redox level imposed by one atmosphere of H_2 (g) in equilibrium with an aqueous system, while an aqueous system in equilibrium with one atmosphere of $O_2(g)$ has a pe + pH value of 20.8 (Lindsay, 1991). Most well-aerated soils have a redox (pe + pH) value of 12 to 16, while submerged paddy soils have a value of 4.0. Figure 13.3 shows that if soil-Fe controls Fe solubility, pe + pH would have to be <10 (at pH 7) or 8 (at pH 8) in order for Fe^{2+} activity to exceed the 10^{-8} M level required by most crop plants.

---

* If an oxidation-reduction is written as

$$Fe^{3+}OOH + e^- + 3H^+ = Fe^{2+} + 2 H_2O, \text{ then it can be generalized as}$$

$$A(OX) + B(e^-) + C(H^+) = D(red) + E(H_2O)$$

where A, B, C, D, and E are reaction coefficients for oxidized species (OX), electron (e⁻), proton (H^+), reduced species (red), and water, respectively. This can be expressed as:

$$K = (red)^D (H_2O)^E / (OX)^A (e^-)^B (H^+)^C$$

where K is the equilibrium constant. Taking the log of both sides of this equation we obtain

$$\log K = \log(red)^D \Big/ \log(OX)A + \log \frac{1}{(e^-)} B + \log \frac{1}{(H^+)^C}$$

$$\text{or } \log K = D \log(red) - A \log(OX) + B(pe) + C(pH)$$

where pe and pH are defined as -log of electron activity and hydrogen ion activity, respectively.
    For one-electronc transfer (B=1) coupled with one-proton consumption (C=1) and when D = A and (red) = (OX),

$$pe + pH = \log K$$

For details see Bartlett and James, (1993).

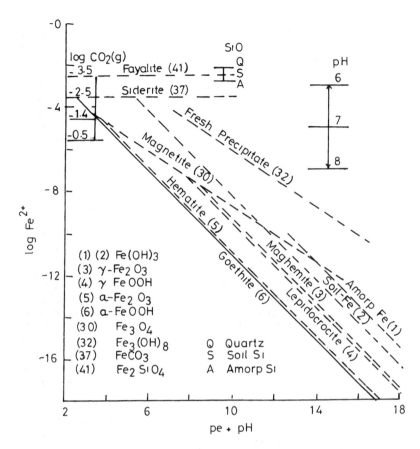

Figure 13.3. Effect of redox level, CO_2 (g), and silica on Fe^{2+} maintained by various Fe minerals at pH 7, showing shifts for other pH values. (From Lindsay, 1991. *Micronutrients in Agriculture*, J.J. Mortvedt, F.R. Cox, L.M. Shuman, and R.M. Welch, Eds. pp. 89–112. With permission of SSSA.)

The availability of Fe in soils therefore very much depends upon the availability of electrons. Thus the release of electrons near actively absorbing roots is considered a major mechanism by which Fe-stressed plants are able to increase the solubility and availability of Fe to meet their nutritional needs (Schwab and Lindsay, 1989). In addition to the release of electrons, plant roots also produce other Fe-reducing compounds and accumulate organic acids such as citrate to make soil-Fe more readily available to them. Monocots usually do not produce or excrete as much reducing compounds or organic acids as dicots. Even in the same species there are considerable differences between the cultivars, and breeding for Fe-deficiency–tolerant cultivars is being attempted (Pierson et al., 1984).

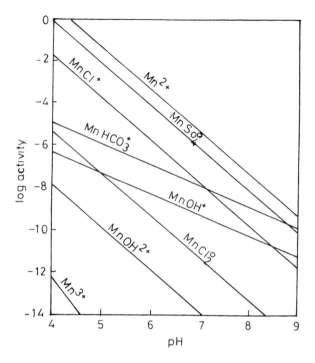

Figure 13.4. Solution species of Mn in equilibrium with manganite and pyrolusite at pe + pH 16.1, when Cl⁻ and SO_4^{2-} are at 10^{-3} M and CO_2 is at $10^{-4.52}$ MPa ($10^{-3.52}$ atm). (From Lindsay, 1991. *Micronutrients in Agriculture*, J.J. Mortvedt, F.R. Cox, L.M. Shuman, and R.M. Welch, Eds. pp. 89–112. With permission of SSSA.)

In contrast to Fe, there are not many hydroxylated species of Mn in soil solution; the most abundant soil solution species is Mn^{2+} (Figure 13.4). The important ion pairs are $MnSO_4$ and $MnCl^+$ at pH values below 7, and only at pH values above 8 would the ion pairs $MnOH^+$ and $MnHCO_3^+$ assume importance. A tenfold increase in CO_2 (g) would raise the $MnHCOS_3^+$ species by one log unit. In acid soils, as well as in submerged paddy soils, Mn solubility can increase to the toxicity level. Liming these soils raises pH and eliminates Mn toxicity.

13.3. FACTORS AFFECTING IRON AND MANGANESE AVAILABILITY

In addition to pH and soil water or aeration as measured by redox, soil organic matter and interactions with other ions in soil solution also affect the availability of Fe and Mn.

13.3.1. Organic Matter

Organic materials added to soil produce chelating agents such as simple aliphatic acids, hydroxamate siderophores, phenols and phenolic acids, complex polymeric phenols, and components of stable humus such as humic and fulvic acids (Stevenson, 1991). These chelating agents make more Fe and Mn available to plants. Hydroxamate siderophores play an important role in the Fe nutrition of plants growing on calcareous soils. These substances are produced by soil bacteria and fungi, including ectomycorrhizal fungi (Powell et al., 1980; Jukrevitch et al., 1988). Mandal (1961) found that the application of rice straw to paddy fields increased soil solution Fe^{2+} concentration from 0 to 132 mg L^{-1} and exchangeable Fe^{2+} from 20 to 200 mg kg^{-1} soil after 4 weeks of submergence. Similarly, the application of 10% peat moss along with 54 mg kg^{-1} Mn to Norfolk sandy loam increased acid-extractable Mn from 48 to 53.9 mg kg^{-1} (Sanchez et al., 1959).

13.3.2. Interaction with Other Nutrients

Excesses of essential nutrients such as P, Cu, Zn, and Mo adversely affect the availability of Fe. Phosphorus application at high rates aggravates Fe and Mn deficiency through some poorly defined inactivation reaction.

Application of NH_4^+-containing or producing fertilizers are likely to result in less Fe or Mn deficiency, while application of NO_3^--containing fertilizers is likely to enhance the deficiency of Fe and Mn. Neutral chloride-containing salts when applied also increase Mn availability.

13.4. SOIL TESTS FOR IRON AND MANGANESE

Because only a small fraction of Fe and Mn present in soil is available for plant growth, a number of chemical extracts have been evaluated for estimating available soil Fe and Mn concentrations. These are listed in Table 13.1. As expected, critical values for different soil tests and different crops using the same soil test will differ. The predictability of Fe and Mn soil tests (or for any other nutrient) will thus depend much upon the thoroughness with which soil test results are correlated with greenhouse or field crop response data.

Efforts have been made to use a common chemical extractant for a number of micronutrients. For example, the DTPA-TEA soil test was developed (Lindsay and Norvell, 1978) for simultaneous estimation of available Fe, Mn, Cu, and Zn in near-neutral and calcareous soils.

Attempts have also been made to fractionate soil Mn forms such as readily soluble Mn, weakly adsorbed Mn, carbonate bound Mn (calcareous soils), specifically adsorbed Mn (noncalcareous soils), and oxide-Mn (Warden and Reisenauer, 1991). Whether such a detailed study will help predict crop responses to Mn fertilization is yet to be seen.

Table 13.1 Critical Levels of Fe and Mn Soil Tests

Crop	Soils	Extractant	Critical level (mg element kg⁻¹ soil)	
		Fe		
Sorghum	35 calcareous soils	DPTA-TEA	4.5	
Sorghum	40 soils; 35 calcareous and 5 noncalcareous	NH_4HCO_3-DTPA	4.8	
		Mn		
Soybean	25 soils; pH 5.7–7.4	0.033 M H_3PO_4 1 M $NH_4H_2PO_4$	20	
Soybean	17 soils; pH 5.2–7.1	Mehlich I	5.2	(pH 6.0)
	30 soils; pH 5.1–6.9	Mehlich I	4.6	(pH 6.0)
	38 soils; pH 5.5–7.1	Mehlich I	4.7	(pH 6.0)
Soybean	2 soils; pH 6.2–7.0	DPTA-TEA	0.22	
		NH_4HCO_3-DTPA	0.40	
Corn	1 soil; pH 6.4–7.2	Mehlich III	3	(pH 6.4)

Adapted from Martens and Lindsay (1990).

Crops differ in their sensitivity to Fe and Mn. Their classification as sensitive, moderately tolerant, and tolerant is shown in Table 13.2 for Fe and Table 13.3 for Mn (Mortvedt, 1980).

13.5. DEFICIENCY SYMPTOMS OF IRON AND MANGANESE

Iron is an immobile nutrient in plants. Therefore its deficiency symptoms show up first in the young leaves of plants. Because 90% of the iron in leaves occurs in chloroplasts and mitochondria, deficiency of Fe results in loss of greenness (chlorophyll) in plants; leaves turn pale green and develop interveinal chlorosis. In severe deficiency the young leaves turn entirely white (Figure 13.5). Such symptoms are distinctly seen in sorghum, especially when growing on neutral to alkaline soils. Sorghum is an excellent indicator plant for Fe deficiency. Rice nurseries on neutral to alkaline soils and direct-seeded rice crops in early growth stages on similar soils may also show similar symptoms, that is, patches of plants with white young leaves. Iron deficiency of soybeans has become quite common in the western Corn Belt of the United States, but the severity of the deficiency varies with variety or cultivar. Iron deficiency is also reported in citrus orchards and blueberry fields. These deficiency symptoms are often referred to as "iron chlorosis."

Manganese is also an immobile nutrient in plants, and deficiency symptoms show up in younger leaves as in the case of Fe. In some wide-leaved plants such as corn or soybean (Figure 13.6), interveinal chlorosis, similar to

Table 13.2 Sensitivity of Crops to Low Levels of
Available Iron in Soil[a]

Sensitive	Moderately tolerant	Tolerant
Berries	Alfalfa	Alfalfa
Citrus	Barley	Barley
Field beans	Corn	Corn
Flax	Cotton	Cotton
Forage sorghum	Field beans	Flax
Fruit trees	Field peas	Grasses
Grain sorghum	Flax	Millet
Grapes	Forage legumes	Oats
Mint	Fruit trees	Potatoes
Ornamentals	Grain sorghum	Rice
Peanuts	Grasses	Soybeans
Soybeans	Oats	Sugar beets
Sudangrass	Orchard grass	Vegetables
Vegetables	Ornamentals	Wheat
Walnuts	Rice	
	Soybeans	
	Vegetables	
	Wheat	

[a] Some crops are listed under two or three categories because
of variations in soil and growing conditions and the different
responses of varieties of a given crop. (From Mortvedt,
1980. Farm Chem. 143(12):43. With permission of Meister
Publishing Co., Willoughby, OH.)

that in the case of Fe-deficiency, is seen. In other crops symptoms are different
and have been described under specific terms such as gray speck of oats, marsh
spot of peas, and speckled yellow of sugar beets.

13.6. TOXICITY SYMPTOMS OF IRON AND MANGANESE

Acid soils having large amounts of Mn may result in Mn toxicity in plants.
Crinkle leaf of cotton observed on highly acid ultisols in the southern United
States is due to Mn toxicity (Figure 13.7). Other crops such as soybeans,
tobacco, and canola on such soils may also show Mn toxicity symptoms. Mn
toxicity is also reported from rice paddies in acidic soils. On such soils Fe-
toxicity may also occur.

13.7. IRON AND MANGANESE FERTILIZERS

Soil application of Fe fertilizer is of limited help in rectifying an iron
deficiency because applied Fe^{2+} is rapidly oxidized. Foliar application to field

**Table 13.3 Sensitivity of Crops to Low Levels of
Available Manganese in Soil[a]**

Sensitive	Moderately tolerant	Tolerant
Alfalfa	Barley	Barley
Citrus	Corn	Corn
Fruit trees	Cotton	Cotton
Oats	Field beans	Field beans
Onions	Fruit trees	Fruit trees
Potatoes	Oats	Rice
Soybeans	Potatoes	Rye
Sugar beets	Rice	Soybeans
Wheat	Rye	Vegetables
	Soybeans	Wheat
	Vegetables	
	Wheat	

[a] Some crops are listed under two or three categories
because of variation in soil and growing conditions and
different responses of varieties of a given crop. (From
Mortvedt, 1980. Farm Chem. 143(12):43. With permission of Meister Publishing Co., Willoughby, OH.)

crops and injection of Fe^{2+}-salts directly into trunks and limbs of fruit trees
is preferred. In field crops a single application or several spray applications
(at weekly or fortnightly intervals) of 2 to 3% solution of ferrous sulfate
(650 to 700 L ha^{-1}) is generally required. A list of some iron fertilizers is
given in Table 13.4. Other than ferrous sulfate, synthetic iron chelates are
quite popular. Iron chelates can be applied to soils; HEDTA chelates are good
for acidic soils, EDTA chelates for neutral soils, and EDDHA chelates for
alkaline soils.

Manganese fertilizers (Table 13.5) can be applied to the soil or directly
to the crop as a foliar spray. Soil application of manganese sulfate may range
from 5 to 25 kg ha^{-1}, depending upon the soil and crop. For foliar application
generally a 0.2 to 0.5% solution of manganese sulfate is used.

Figure 13.5. Iron deficiency in maize (top) and rice (bottom). (From Yoshida, 1981. *Fundamentals of Rice Crop Science*, p. 269. With permission of International Rice Research Institute, Los Bancos, Philippines.) See Plate 5 following p. 170.

Figure 13.6. Manganese deficiency in maize. (From *Corn Field Manual*, J.R. Simplot Company Mineral & Chemical Division, Pocatello, ID, ©1984. With permission.) See Plate 6.

Figure 13.7. Iron toxicity in rice. (From Yoshida, 1981. *Fundamentals of Rice Crop Science*, p. 269. With permission of International Rice Research Institute, Los Baños, Philippines.) See Plate 7 following p. 170.

Table 13.4 Iron Fertilizers

Source	Formula	Percent Fe (approx.)
Ferrous sulfate	$FeSO_4.7H_2O$	19
Ferric sulfate	$Fe_2(SO_4)_3.4H_2O$	23
Ferrous oxide	FeO	77
Ferric oxide	Fe_2O_3	69
Ferrous ammonium phosphate	$Fe(NH_4)PO_4.H_2O$	29
Ferrous ammonium sulfate	$(NH_4)_2SO_4.FeSO_4.6H_2O$	14
Iron frits	Varies	Varies
Iron ammonium polyphosphate	$Fe(NH_4)PO_4.H_2O$	22
Iron-sul	Mixture $FeO(OH)$, $KFe_3(OH)_6(SO_4)_2$, FeS_2, and $CuFeS_2$	20
Iron chelates	NaFeEDTA	5–14
	NaFeHEDTA	5–9
	NaFeEDDHA	6
	NaFeDTPA	10
Iron polyflavonoids	—	9–10
Iron ligninsulfonates	—	5–8
Iron methoxyphenylpropane	FeMPP	5

From Murphy and Walsh, 1972. *Micronutrients in Agriculture*, J.J. Mortvedt, P.M. Giordano, and W.L. Lindsay, Eds. pp. 347–387. With permission of the Soil Science Society of America, Madison, WI.

Table 13.5 Manganese Fertilizers

Source	Formula	Percent Fe (approx.)
Manganese sulfate	$MnSO_4.4 H_2O$	26–28
Manganous oxide	MnO	41–68
Manganese carbonate	$MnCO_3$	31
Manganese chloride	$MnCl_2$	17
Manganese oxide	MnO_2	63
Manganese frits	Fritted glass	10–25
Natural organic complexes	—	5–9
Synthetic chelates	MnEDTA	5–12
Manganese methoxyphenylpropane	MnMPP	10–12

From Murphy and Walsh, 1972. *Micronutrients in Agriculture*, J.J. Mortvedt, P.M. Giordano, and W.L. Lindsay, Eds. pp. 347–387. With permission of the Soil Science Society of America, Madison, WI.

REFERENCES

Baes, C.F. and R.E. Mesmer. 1976. *The Hydrolysis of Cations*, John Wiley & Sons, New York.

Bartlett, R.J. and B.R. James. 1993. Redox chemistry of soils. Adv. Agron. 50:151–208.

Jurkevitch, E., Y. Hadar, and Y. Chen. 1988. Involvement of bacterial siderosphores in the remedy of lime induced chlorosis in peanut. Soil Sci. Soc. Am. J. 52:1032–1037.

Lindsay, W.L. 1988. Solubility and redox equilibria in iron compounds in soils, in *Iron in Soils and Clay Minerals*, J.W. Stucki, B.A. Goodman, and V.V. Schwertmann, Eds., D. Reidel, Dordrecht, Netherlands, pp. 37–62.

Lindsay, W.L. 1991. Inorganic equilibria affecting micronutrients in soils, in *Micronutrients in Agriculture*, J.J. Mortvedt, F.R. Cox, L.M. Shuman, and R.M. Welch, Eds., Soil Sci. Soc. Am., Madison, WI, Book Ser. No. 4, pp. 89–112.

Lindsay, W.L. and W.A. Norvell. 1978. Development of a DTPA soil test for zinc, iron, manganese and copper. Soil Sci. Soc. Am. J. 42:421–428.

Loeppert, R.H. and C.T. Hallmark. 1985. Indigenous soil properties influencing the availability of iron in calcareous soils. Soil Sci. Soc. Am. J. 49:597–603.

Mandal, L.N. 1961. Transformations of iron and manganese in water-logged rice soils. Soil Sci. 91:121–126.

Martens, D.C. and W.L. Lindsay. 1990. Testing soils for copper, iron, manganese and zinc, in *Soil Testing and Plant Analysis*, 3rd ed., R.L. Westermann, Ed., Soil Sci. Soc. Am., Madison, WI, pp. 229–264.

Mortvedt, J.J. 1980. Do you really know fertilizers? IV. Iron, manganese and molybdenum. Farm Chem. 143(12):42–47.

Murphy, L.S. and L.M. Walsh. 1972. Correction of micronutrient deficiencies with fertilizers, in *Micronutrients in Agriculture*, J.J. Mortvedt, P.M. Giordano, and W.L. Lindsay, Eds., Soil Science Society of America, Madison, Wisconsin, pp. 347–387.

Pierson, E.E., R.B. Clark, J.W. Maranville, and D.P. Coyne. 1984. Plant genotype differences to ferrous and total iron in emerging leaves. I. Sorghum and maize. J. Plant Nutr. 7:371–387.

Powell, P.E., G.R. Cline, C.P.P. Redi, and P.J. Szaniszlo. 1980. Occurrence of hydroxamate siderosphore iron chelators in soils. Nature (London) 287:833–834.

Sanchez, C. and E.J. Kamprath. 1959. The effect of liming and organic matter content on the availability of native and applied manganese. Soil Sci. Soc. Am. Proc. 23:302–304.

Schwab, A.P. and W.L. Lindsay. 1989. A computer simulation of Fe (III) and Fe (II) complexation in nutrient solution. II. Experimental. Soil Sci. Soc. Am. J. 53:34–38.

Stevenson, F.J. 1991. Organic matter—nutrient reactions in soil, in *Micronutrients in Agriculture*, J.J. Mortvedt, F.R. Cox, L.M. Shuman, and R.M. Welch, Eds., Soil Sci. Soc. Am., Madison, WI, pp. 145–186.

Warden, B.T. and H.M. Reisenauer. 1991. Fractionation of soil manganese forms important to plant availability. Soil Sci. Soc. Am. J. 55:345–349.

Yoshida, S. 1981. *Fundamentals of Rice Crop Science*, Int. Rice Research Institute, Los Baños, Philippines, p. 269.

14   COPPER AND ZINC

Copper, like Fe, participates in oxidation-reduction processes in plants. Copper is associated with enzymes that can hydroxylate monophenols, oxidizing them to create complex polymers such as lignin and melanin; detoxify superoxides; oxidize amines; terminate electron transfer chains; and act generally as cytoplasmic oxidases (Tisdale et al., 1985).

Newly reclaimed acid histosols (organic soils) are invariably Cu deficient, hence the name reclamation disease. In the United States Cu deficiency is most common in Florida, Wisconsin, Michigan, and New York, where vegetable and fruit crops are grown. Similarly, in Canada the Cu deficiency has been reported from Manitoba, Alberta, Saskatchewan, and Prince Edward Island. Responses of cereals to Cu have been reported from Australia (Robson et al., 1984), the United States (Varvel, 1983), Scotland (Reith, 1968), and Canada (Karamanos et al., 1986).

Zinc is a transition element, but has no impaired electrons; both the third and fourth orbitals are filled. Zinc ion (Zn^{2+}) is formed by the loss of the 4S electrons, and reactions tend to be similar to those of Ca^{2+} (Harter, 1991). Zinc is involved in a diversity of enzymatic activities such as auxin metabolism, dehydrogenases, phosphodiesterase, and the promotion of synthesis of cytochrome C. Zinc deficiencies are more prevalent than those of Cu and are virtually global in nature. In the United States and India most states now recognize the need for supplemental Zn for one or more crops. Analysis of about 40,000 soil samples from different parts of India showed 50% of the samples to be deficient in plant available Zn (Katyal and Sharma, 1979). Large areas of Zn deficiency have been identified in Canada, Europe, Great Britain, Australia, New Zealand, Central and South Africa, and Brazil. Acidic, sandy soils low in total Zn, calcareous soils, soils heavily fertilized with phosphorus, and subsoils exposed by land leveling operations or by wind and/or water erosion are prone to zinc deficiency.

While the chemistry in soils and the functions in plants of Cu and Zn differ greatly, there are several similarities:

1. Both are metals.
2. Both are taken up by plants as cations in +2 valence, as Cu^{2+} and Zn^{2+}.

3. Both Cu and Zn have a strong tendency to combine with S and occur as sulfides in the lithosphere.
4. Both Cu and Zn have three ionic forms in soil solution with possible valencies of +2, +1, and 0.

14.1. AMOUNTS IN SOIL

The Cu content in soils ranges from nearly 5 to 60 mg kg^{-1}, although both lower (<2 mg kg^{-1}) and higher values are not uncommon (Stevenson, 1986). The average value of Cu in soils is about 9 to 10 mg kg^{-1}. The most widely occurring copper mineral is chalcopyrite ($CuFeS_2$). Other Cu-bearing minerals are chalcocite (Cu_2S) and bornite ($CuFeS_4$).

Total Zn content in soils ranges from 10 to 300 mg kg^{-1} with an average of 50 mg kg^{-1}. The Zn content in various rocks varies from 4 to 100 mg kg^{-1}. The general values (mg kg^{-1}) are basalt 100, shale 45, granite 10, limestone 4, and sandstone 30 mg kg^{-1} (Tisdale et al., 1985; Stevenson, 1986). The most important minerals are sphalerite (ZnS), smithsonite ($ZnCO_3$), and hemimorphite ($Zn_4(OH)_2Si_2O_7.H_2O$).

14.2. FORMS OF COPPER AND ZINC IN SOILS

Both Cu and Zn occur in soils in the following forms:

1. As a structural component of clay minerals. Zinc can substitute for Mg in the crystal lattice. A large part of the Zn present in soil exists as a component of augite, hornblende, and biotite, which are known as ferromagnesium minerals.
2. Adsorbed on clay mineral surfaces—partly as exchangeable cations.
3. Adsorbed on the oxides, hydroxides, and oxyhydroxides of Fe and Al. Some of the Cu so adsorbed may be occluded by oxide minerals because Cu readily coprecipitates with Fe and Al. Zinc is also adsorbed on carbonates.
4. Complexed with soil organic matter.
5. Present as Cu^{2+} and Zn^{2+} in soil solution.

Different forms of Cu or Zn tend to maintain an equilibrium in soil.

14.2.1. Copper and Zinc in Soil Solution

The concentrations of both Cu^{2+} and Zn^{2+} in soil solution remain very low, approximately 0.5 to 70 mg Mg^{-1} (ppb). At any time the concentration of Zn^{2+} in the soil solution is generally greater than that of Cu^{2+}. Plant uptake of Cu^{2+} is mostly by root interception, while that of Zn^{2+} is by diffusion.

Both Cu and Zn are present in three ionic forms—M^{2+}, $M(OH)^+$, and $M(OH)_2^0$, where M stands for Cu and Zn. At pH <6.0 nearly all Cu and Zn are present as Cu^{2+} or Zn^{2+}. As pH approaches neutrality, nearly half the Cu in soil solution is hydrolyzed to $Cu(OH)_2^0$, and the change to this form is 92% at pH 8. This change in form reduces Cu availability to plants. In contrast $Zn(OH)_2^0$ begins to appear only at pH 8. Reduction in the divalent form of Zn^{2+} to the hydrolyzed form is slower for Zn. At pH 7 about 83% is still present as Zn^{2+} and 31% at pH 8. At pH 8 about two-thirds of Zn may be present as $Zn(OH)^+$, which is less available to plants.

14.2.2. Copper and Zinc Adsorbed onto Clay Minerals

Copper can be sorbed by clay minerals at lower activities than those required for Zn. Thus it is suggested that Cu^{2+} is retained by specific sorption sites, while Zn^{2+} is retained on nonspecific sites. Although quite divergent data are available, 2:1 layer silicates in general adsorb more Cu^{2+} and Zn^{2+} than 1:1 layer silicates (Farrah, et al., 1980); Krishnasamy et al., 1985).

14.2.3. Copper and Zinc Adsorbed onto Fe and Al Hydrous Oxides

Hydrous oxides of Fe and Al play an important role in the retention of Cu and Zn and their concentration in soil solution. Removal of hydrous oxides of Fe and Al can significantly reduce the Cu^{2+} and Zn^{2+} retention capacity of soil (Cavallrao and McBride, 1984). Also since Fe and Al oxides have a variable charge, retention of Cu^{2+} and Zn^{2+} by these oxides depends very much upon soil pH. Zn^{2+} may be adsorbed by oxides specifically (irreversibly) or nonspecifically. The specific or irreverisble sorption mechanism proposed by Quirk and Posner (1975) is shown below:

14.2.4. Copper and Zinc Adsorbed by Organic Matter

Soil organic matter also plays an important role in the retention of both Cu and Zn, more so for Cu than for Zn. Carboxylate and phenolic group and N lone-pair electrons are the dominant retention sites for Cu^{2+} (Harter, 1991).

Table 14.1 Effect of pH on Solution Composition of Cu and Zn, Expressed as Percent in Solution

pH	4	5	6	7	8
Cu^{2+}	100	100	96	33	1
$CuOH^+$			2	7	1
$Cu(OH)_2^0$			2	56	92
Zn^{2+}	100	100	98	83	31
$ZnOH^+$			2	17	64
$Zn(OH)_2^0$					5

From Harter, 1983. Soil Sci. Soc. Am. J. 47:49. With permission of SSSA.

14.3. FACTORS AFFECTING THE AVAILABILITY OF COPPER AND ZINC

The availability of Cu and Zn to plants is affected by a number of factors such as pH, kind and amount of organic matter, interaction with other elements in soil solution, fertilizer practices, soil amendments, flooding (as in rice culture), environmental factors, and plant factors.

14.3.1. Soil pH

The influence of soil pH on the solubility and ionic species of Cu and Zn in soils has already been discussed. In general, increases in soil pH above 6.0 decreases the availability of Cu and Zn due to the following:

1. A change in the hydrolysis status—for example, at pH 8 Cu $(OH)_2^0$ and Zn $(OH)^{+1}$ are the dominant ionic species (Table 14.1), which have less solubility (activity) than Cu^{2+} and Zn^{2+}.
2. Increased sorption on Al and Fe oxides and hydroxides, clay minerals, and soil organic matter due to increased pH-dependent charge.
3. Reduced competition with H^+ for the adsorption sites.
4. A change in the quantity and nature of organic binding chemicals; for example, zinc complexing by humic acids increases with rising pH.

14.3.2. Interaction With Other Elements in Soil

High concentrations of Zn, Al, P, and Fe in soil solution restrict Cu absorption by plants. Similarly, Cu, Fe, and Mn inhibit uptake of Zn (Kausar, et al., 1976; Giordano et al., 1974). This is possibly due to competition for the carrier sites on roots.

The most frequently reported and researched interaction is P x Zn, namely, phosphorus-induced Zn deficiency. Application of high rates of P enhances the adsorption of Zn in some soils, especially those rich in Fe and Al oxides and hydroxides (Saeed and Fox, 1979). This could be due to the additional negative charge or complexation sites created due to sorption of PO_4^{3-} into oxide surfaces (Bolland et al., 1977). Furthermore, physiological research has brought out that high P rates increase the amount of Zn in the ethanol-soluble and pectate fraction of root cell walls, and this binding of Zn results in reduced amounts of Zn available for transport to the plant shoot (Youngdahl et al., 1977). An alternative hypothesis suggests that what is generally considered a Zn deficiency is in reality P-toxicity due to higher uptake of P (Webb and Loneragan, 1988). Whatever the explanation, large amounts of data (Adams, 1980) suggest the presence of a negative P x Zn interaction under field conditions.

14.3.3. Fertilizer Practices

Application of high rates of NPK fertilizers has aggravated Cu and Zn deficiencies. In addition to P x Zn interaction effects, a number of other factors listed below are responsible for this:

1. Increased fertilization results in increased yields and thus increased demands for Cu and Zn (this would apply for other nutrients as well).
2. Increased fertilization results in greater depletion of Cu and Zn (also other plant nutrients) from soils and thus increases the chances of their deficiencies occurring.
3. Addition of large amounts of N-fertilizers can increase soil acidity, which may increase Al and Fe levels in soils and thereby depress the absorption of Cu and Zn by plant roots.

14.3.4. Soil Amendments

The application of lime ($CaCO_3$) to acidic soils and gypsum ($CaSO_4.5H_2O$) on sodic soils affects the availability of these nutrients. When lime is added to acidic soils, or in calcareous soils where free $CaCO_3$ is present, zinc can be adsorbed on $CaCO_3$ particles, which may temporarily reduce Zn availability. In the long run, however, $ZnCO_3$ becomes available to plants because its solubility is too high to persist in soils.

The application of gypsum on sodic soils increases the availability of Zn (1) because of slight lowering of pH and (2) because the highly mobile $ZnSO_4^0$ (an important species of Zn in soils) contributes significantly to soil solution Zn.

14.3.5. Flooding or Submergence

Flooding or submergence of fields, as practiced in paddy rice culture results in increased pH on acidic soils and therefore is likely to reduce Zn availability (Mikkelsen and Brandon, 1975). The reverse could be true on alkaline or sodic soils. The formation of ZnS under the anaerobic conditions created is also responsible for decreased Zn availability (Ponnamperuma, 1972).

14.3.6. Weather Conditions

In cooler regions the availability of Cu and Zn increases with an increase in temperature (Moraghan and Mascagnie, 1991). The severity of Zn deficiency increases under reduced or intermediate light (Edwards and Kamprath, 1974). Thus Zn deficiency symptoms appear when days are cool and cloudy.

14.3.7. Plant Factors

There are genotypic differences in plant species in respect of sensitivity to Cu or Zn deficiency. Among small grains, rye has exceptional tolerance to Cu deficiency. The usual order of sensitivity of the small grains to Cu deficiency in the field is wheat > barley > oats and rye (Tisdale et al., 1985). The sensitivity of different crops to Cu and Zn are given in Tables 14.2 and 14.3.

14.4. SOIL TESTS FOR COPPER AND ZINC

14.4.1. Copper

Two extractants tested and used for upland crops are 0.05 M EDTA (pH 7.0) (Reith, 1968) and Mehlich-Bowling (0.5 M HCl + 0.016 M AlCl$_3$) (Makarin and Cox, 1983). The critical value for barley and oat for the 0.05 M EDTA extractable Cu is 1.1 mg kg^{-1}, while that for soybean and wheat using Mehlich-Bowling extract is 0.7 mg Cu kg^{-1} soil.

14.4.2. Zinc

There are four extractants generally used for determining available Zn in soils, namely, 0.1 M HCl (Wear and Sommer, 1948), EDTA-(NH$_4$)$_2$CO$_3$ (Trierweiler and Lindsay, 1969), Dithizone—NH$_4$OAC (Tierweiler and Lindsay, 1969), and DTPA-TEA (Brown et al., 1971). The first proposed extractant was 0.1 M HCl, and it is now used the least. Dithizone—NH$_4$OAC extraction of zinc involves the use of CCl$_4$, which is hazardous to human health and is therefore less adaptable to routine use, yet this method has been widely used. Dithizone—NH$_4$OAC and DTPA-TEA were equally

Table 14.2 Relative Sensitivity of Crops to Cu Deficiency

Low	Medium	High
Barley	Broccoli	Alfalfa
Beans	Cabbage	Cauliflower
Blueberry	Carrot	Celery
Cucumber	Clover	Sugarbeet
Corn	Lettuce	Turnip
Sudan Grass	Radish	
Oat	Spinach	
Onion	Sweet corn	
Pea	Tomato	
Potato		
Rye		
Sorghum		
Soybean		
Wheat		

From Martens and Westermann. 1991. *Micronutrients in Agriculture*, 2nd ed., J.J. Mortvedt, F.R. Cox, L.M. Shuman, and R.M. Welch, Eds. With permission of SSSA.

Table 14.3 Sensitivity of Crops to Low Levels of Available Zinc in Soils

Sensitive	Moderately tolerant	Tolerant
Corn	Grain sorghum	Alfalfa
	Clover	Barley
Field beans	Potatoes	
	Forage	Oats
	Sorghum	Millet
Sweet corn	Soybeans	Rye
	Sugar beets	Wheat
Rice[a]	Sudan grass	Grasses

[a] Added by authors.

From Rehm and Penas, 1982. NebGuide, G82-596. With permission of the University of Nebraska, Lincoln.

effective in the separation of 92 California soils into Zn-sufficient and Zn-deficient groups for sweet corn production (Brown et al., 1971). The critical values for DTPA-TEA are from 0.5 to 0.8 mg Zn kg^{-1} soil for corn (Brown et al., 1971; Lindsay and Norvell, 1978), 0.48 mg Zn kg^{-1} soil for green gram (Gupta and Mittal, 1981), and 0.86 mg Zn kg^{-1} soil for rice (Singh and Takkar, 1981).

14.5. DEFICIENCY SYMPTOMS IN PLANTS

14.5.1. Copper

In cereals such as wheat and oats receiving adequate Cu supply the Cu concentration at boot stage may vary from 5 to 21 mg kg^{-1}, while in forage alfalfa and timothy (*Phleum pratense* L.) it may vary from 9 to 54 mg kg^{-1} dry matter (Gupta, 1989a). Deficiencies are likely when Cu concentration in plant dry matter falls below 4 mg kg^{-1} (Tisdale et al., 1985). On the other hand, Cu concentration as high as 55 mg kg^{-1} in timothy and wheat exhibited no toxicity.

Symptoms of Cu deficiency appear first at the top of the plant with the youngest leaves becoming yellow and pale as the deficiency advances. Eventually, dead tissue appears along the tips and edges of leaves as in the case of K. In vegetables the leaves lack turgor and develop a bluish green cast (Tisdale et al., 1985). Copper deficiency symptoms in barley and wheat are shown in Figure 14.1. Crops most susceptible to Cu deficiency are alfalfa, wheat, barley, oats, and onions.

14.5.2. Zinc

In cereals such as wheat and barley receiving adequate Zn supply, plant concentrations at boot stage may vary from 20 to 123 mg kg^{-1} dry matter (Gupta, 1989b) or even higher. Zinc deficiency in plants may be expected when the plant concentration is less than 20 mg kg^{-1} dry matter, while Zn toxicity can occur when concentrations exceed 400 mg kg^{-1} dry matter (Tisdale et al., 1985).

Corn and beans in field crops and citrus in orchard crops are sensitive to zinc deficiency. Experience in Asian countries has shown that rice is also sensitive to Zn deficiency.

Since Zn is highly immobile in plants, its deficiency symptoms are seen on the growing points and young leaves. In corn the young leaves become yellow to white (sometimes stripped) and the plant has stunted growth; therefore the name "white bud" (Figure 14.2). Zinc deficiency in rice fields is indicated by areas of chocolate-brown, burned-up plants, hence the name "Khaira" disease. In citrus the deficiency is indicated by a cluster of leaves (rosette) at the top of mainly bare branches, referred to as "mottle leaf" or "frenching." Some other zinc deficiency diseases are "little leaf" in cotton and "fern leaf" in russet Burbank potatoes.

Crops sensitive, tolerant, and moderately tolerant to zinc deficiency are listed in Table 14.3.

14.6. COPPER AND ZINC FERTILIZERS

A list of Cu and Zn fertilizers is given in Table 14.4. The most popular and widely used source of Cu is $CuSO_4.5H_2O$ and that of Zn is $ZnSO_4$. Generally, a dose of 10 to 25 kg ha^{-1} $CuSO_4$ or $ZnSO_4$ is recommended on Cu/Zn deficient soils, depending upon soil texture; the heavier the texture, the

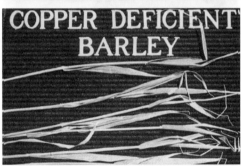

Figure 14.1. Copper deficiency in wheat (above) and barley. (From *Small Grains Field Manual*, pp. 11–12, J.R. Simplot Co. Minerals & Chemical Division, Pocatello, ID, ©1984. With permission.) See Plate 8 following p. 170.

greater the dose. In general, soil application of Cu, as well as Zn, is considered more effective. When a deficiency is observed in a standing crop, foliar applications can also be made. A 0.5% w/v solution of $CuSO_4$ or $ZnSO_4$ is recommended with a small amount of slaked lime (0.5 kg 100 l^{-1}); this practice prevents scorching of leaves. Dosages are much less when chelates are used, and these are preferably used for foliar application. Dipping of rice seedlings or cut potato-seed pieces in a 2%-$ZnSO_4$ or -ZnO suspension or slurry is an economical way of applying Zn to crops on Zn-deficient soils; only 2 to 5 kg ha^{-1} $ZnSO_4$ or ZnO are required with this method.

Figure 14.2. Zinc deficiency in maize (corn). (From *Corn Field Manual*, pp. 9–10, J.R. Simplot Co. Minerals & Chemical Division, Pocatello ID, ©1984. With permission.) See Plate 9 following p. 170.

Table 14.4 Copper and Zinc Fertilizers

Source	Formula	Cu or Zn (%)
Copper		
Copper sulfate monohydrate	$CuSO_4.H_2O$	35
Basic copper sulfates	$CuSO_4.3Cu(OH)_2$	13–53
Copper ammonium phosphate	$Cu(NH_4)PO_4.H_2O$	32
Copper chelates	$Na_2CUEDTA$	13
	NaCuHEDTA	9
Copper-sulfur frits		0.5–20
Zinc		
Zinc sulfate monohydrate	$ZnSO_4.H_2O$	35
Zinc sulfate heptahydrate	$ZnSO_4.H_2O$	23
Basic zinc sulfate	$ZnSO_4.6Zn(OH)_2$	55
Zinc oxide	Zno	78
Zinc phosphate	$Zn_3(PO_4)_2$	51
Zinc chelates	$Na_2ZnEDTA$	14
	NaZnNTA	13
	NaZnHEDTA	9
Zinc frits		Variable

From Murphy and Walsh, 1972. *Micronutrients in Agriculture*, J.J. Mortvedt, P.M. Giordano, and W.L. Lindsay, Eds., p. 351. With permission of SSSA.

REFERENCES

Adams, F. 1980. Interaction of phosphorus with other elements in soils and plants, in *The Role of Phosphorus in Agriculture*, F.E. Khasawneh, E.C. Sample, and E.J. Kamprath, Eds., Am. Soc. Agron. Soil Sci. Soc. Am., Madison, WI, pp. 655–680.

Bolland, M.D.A., A.M. Posner, and J.P. Quirk. 1977. Zinc adsorption by geothite in the absence and presence of phosphate. Aust. J. Soil Res. 15:279–286.

Brown, A.L., J. Quick, and J.L. Eddings. 1971. A comparison of analytical methods for soil zinc. Soil Sci. Soc. Am. Proc. 35:105–107.

Cavallaro, N. and M.B. McBride. 1984. Zinc and copper sorption and fixation by an acid soil clay: effect of selective dissolutions. Soil Sci. Soc. Am. J. 48:1050–1054.

Edwards, J.H. and E.J. Kamprath. 1974. Zinc accumulation by corn as influenced by phosphorus, temperature and light intensity. Agron. J. 66:479–482.

Farrah, H.D., D. Hatton, and W.F. Pickering. 1980. The affinity of metal ions for clay surfaces. Chem. Geol 28:55–68.

Giordano, P.M., J.C. Noggle, and J.J. Mortvedt. 1974. Zinc uptake by rice as affected by metabolic inhibitors and competing cations. Plant Soil 41:637–646.

Gupta, U.C. 1989a. Copper nutrition of cereals and forages grown in Prince Edward Island. J. Plant Nutr. 12:53–64.

Gupta, U.C. 1989b. Effect of zinc fertilization on plant zinc concentration of forages and cereals. Can. J. Soil Sci. 69:473–479.

Gupta, V.K. and S.B. Mittal. 1981. Evaluation of chemical methods for estimating available zinc and response of green grass (Phaseolus aureus Roxb.) to applied zinc in non-calcareous soils. Plant Soil 63:477–484.

Harter, R.D. 1983. Effect of soil pH on adsorption of lead, copper zinc, and nickel. Soil Sci. Soc. Am. J. 47:47–51.

Harter, R.D. 1991. Micronutrient adsorption-desorption reactions in soils, in *Micronutrients in Agriculture*, J.J. Mortvedt, F.R. Cox, L.M. Shuman, and R.M. Welch, Eds., Soil Sci. Soc. Am. Madison, WI, Book Ser. No. 4, pp. 59–87.

Karamanos, R.E., G.A. Kruger, and J.W.B. Stewart. 1986. Copper deficiency in cereal and oilseed crops in northern Canadian prairie soils. Agron. J. 78:317–323.

Katyal, J.C. and B.D. Sharma. 1979. Role of micronutrients in crop production. Fert. News 24(9):33–50.

Kausar, M.A., F.M. Chandhary, A. Rashid, A. Latif, and S.M. Alam. 1976. Micronutrient availability to cereals from calcareous soils. I. Comparative Zn and Cu deficiency and their mutual interaction in rice and wheat. Plant Soil. 45:397–410.

Krishnasamy, R., K.K. Krishnamoorthy, and T.S. Manickam. 1985. Zinc adsorption isotherms for soil clays. Clay Res. 4:92–95.

Lindsay, W.L. and W.A. Norvell. 1978. Development of a DTPA soil test for zinc, iron, manganese and copper. Soil Sci. Soc. Am. J. 42:421–428.

Makarim, A.K. and F.R. Cox. 1983. Evaluation of the need for copper with several soil extractants. Agron. J. 75:493–496.

Martens, D.C. and D.T. Westermann. 1991. Fertilizer applications for correcting micronutrient deficiencies, in *Micronutrients in Agriculture*, 2nd ed., J.J. Mortvedt, F.R. Cox, L.M. Shuman, and R.M. Welch, Eds., Soil Sci. Soc. Am., Madison, WI, Book series no. 4, pp. 549–592.

Mikkelson, D.S. and D.M. Brandon. 1975. Zinc deficiency in California rice. California Agric. 29(9):8–9.

Moraghan, J.T. and H.J. Mascagni, Jr. 1991. Environmental and soil factors affecting micronutrient deficiencies and toxicites, in *Micronutrients in Agriculture*, 2nd ed., J.J. Mortvedt, F.R. Cox, L.M. Schuman, and R.M. Welch, Eds., Soil Sci. Soc. Am., Madison, WI, Book series No. 4, pp. 371–425.

Murphy, L.S. and L.M. Walsh. 1972. Correction of micronutrient deficiencies with fertilizers, in *Micronutrients in Agriculture*, J.J. Mortvedt, P.M. Giordano, and W.L. Lindsay, Eds., Soil Sci. Soc. Am., Madison, WI, pp. 347–387.

Ponnamperuma, F.N. 1972. Chemistry of submerged soils. Adv. Agron. 24:29–95.

Quirk, J.P. and A.M. Posner. 1975. Trace element adsorption by soil minerals, in *Trace Elements in Soil-Plant-Animal Systems*, D.J.D. Micholas and A.R. Egan, Eds., Academic Press, New York, pp. 95–107.

Rehm, G.W. and E.J. Penas. 1982. Use and management of micronutrient fertilizers in Nebraska. NebGuide G82-596.

Reith, J.W.S. 1968. Copper deficiency in crops in northeast Scotland. J. Agric. Sci. Camb. 70:39–45.

Robson, A.D., J.F. Loneregan, J.W. Gartell, and K. Snowball. 1984. Diagnosis of copper deficiency in wheat by plant analysis. Aust. J. Agric. Res. 35:347–358.

Saeed, M. and R.L. Fox. 1979. Influence of phosphate fertilization on zinc adsorption by tropical soils. Soil Sci. Soc. Am. J. 43:683–686.

Singh, H.G. and P.N. Takkar. 1981. Evaluation of efficient soil test methods for Zn and their critical values in salt-affected soils for rice. Commu. Soil Sci. Plant Anal. 12:383–406.

Stevenson, F.J. 1986. *Cycles of Soil*, John Wiley & Sons, New York, p. 380.

Tisdale, S.L., W.L. Nelson, and J.D. Beaton. 1985. *Soil Fertility and Fertilizers*, 4th ed., Macmillan, New York, p. 754.

Trierweiler, J.F. and W.L. Lindsay. 1969. EDTA-ammonium carbonate soil test for zinc. Soil Sci. Soc. Am. Proc. 33:49–54.

Varvel, G.E. 1983. Effect of banded and broadcast placement of Cu fertilizers on correction of Cu deficiency. Agron. J. 75:99–101.

Wear, J.I. and A.L. Sommer. 1948. Acid extractable zinc of soils in relation to the occurrence of zinc deficiency symptoms of corn: a method of analysis. Soil Sci. Soc. Am. Proc. 12:143–144.

Webb, M.J. and J.F. Loneragan. 1988. Effect of zinc deficiency on growth, phosphorus concentration and phosphorus toxicity of wheat plants. Soil Sci. Soc. Am. J. 52:1676–1680.

Youngdahl, L.J., L.V. Svec, W.C. Leibhardt, and M.R. Teel. 1977. Changes in the zinc-65 distribution in corn root tissue with a phosphorus variable. Crop. Sci. 17:66–69.

15   BORON AND MOLYBDENUM

Boron and molybdenum are the two micronutrients that are taken up by plant as anions. However, their chemistry in soils is quite different, and therefore each nutrient is separately discussed.

15.1. BORON

The total concentration of boron in most soils varies between 2 and 200 mg kg^{-1}, and less than 5% is generally available to plants (Tisdale et al., 1985). Boron-containing minerals in soils are tourmaline, axenite, ulexite, colemanite, and kermite; tourmaline is the most important. Boron-containing minerals are quite resistant to weathering, and most plant-available B comes from the decomposition of soil organic matter and from B adsorbed and precipitated onto the surface of soil particles.

Boron is highly mobile in soils (in contrast to being highly immobile in plants). Consequently, both B deficiencies and toxicities are of concern. Soils of humid regions such as sandy podzols, vertisols, alluvial soils, and organic soils have low amounts of plant-available B due to leaching of B. Boron deficiencies have been reported from many countries having such soils, namely, the United States, Canada, England, New Zealand, India, and Nigeria (Gupta, 1985). In the United States some soils in the Atlantic coastal plain, the Pacific coastal area, the Pacific Northwest, and northern Michigan, Wisconsin, and Minnesota appear to be low in B.

Boron, like sodium and chloride, is soluble and tends to accumulate where salts accumulate. Thus B may be found in toxic levels in salinic and sodic soils, in low lying areas with impeded drainage, and in areas with a shallow water table. Irrigation water high in B content is a major cause of B toxicity. Overall in nature, B toxicity is not as widespread as B deficiency.

15.1.1. Forms of Boron in Soils

In addition to being present as a component of minerals and rocks, B exists as (1) adsorbed onto surfaces of clay minerals and hydroxides of Al

283

and Fe; (2) complexed with soil organic matter; and (3) in the soil solution as soluble B, either as free, nonionized H_3BO_3 or as ionized $B(OH)_4^-$ forms.

1. Adsorbed B: There are three possible mechanisms for interlayer adsorption of B: (a) H bonding of $B(OH)_3$ with the oxygen-rich interlayer of clay platelets; (b) borate ion ionically bound to materials adsorbed onto the interlayer surfaces (e.g., hydroxyl-Al material); and (c) interlayer material that could move out to external sites or into solution and then react with borate ions (Evans and Sparks, 1983). Thus Fe and Al, present as interlayer materials in various forms, seem necessary for B sorption on clays (Gupta, 1985). On a weight basis illite is most reactive and kaolinite least reactive. Chain silicate minerals such as olivine and augite adsorb more B than the micaceous layer silicates muscovite, vermiculite, and biotite. At a high pH, Al and Fe hydroxides and oxides, through ligand exchange, provide the main mechanism for bonding B (Bingham et al., 1970).
2. B complexed with organic matter: The exact mechanism of B-organic matter complex formation are not known, but some studies suggest basic acid condensation with diol groups associated with carboxylic acids.
3. Soluble B: Boron dissolved in soil solution is present as undissociated boric acid (H_3BO_3) or as soluble species $B(OH)_4^-$ (which is $H_2BO_3^- + H_2O$). $H_2BO_3^-$ is of importance in soils having a pH of 9.0 and above. Undissociated H_3BO_3 is considered the form in which most B is taken up by the plants (Bingham et al., 1970). The presence of different species of B as affected by pH is shown in Figure 15.1. The horizontal line H_3BO_3 passes through the solubility data for Swedish soils and can be designated as soil-B line, representing a mean level of approximately $10^{-5.5}$ M B. The bulk of B is taken up by plants by the mechanism of mass flow.

15.1.2. Factors Affecting Boron Availability

As brought out in the discussion so far, soils having a relatively high content of micaceous clay minerals (such as illite) and organic matter contain greater amounts of total and available B. On the other hand, coarse, well-drained, low-organic-matter, sandy soils usually have less total and available B.

Soil pH and liming; interactions with Ca, K, and other nutrients; and weather factors influence B availability to plants.

15.1.2.1. Soil pH and Liming

Liming strongly acidic soils frequently induces at least a temporary B deficiency in susceptible plants, which is believed to be caused by increased

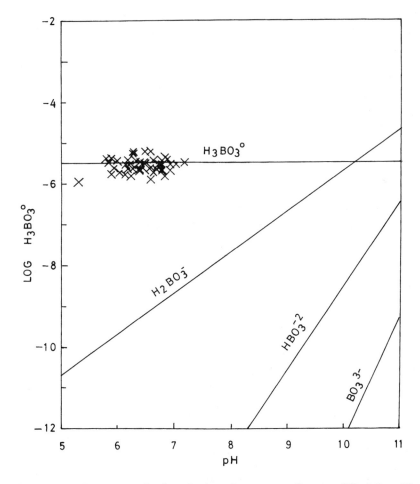

Figure 15.1. Boron species in solution, drawn to reflect equilibrium with the average soluble-B levels in 49 soils in Sweden. (From Lindsay, 1991. *Micronutrients in Agriculture*, J.J. Mortvedt, Ed., p. 107. With permission of SSSA.)

adsorption of B on freshly formed $Al(OH)_3$ (Tisdale et al., 1985). As a corollary, moderate liming can be used as a corrective treatment on soils containing excess B.

15.1.2.2. Interactions with Other Nutrients

The most well known of the B interactions with other nutrients is that with Ca, as seen in liming. This effect is probably related to the Ca:B ratio in plants. For example Ca:B ratios of 10 to 45 were toxic to barley, a ratio of 180 was optimal, and ratios greater than 697 may produce B deficiency.

However, Gupta and MacLeod (1981) using $CaCO_3$ and $CaSO_4$ as sources of Ca, suggested that reduced B concentration in plants was a pH rather than a direct Ca effect.

High levels of K accentuate B deficiency, as well as toxicity symptoms, mainly by affecting Ca concentration. For example, Patel (1967) showed that B deficiency symptoms of tobacco increased and toxicity symptoms decreased with increasing Ca:B and K:B ratios.

Liberal N applications decrease the severity of B toxicity symptoms in citrus and in cereals (Gupta, 1985).

15.1.2.3. Environmental Factors

Environmental factors can influence B availability to plants. For example, in areas of adequate precipitation, B deficiency is observed only in dry seasons or in late summer when water availability is low. Plants can tolerate higher B concentrations, without experiencing harmful effects, during cool weather than during warm, humid conditions (Eaton, 1935). B toxicity can be exacerbated by high light intensity.

15.1.3. Soil Tests for Boron

The most widely used soil test for B is the hot water extraction method as proposed by Berger and Truog (1939). Other methods, including variations in the hot water extraction (Gupta, 1993a), are 0.05 M HCl, 0.01 M $CaCl_2$ + 0.5 M Mannitol, hot 0.02 M Ca Cl_2, and Morgan's reagent.

The boron requirement for most crops is usually met when soil contains 0.5 to 1.0 mg kg^{-1} of hot water–soluble B, and levels above 5 mg kg^{-1} are likely to be toxic (Ponnamperuma et al., 1981). Deficiency, sufficiency, and toxic B levels in crop tissues are given in Table 15.1. In general, crucifers and legumes have a much greater need for B than the cereals. Nevertheless, large differences exist between cultivars of the same species in respect to B requirements.

15.1.4. Boron Deficiency Symptoms in Plants

Boron plays an essential role in the development and growth of new cells in the meristematic tissue and imparts stability to the pollen tubes. It is also involved in the germination and growth of pollen. Boron also facilitates the translocation of sugar through the phloem.

Boron is one of the least mobile of the micronutrients in plants and is not readily translocated from old to young plant parts. The first symptoms of B deficiency therefore appear in the growing points and meristematic tissue—the stem tips, root tips, new leaves, and flower buds. Boron deficiency symptoms include thickened, cracked, and wilted leaves, petioles, and stems, as well as

Table 15.1 Deficiency, Sufficiency, and Toxicity Levels of Boron on Crops

Crop	Part of plant tissue sampled	Deficient	Sufficient	Toxic
		mg B kg⁻¹ dry matter		

$$\text{mg B kg}^{-1}\text{ dry matter}$$

Crop	Part of plant tissue sampled	Deficient	Sufficient	Toxic
A. Crops responsive to B application; may tolerate high B levels				
Alfalfa (*Medicago sativa* L.)	Whole tops at early bloom	<15	20–40	200
Snap beans (*Phaseolus vulgaris* L.)	Whole tops at early bloom stage	<12	42	>125
Birdsfoot trefoil (*Lotus corniculatus* L.)	Whole tops at bud stage	14	30–45	>68
Brussel sprouts (*Brassica oleracea* var. *gemmifera* Zenker)	Leaf tissue when sprouts begin to form	6–10	13–101	—
Carrots (*Daucus carota* L.)	Mature leaf lamina	<16	32–103	175–307
Celery (*Apium graveslens* L.)	Leaflets	20	68–432	720
Cucumber (*Cucumus sativus* L.)	Mature leaves from center of stems 2 weeks after first picking	<20	40–120	>300
Red clover (*Trifolium pratens* L.)	Whole tops at bud stage	12–20	21–45	>59
Rutabaga (*Brassica napobrassica*, Mill)	Leaf tissue at harvest	20–38	38–140	>250
Sugarbeets (*Beta vulgaris* L.)	Middle fully developed leaf without stem taken at end of June/early July	<20	31–200	>800
Tomatoes (*Lycopersicon eselentum* Mill)	Mature young leaves from top of 63-day-old plants	<10	30–75	>200
Spanish peanuts (*Arachis hypogea* L.)	Young leaf tissue from 30-day-old plants	—	54–65	>250

Table 15.1 Deficiency, Sufficiency, and Toxicity Levels of Boron on Crops (Continued)

Crop	Part of plant tissue sampled	Deficient	Sufficient	Toxic
		mg B kg⁻¹ dry matter		
		$mg\ B\ kg^{-1}$ dry matter		

B. Crops less responsive to B application; less tolerant to high B levels

Crop	Part of plant tissue sampled	Deficient	Sufficient	Toxic
Corn (Zea mays L.)	Total above-ground plant material at vegetative stage until ear formation	<9	15–90	>100
Oats (Avena sativa L.)	Boot stage tissue	3.5–5.6	14–24	>50
Soybean (Glycine max)	Mature trifoliate leaves at early bloom	9–10	—	63
Barley (Hordeum vulgare L.)	Boot stage tissue	7.1–8.6	21	>46
Wheat (Triticum aestivum L.)	Boot stage tissue	2.1–5.0	8	>16

Adapted from Gupta (1993b).

A

B

Figure 15.2. Boron deficiency symptoms. (A) Twisted, deformed ears with many rows missing and poorly filled kernels. (B) Deformed ears with very few kernels forming. (From *Corn Field Manual*, J.R. Simplot Company Minerals & Chemistry Division, Pocatello, ID, ©1984. With permission.) See Plate 10.

discoloration, cracking, or rotting of fruits, tubers, or roots (Figure 15.2). In cotton square or boll shedding may take place due to B deficiency.

15.1.5. Boron Fertilizers

Boron is applied to both soils and foliage. When applied to soil, B-fertilizers should be uniformly banded or broadcast and incorporated into the

Table 15.2 Commonly Used Boron Fertilizers

Fertilizer	Chemical formula	B content (%)
Borax	$Na_2B_4O_7.10H_2O$	11
Boric acid	H_3BO_3	17
Colemanite	$Ca_2B_6O_{11}.5H_2O$	10–18
Sodium tetraborate		
Fertilizer borate 68	$Na_2B_4O_7.5H_2O$	14–15
Fertilizer borate 68	$Na_2B_4O_7$	21
Solubor	$Na_2B_4O_7{}^{15}H_2O+$	20–21
	$Na_2B_{10}O_{16}.10H_2O$	
Boron frits	Complex borosilicates	2–11

From Tisdale et al. 1993. *Soil Fertility and Fertilizers*, 5th ed., p. 342.
With permission of Prentice-Hall, Inc., Upper Saddle River, NJ.

soil. Foliar application of B is practiced in orchards and also in field crops such as cotton, which need insecticidal sprays; B can be easily mixed with insecticides. The method of application has an important bearing on the rate at which B is applied. A dose of 0.5 to 1 kg B ha^{-1} is recommended for soil application; it may be increased when B is broadcast. For foliar application a dose of 0.1 to 0.5 kg ha^{-1} is generally recommended. Some commonly used B fertilizers are listed in Table 15.2

15.2. MOLYBDENUM

Molybdenum content in soils ranges from 0.2 to 5 mg kg^{-1} and averages about 2 mg kg^{-1} (Tisdale et al., 1985). The essentiality of Mo was established by Arnon and Stout (1939), while its role in N_2-fixation by *Azotobacter chroococcum* was reported by Bortels (1930). Mulder (1948) showed that Mo was also essential for symbiotic N_2-fixation by Rhizobium.

For field crops, response to Mo application was first reported by Anderson (1942) who applied 1 kg ha^{-1} ammonium molybdate to subterranean clover (*Trifolium subterraneum* L.)–perennial rye grass (*Lolium perenne* L.)–*Phalaris aquatica* L. pastures in South Australia. Molybdenum then became an essential component of fertilization packages in Australia. Response of wheat and oats to Mo was later reported from Western Australia (Gartell, 1966). Responses of crops to Mo are closely related to soil properties, and, consequently, there are established geographical patterns of deficiency and of excess. Large areas of North America, Australia, New Zealand, and probably eastern Europe are potentially deficient in Mo (Gupta and Lipsett, 1981). Mo deficiency is expected on well-drained, leached-acid soils and on some sandy soils. In the United States, responses to Mo have been obtained in the Atlantic and Gulf coasts, California, the Pacific Northwest, Nebraska, and the states bordering the Great Lakes.

It may be pointed out that Mo is needed in very small amounts (a few mg ha^{-1}), and often seed reserves of Mo are sufficient to take care of the crop needs. For example, Weir and Hudson (1966) observed that Mo deficiency symptoms in maize were unlikely even in low-Mo soils when the Mo content in seed was >0.08 mg kg^{-1}, but were likely for seed Mo concentrations of <0.02 mg kg^{-1}.

15.2.1. Forms of Molybdenum in Soils

Like any other nutrient, Mo in soils is present in the crystal lattice of primary and secondary minerals, is bound to Al and Fe hydroxides and oxides, complexed with soil organic matter, held as an exchangeable anion, and dissolved in soil solution.

In soil solution Mo is present the following ionic species: MoO_4^{2-}, $HMoO_4^-$, and $H_2MoO_4^0$. The concentration of these three species is highly pH dependent. As can be seen in Figure 15.3, MoO_4^{-2} is the most prevalent species, followed by $HMoO_4^{-1}$. The availability (solubility) of both these species and thus of Mo increases with pH. At pH 6.5 the concentration of MoO_4^{-2} is $10^{-7.5}$ M. The concentration of Mo in soil solution is generally 2 to 8 mg Mg^{-1} (2 parts per billion). When the concentration is less than 4 mg Mg^{-1}, diffusion is the main mechanism of Mo uptake by plants. When the concentration exceeds 4 mg Mg^{-1}, considerable Mo is transported to plant roots by mass flow.

15.2.2. Factors Affecting Molybdenum Availability in Soils

Soil pH, content of Fe and Al hydroxides and oxides, interaction with other ions in soil solution, and environmental factors affect Mo availability in soils.

15.2.2.1. Soil pH and Liming

The effect of pH on the increased availability of MoO_4^{-2} and $MHoO_4^{-1}$ ions has already been discussed. The general relationship between soil particles, Mo, and pH can be written as follows:

$$Soil + MoO_4^{-2} \rightarrow Soil - MoO_4^- + OH^-$$

Thus, in general, there is a tenfold increase in MoO_4^{-2} for each unit increase in soil pH.

Liming will improve Mo availability because pH is increased. On the other hand, acid-forming fertilizers such as ammonium sulfate are likely to decrease Mo availability. Large and continuous application of such fertilizers will increase Mo deficiency.

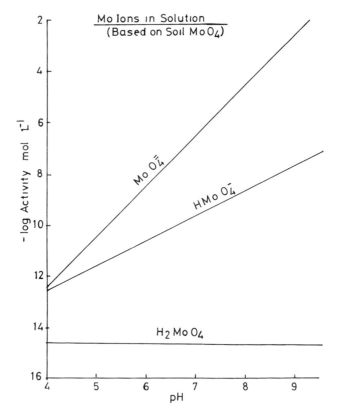

Figure 15.3. Solution species of molybdenum shown in equilibrium with soil-Mo, which has a value of $10^{-7.5}$ M at pH 6.5. (From Lindsay, 1991. *Micronutrients in Agriculture*, J.J. Mortvedt, Ed., p. 110. With permission of SSSA.)

15.2.2.2. Effects of Fe and Al Oxides

Fe oxides (and Al oxides to a lesser degree) can sorb large quantities of MoO_2^{-2}, particularly under acidic conditions when these oxides acquire a positive charge. For example, in a study by Jones (1956) adsorption of Mo by Fe_2O_3 in a solution containing 100 mg Mo decreased from 98 μg at pH 7 to 22 μg at pH 9 after shaking with 100 mg of amorphous Fe_2O_3. Similarly, the adsorption maximum of MoO_4^{-2} on hematite was reduced by 80% if pH was changed from 4 to 7.75 (Reyes and Jurinak, 1967).

15.2.2.3. Interaction with Other Ions in Soil Solution

Of the anions, phosphate increases and sulfate decreases Mo uptake by plants. The increase in Mo uptake by plants due to phosphate may result from increased release of adsorbed MoO_4^{2-} or from the formation of a complex phosphomolybdate ion, which is more soluble and adsorbed more readily by plants (Barshad, 1951). The inhibitory antagonistic effects of SO_4^{2-} on Mo content have been suggested to occur primarily during the absorption process, with some antagonistic mechanism involved during translocation from roots to shoots (Gupta and Lipsett, 1981).

Of the cations, Mg has been reported to increase Mo uptake, while Cu and Mn have antagonistic effects (Tisdale, et al., 1985).

15.2.2.4. Environmental Effects

Molybdenum deficiency is more severe under dry weather conditions, which reduce soil water content and reduce the mobility of Mo in soil solution.

15.2.3. Soil Test for Molybdenum

The acid ammonium oxalate (AAO) procedure proposed by Grigg (1953) is the most commonly used soil extractant for Mo, and a level of >0.2 mg kg^{-1} AAO extractable Mo is identified as adequate. The other extractants proposed are ammonium acetate-EDTA and 1 M $(NH_4)_2$ CO_3 at pH 9.0 for alkaline soils. Anion-exchange resin (Dowey I × 4) has also been suggested as an extractant for Mo. None of these reactants has been very successful in predicting Mo needs of soils or in predicting the response of crops to Mo.

15.2.4. Molybdenum Deficiency Symptoms in Plants

The symptoms of Mo deficiency are closely related to N metabolism since most of the Mo in plants is concentrated in the nitrate reductase enzyme. Deficient plants are suffering essentially from a shortage of protein due to the failure of the initial process of nitrate reduction. These specific symptoms commonly involve deformation of leaves, as in "whiptail" of cauliflower (Figure 15.4). In wheat the symptoms are yellowing of older leaves and, in acute deficiency, the presence of empty heads. In clovers and legumes also yellowing of leaves is the most common symptom. Deficiency and sufficiency levels of Mo for some crops are given in Table 15.3.

15.2.5. Molybdenum Toxicity to Animals

Molybdenum toxicity in plants has been seldom reported and generally is not a matter of concern. However, excessive amounts of Mo (20 to 30 mg kg^{-1} dry matter) in forage can be toxic to animals feeding on such forage. This toxicity disease in animals is known as molybdenosis and is due to Mo-Cu imbalance. This disease is also known as "teart" in England and "peat scours" in New Zealand. For molybdenosis the Cu:Mo ratio in animal feeds from

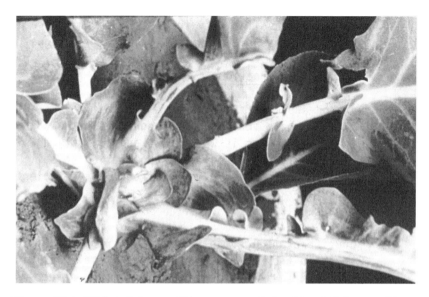

Figure 15.4. Whiptail in cauliflower caused by Mo deficiency. (From Katyal, J.C. and N.S. Randhawa. 1983. *Micronutrients*, FAO Fertilizer and Plant Nutrition Bulletin 7, Food and Agriculture Organization, Rome. With permission.) See Plate 11 following p. 170.

western Canada was found to be 2, whereas in some English pastures the ratio was reported to be closer to 4 (Gupta and Lipsett, 1981). Inflicted animals have bone malformation and stunted growth. Addition of Cu to the diet can cure the disease.

Molybdenosis has been reported from western areas of the United States, western Canada, England, and New Zealand. In some instances the problem is associated with soils developed from marine shales.

15.2.6. Molybdenum Fertilizers

The commonly used Mo fertilizers are ammonium molybdate $(NH_4)_6$ $(Mo_7O_{24} \cdot 2H_2O$, containing 54% Mo); sodium molybdate $(Na_2 MoO_4 \cdot 2H_2O$, containing 39% Mo); molybdenum trioxide $(MoO_3$, containing 66% Mo); and molybdenum frits (1 to 30% Mo).

Molybdenum may be applied to soil or foliage or put on seed prior to sowing. Coating or soaking seeds with Mo is the easiest method and involves the least amount of Mo fertilizers. When applied to soil, the doses vary from 35 to 350 g ha^{-1} depending on the soil and crop.

Table 15.3 Deficient and Sufficient Levels of Molybdenum in Plants

Plant	Part of plant tissue samples	Mo in dry matter (mg kg^{-1}) Deficient	Mo in dry matter (mg kg^{-1}) Sufficient
Alfalfa (*Medicago sativa* L.)	Leaves at 10% bloom	0–26–0.28	0.34
Barley (*Hordeum vulgare* L.)	Blades 8 weeks old	—	0.03–0.07
Beans (*Phaseolus vulgaris*)	Tops 8 weeks old	—	0.4
Beets (*Beta vulgaris* L.)	Tops 8 weeks old	0.05	0.62
Broccoli (*Brassica oleracea* var. *italica* Plenck.)	Tops 8 weeks old	0.04	
Brussels spouts (*Brassica oleracea* var. *gemmifera* Zenker)	Whole plants when sprouts begin to form	<0.08	0.16
Cabbage (*Brassica oleracea* var. *capitata* L.)	Above-ground portion of plants at appearance of a curd	<0.26	0.68–1.49
Cauliflower (*Brassica oleracea* var. *botrytis* L.)	Whole plants before the appearance of curd	<0.11	0.56
Corn (*Zea mays* L.)	At tassel, middle of first leaf opposite and below the lower ear	<0.1	>0.2
Lettuce (*Lactuca sativa* L.)	Leaves	0.06	0.08–0.14
Pasture grass (*Graminae*)	First cut at first bloom	—	0.2–0.7
Red clover (*Trifolium pratense* L.)	Total above-ground plants at bloom	<0.15	0.3–1.59
	Whole plants at bud stage	0.1–0.2	0.45

Table 15.3 Deficient and Sufficient Levels of Molybdenum in Plants (Continued)

Plant	Part of plant tissue samples	Mo in dry matter (mg kg⁻¹)	
		Deficient	Sufficient
Spinach (*Spinacea oleracea* L.)	Whole tops at normal maturity	—	0.15–1.09
Sugar beets (*Beta vulgaris* L.)	Blades shortly after symptoms appear	0.01–0.15	0.2–20.0
Temperate pasture legumes	Plant shoots	—	>0.1
Timothy (*Phleum pratense* L.)	Whole tops at prebloom, head fully emerged from the panicle	0.11	—
Soybeans	Plants when 26–28 cm high	0.19	—
Tobacco (*Nicotiana tabacum* L.)	Leaves 8 weeks old	—	1.08
Tomatoes (*Lycopersicon esculentum* Mill)	Leaves 8 weeks old	0.13	0.68
Tropical pasture legumes in mixture with (*Panicum maximum* cv Gatton)	Plant shoots	—	>0.02
Wheat (*Triticum aestivum* L.)	Whole tops at boot stage	—	0.09–0.18
	Grain	—	0.16–0.20

Adapted from Gupta and Lipsett (1981).

REFERENCES

Anderson, A.J. 1942. Molybdenum deficiency on a south Australian ironstone soil. J. Aust. Inst. Agric. Sci. 8:73–75.

Arnon, D.I. and P.R. Stout. 1939. Molybdenum as an essential element for higher plants. Plant Physiol. 14:599–602.

Barshad, I. 1951. Factors affecting the molybdenum content of pasture plants. I. Nature of soil Mo, growth of plants and soil pH. Soil Sci. 71:387–398.

Berger, K.C. and E. Truog. 1939. Boron determination in soils and plants. Ind. Eng. Chem. Anal. Ed. 11:540–545.

Bingham, F.T., A. Elseewi, and J.J. Oertli. 1970. Characteristics of boron adsorption by excised barley roots. Soil Sci. Soc. Am. Proc. 34:613–617.

Bortels, H. 1930. Molybdan als Katalysator bei der biologischen stickstoffbindung. Arch. Mikrobiol. 1:333–342.

Eaton, F.M. 1935. Boron in soils and irrigation waters and its effect on plants with particular reference to the San Joaquin Valley of California. USDA Tech. Bull. No. 448.

Evans and Sparks. 1983. On the chemistry and mineralogy of boron in pure and mixed systems. A review. Commu. Soil Sci. Plant Anal. 14:827–846.

Gartell, J.W. 1966. Field response of cereals to molybdenum. Nature (London) 209:1050.

Grigg, J.L. 1953. Determination of available molybdenum of soils. N.Z. J. Sci. Technol. 34:405–414.

Gupta, U.C. 1985. Boron toxicity and efficiency: a review. Can. J. Soil Sci. 65:381–409.

Gupta, U.C. 1993a. Boron, molybdenum and selenium, in Soil *Sampling and Methods of Analysis*, M.R. Carter, Ed., CRC Lewis, Boca Raton, FL, pp. 91–99.

Gupta, U.C. 1993b. Deficiency, sufficiency and toxicity levels of boron in crops, in *Boron and Its Role in Crop Production*, U.C. Gupta, Ed., CRC Press, Boca Raton, FL, pp. 137–145.

Gupta, U.C. and J. Lipsett. 1981. Molybdenum in soils, plants and animals. Adv. Agron. 34:73–115.

Gupta, U.C. and J.A. Macleod. 1981. Plant and soil boron as influenced by soil pH and calcium sources on podzol soils. Soil Sci. 131:20–25.

Jones, L.H.P. 1956. Interaction of molybdenum and iron in soils. Science 123:1116.

Lindsay, W.L. 1991. Inorganic equilibria affecting micronutrients in soil, in *Micronutrients in Agriculture*, J.J. Mortvedt, F.R. Cox, L.M. Shuman, and R.M. Welch, Eds., Soil Sci. Soc. Am. Madison, WI. Book Ser. No. 4, pp. 89–112.

Mulder, E.G. 1948. Importance of molybdenum in the nitrogen metabolism of microorganisms and higher plants. Plant Soil 1:94–119.

Patel, N.K. 1967. Effect of various Ca-B and K-B ratios on the growth and chemical composition of aromatic strain of Bidi tobacco. J. Indian Soc. Soil Sci. 14:241–251.

Ponnamperuma, F.N., M.T. Clayton, and R.S. Lantin. 1981. Dilute hydrochloric acid as an extractant for available zinc, copper and boron in rice soils. Plant Soil 61:297–310.

Reyes, E.D. and J.J. Jurinak. 1967. A mechanisms of molybdate absorption on Fe_2O_3. Soil Sci. Soc. Am. Proc. 31:637–641.

Tisdale, S.L., W.L. Nelson, and J.D. Beaton. 1985. *Soil Fertility and Fertilizers*, 4th ed., MacMillan, New York, p. 754.

Weir, R.G. and A. Hudson. 1966. Molybdenum deficiency in maize in relation to seed reserves. Aust. J. Exp. Agric. Anim. Husb. 6:35–41.

16   CHLORINE

Because of the ubiquitous presence of Cl in soil and water, it did not receive much attention as a plant nutrient until 1954 when Broyer et al. (1954) offered convincing evidence for its essentiality. Chlorine as an element is a gas and is taken up by plants as the chloride ion (Cl^-). This is the form involved in most reactions in soil. In further discussion therefore chloride (Cl^-) is generally used.

Chloride has biochemical, as well as osmoregulatory, functions in plants. It is involved in the splitting of water molecules in photosystem II of photosynthesis (Izawa et al., 1969). Several enzymes such as ATPase, alpha-amylase, and asparagine synthetase require Cl^- for stimulation or activation. As regards osmoregulatory functions, when solutes such as Cl^- accumulate within a cell, a water potential gradient is developed across the cell wall causing more water to enter the cell. This leads to increased cell turgidity. If the cell of concern is a stomata, this process leads to stomatal opening when water moves into guard cells. Because Cl^- is very mobile and is tolerated at high concentrations, it is ideally suited to maintain charge balance when cations such as K^+ move across cell membranes (Fixen, 1993). The chloride requirement of plants for biochemical functions is hardly more than 100 mg kg^{-1} (plant dry matter). However, it is usually present in much higher concentrations (2000 to 20,000 mg kg^{-1}), suggesting a role in functions other than those of a biochemical nature.

16.1. CHLORINE IN SOILS

Chlorine occurs in soils as soluble salts such as NaCl, $CaCl_2$, and $MgCl_2$. Most Cl in soils has originated from salts trapped in parent material, marine aerosols, and volcanic emissions. The amount of Cl^- in soil solution ranges from 0.5 to 6×10^3 mg kg^{-1} soil; Cl^- is often the dominant anion in extracts of several saline soils.

16.2. ADDITION OF CHLORINE TO SOILS

Rain and irrigation waters, plus HCl released during volcanic eruptions and from marine aerosols, continuously add Cl to soil. Annual deposits in rainwater vary from 15 to 40 kg ha^{-1} in the inlands and may be 100 kg ha^{-1} or more near a seacoast. Actual quantities of Cl in marine aerosols and rain water near a seacoast depends upon the foam formation on the top of waves, the velocity of wind sweeping inland from the sea, temperature, topography of the coastal region, and the amount, frequency, and intensity of rainfall. Salty droplets or dry salt dust may be whirled to great heights by strong air currents and carried over large distances toward the inlands (Tisdale et al., 1985). Irrigation water can also add large amounts of Cl to soil.

16.3. TESTING SOILS FOR CHLORINE DEFICIENCY

Water-soluble Cl in the soil to a depth of 60 cm is a good indicator of Cl availability (James et al., 1970). In one of the most extensive evaluations of soil Cl, field trials were conducted at 36 locations in 5 years across South Dakota using spring-wheat cultivars known to be Cl responsive. A critical level of 43 kg Cl ha^{-1} divided 83% of the sites into low Cl responsive or high Cl nonresponsive quadrants (Figure 16.1) (Fixen, 1987). In general, the magnitude of response to Cl is on the average more limited than responses to macronutrients.

16.4. CHLORINE DEFICIENCY SYMPTOMS

As in the case of other elements, deficiency symptoms of Cl differ from crop to crop. The common symptoms characteristic of Cl deficiency are wilting of the leaf blade tips followed by chlorosis, bronzing, and necrosis. Restricted root growth with stubby, club-tipped laterals is also characteristic of Cl deficiency. In barley, leaves may remain wrapped in tubular form longer than normal, have slower growth, and become more fragile than normal leaves. In potatoes, leaves become lighter green in color and give a pebbled appearance (vertical protrusions on the upper side of leaflets). Coconut palm (*Cocus mucifera* L.) trees deficient in Cl have older leaves with yellowing and/or orange mottling and dried up leaf tips and edges (von Uexkull and Sanders, 1986).

16.5. CHLORINE TOXICITY SYMPTOMS

Excess Cl in soil and its toxic effects on plants have received more attention than Cl deficiency. The primary influence of high Cl concentration in soil solution is increased osmotic pressure of soil water, resulting in reduced availability of water to plants and consequent wilting (Cl-induced drought). Most fruit trees, berry and vine crops, and ornamental shrubs are specifically sensitive to Cl ions and develop leaf-burn symptoms when Cl concentration in plants reaches about 0.5% (on dry matter basis) (Tisdale et al., 1985). Thickening and rolling in tobacco and tomato leaves may occur.

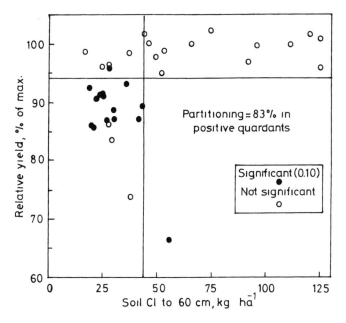

Figure 16.1. Influence of soil Cl content on relative grain yield of spring wheat. (From Fixen, 1987. Crops Soils 39:14–16. With permission.)

16.6. INTERACTIONS WITH OTHER NUTRIENTS

Uptake of Cl is competitively inhibited by NO_3^- and sulfate. Irrigated hard red spring wheat in Montana inoculated with take-all fungus (*Gaeumannomyces graminis* var *tritici*) responded to Cl fertilization when NH_4^+-N was applied and not when NO_3^--N was applied (Engel and Mathre, 1988). Nitrate and Cl⁻ compete with each other for uptake in potatoes, sugarbeets, tomatoes, beans, tobacco, barley, and wheat (Fixen, 1993). For example, in barley the net accumulation of Cl was virtually eliminated in roots and reduced by 40% in shoots when external media (0.5 mol m⁻³ $CaSO_4$ plus 0.5 mol m⁻³ KCl) were supplemented with 0.25 mol m⁻³ $Ca(NO_3)_2$ (Table 16.1) (Glass and Siddiqi, 1985).

Chloride inhibits nitrification in acid soils; concentrations of 46 to 152 mg Cl⁻ kg⁻¹ soil are required for measurable inhibitions of nitrification. Such high concentrations of Cl⁻ can be encountered in saline soils. For example, Christensen and Brett (1985) showed that the NH_4^+-N:NO_3^--N ratio remained above 3:1 for 12 days longer with NH_4Cl than with $(NH_4)_2SO_4$ at pH 5.5. This nitrification-inhibiting property of Cl⁻ was not observed when soil pH was raised to 6.6 by liming.

Chloride appears to interact with P in a complex and largely unknown manner. In some cases P availability is increased by elevating Cl⁻, while in other cases it is not affected or decreases (Fixen, 1993).

Table 16.1 Chloride and Nitrate Concentrations of Roots and Shoots as a Function of Nitrate and Chloride Provision[a]

| Nitrate/Chloride provision in growth medium (mol m⁻³) | | Ion content of tissues (μmol g⁻¹) | | | |
| | | Cl⁻ | | NO₃⁻ | |
KCl	Ca(NO₃)₂	Root[b]	Shoot	Root	Shoot
0.0	0.0	114	21	6	23
0.5	0.0	45	88	3	17
0.5	0.0625	17	57	55	57
0.5	0.125	17	60	58	64
0.5	0.25	15	53	62	70
0.5	0.5	15	52	68	76
0.5	1.0	12	50	73	78
1.0	0.0	52	93	8	21
1.0	0.1	26	73	45	50
1.0	0.5	13	59	73	79
1.0	1.0	11	58	78	89
1.0	2.5	9	46	78	97
1.0	5.0	9	46	81	99
1.0	10.0	8	40	84	107

[a] Nutrients were completely replaced twice daily.

[b] Standard errors of the means were generally within 2% of the means.

Adapted from Glass and Siddiqi (1985).

Application of KCl increases Mn release from soils and its uptake by plants, which may reach toxic levels. In a study in Oregon on a poorly drained soil with pH 4.7 to 5.3 and with appreciable Mn concentrations in its profile, Cl application increased Mn uptake of bush beans (Table 16.2) and sweet corn and resulted in Mn toxicity symptoms in beans (Jackson et al., 1966). Mn toxicity has been offered as a possible explanation for the yield reductions on acidic soils observed from band application of KCl at planting.

16.7. CHLORIDES AND PLANT DISEASES

Chloride application is reported to suppress or reduce the effects of numerous diseases on a variety of crop species. Some of these are listed in Table 16.3.

16.8. CROP RESPONSES TO CHLORIDE FERTILIZATION

Extensive Cl research has been conducted on wheat and barley in the northwestern United States and the Great Plains of North America. Responses in the Great Plains have generally been modest compared with those measured

Table 16.2 The Effect of Additions of Lime and KCl on the Mn Content in Recently Matured Trifoliate Leaves and Yield of Bush Beans (*Phaseolus vulgaris*)

| | No lime | | 6.72 Mg lime ha⁻¹ | |
kg K ha⁻¹	mg Mn kg⁻¹ DM	Grain Mg ha⁻¹	mg Mn kg⁻¹ DM	Grain Mg ha⁻¹
0	798	11.5	528	8.09
55.5	1190	5.8	772	15.7
111	1048	4.6	610	13.8
LSD (P = 0.05)				
Lime at constant K	458	3.7		
K at constant lime	482	4.7		

Adapted from Jackson et al. (1966).

Table 16.3 Plant Disease with Reported Suppression Using Cl Fertilizers

Crop	Diseases
Winter wheat	Take-all root rot, tan spot, stripe rust, Septoria, leaf rust
Spring wheat	Common root rot, tan spot, leaf rust, Septoria
Barley	Common root rot, Fusarium root rot, spot blotch
Durum wheat	Common root rot
Corn	Stalk rot
Pearl millet	Downy mildew
Coconut palm	Gray leaf spot
Potatoes	Hollow heart, brown center
Celery	Fusarium yellows
Rice	Stem rot, sheath blight

Adapted from Fixen (1987).

in the northwestern United States. The results from the Great Plains are shown in Figure 16.2. There appears to be less potential for corn to benefit from Cl fertilization as compared with other cereal crops (Fixen, 1993).

Coconut and oil palms respond markedly to Cl fertilization on low Cl soils, typically occurring at distances greater than 20 to 25 km from the sea (von Uexkull and Sanders, 1986; Ollagnier and Olivin, 1984).

16.9. CHLORIDE FERTILIZERS

Potassium chloride (47% Cl), ammonium chloride (66% Cl), calcium chloride (65% Cl), magnesium chloride (74% Cl), and sodium chloride (66% Cl) are the most common chloride fertilizers.

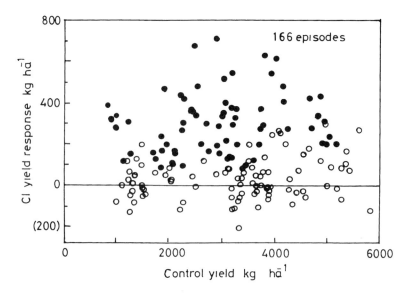

Figure 16.2. Chloride yield response relative to control yields in the Great Plains. ○, Nonsignificant; ●, significant (>67 kg ha⁻¹). Each data point represents a cultivar × site × year episode. (From Engel et al., 1992. Proceedings of the Great Plains Soil Fertility Conference, 4:232–241. With permission of Kansas State University, Manhattan, KS.

REFERENCES

Broyer, T.C., A.B. Carlton, C.M. Johnson, and P.R. Stout. 1954. Chlorine — a micronutrient element for higher plants. Plant Physiol. 29:526–532.

Christensen, N.W. and M. Brett. 1985. Chloride and liming effects on soil nitrogen form and take-all of wheat. Agron. J. 77:157–163.

Engel, R.E. and D.E. Mathre. 1988. Effect of fertilizer nitrogen source and chloride on take-all of irrigated hard red spring wheat. Plant Dis. 72:393–396.

Engel, R.E., H. Woodard, and H.L. Sanders. 1992. A summary of chloride research in the Great Plains, in *Proceedings of the Great Plains Soil Fertility Conference*, J.L. Halvin, Ed., Kansas State University, Manhattan, 4:232–241.

Fixen, P.E. 1987. Chloride fertilization? Recent research gives new answers. Crops Soils 39:14–16.

Fixen, P.E. 1993. Crop responses to chloride. Adv. Agron. 50:107–150.

Glass, A.D.M. and M.Y. Siddiqi. 1985. Nitrate inhibition of chloride influx in barley: implications for a proposed chloride homestat. J. Exp. Botany 36:556–566.

Izawa, S., R.L. Heath, and G. Hind. 1969. The role of chloride ion in photosynthesis. III. The effect of artificial electron donors upon electron transport. Biochem. Biophys. Acta. 180:388–398.

Jackson, T.L., D.T. Westermann, and D.P. Moore. 1966. The effect of chloride and lime on the manganese uptake by bush beans and sweet corn. Soil Sci. Soc. Am. J. 30:70–73.

James, D.W., W.H. Weaver, and H.L. Reeder. 1970. Chloride uptake by potatoes and the effects of potassium chloride, nitrogen and phosphorous fertilization. Soil Sci. 109:48–52.

Ollagnier, M. and J. Olivin. 1984. Effects of potassium nutrition on the productivity of the oil palm. Oleagineux 39:349–363.

Tisdale, S.L., W.L. Nelson, and J.D. Beaton. 1985. *Soil Fertility and Fertilizers*, 4th ed., Macmillan, New York, p. 754.

vonUexkull, H.R. and J.L. Sanders. 1986. Chlorine in the nutrition of palm trees, in Special Bulletin on Chloride and Crop Production No. 2, T.L. Jackson, Ed., Potash and Phosphate Institute, Atlanta.

17   BENEFICIAL ELEMENTS

In addition to the 16 elements (C,H,O; 6 macronutrients—N, P, K, S, Ca, Mg; 7 micronutrients—Fe, Mn, Cu, Zn, B, Mo, Cl) that are considered essential for plant growth, according to the criteria proposed by Arnon and Stout (1939), a number of other elements have been reported to be essential, or at least beneficial, by way of increased growth or improved resistance to diseases or pests for some species. These elements, which include Na, Si, Co, Ni, La, Ce, V, and even Al, are currently considered as beneficial plant nutrients. Recognition of the role of some of these elements resulted from the improvement in research methodology and plant chemical analytical techniques, since some of these elements may be present in concentrations less than a few parts per billion (μg kg^{-1}). More than half the elements in the Periodic Table are known to occur in plant tissues (Asher, 1991). Thus there are a fairly large number of elements about which little is known concerning their roles in the tissue of one or more species; is it merely a coincidence or do they have a specific role?

It may be mentioned that there has been considerably more research on nutrient essentiality in animals than in plants and plant products. Between 1957 and 1973 the essentiality of seven new trace elements (F, Si, V, Cr, Ni, Se, Sn) for warm-blooded animals was established (Schwarz, 1974), and about 19 others are being considered. It is important that plants absorb these elements in adequate quantities, whether they are required for plant growth or not (Asher, 1991). The case for Co in herbage is an example. A brief discussion on a few elements considered beneficial for plants follows.

17.1. SODIUM

Sodium makes up about 2.6% of the earth's crust and is the sixth most abundant element. It is also present as a major component of seawater. Sodium has generally received the attention of the soil scientist because it is a problem element in sodic soils. However, the following findings need consideration.

1. Sodium is required for the conversion of pyruvate to phosphoenol pyruvate in the mesophyll in C_4 plants (Johnson et al., 1988). External Na concentrations $\leq 100 \mu M L^{-1}$ have been shown to be needed for maximum growth of C_4 plants (Brownell, 1968).
2. Sodium is required for assimilation of NO_3^- and conversion to NO_2^- and other intermediates in blue-green alga *Anabaena cylindrica* (Brownell, 1968).
3. Sodium is reported to increase growth of halophytes even when adequate K is present; this has been attributed to increased turgor (Flowers et al., 1977).
4. Some glycophytes, notably sugarbeet (*Beta vulgaris*, L.), respond positively to Na even in the presence of adequate K; Na is also reported to increase sugar concentration (El-Seikh and Ubrich, 1970).
5. Sodium can partly substitute for K in several species, and it is possible that it may be involved in different mechanisms in different species. Two generally reported mechanisms are (a) improved stomatal function (Raghvendra et al., 1976) and (b) osmotic and electrical balance of cells (Asher, 1991).
6. Sodium can overcome the impairment of carbohydrate transport associated with Ca deficiency (Joham and Johanson, 1973).
7. Sodium deficiency of grazing cattle has been reported from several parts of the world (Playne, 1970), and Na concentration in herbage is an important quality factor in relation to animal nutrition.

Thus for halophytes, glycophytes, cultures of blue-green algae, and herbage plants, Na nutrition seems to be important.

17.2. SILICON

Silicon is the second most abundant element in the lithosphere, accounting for about 26% by weight. Silicon is regarded as an essential trace element for the normal growth and development of higher animals since it is involved in the formation of bones and cartilages. Essentiality of Si for the growth of several species of diatom was reported as early as 1933 (King and Davidson, 1933). Regarding the crop plants, the following need to be considered.

1. Silicon has been found necessary for the normal growth of rice (Okawa, 1936, 1937), sugarcane (Fox et al., 1967), barley (Okawa, 1937), and beet (Raleigh, 1939).
2. In rice and sugarcane, application of silicate materials has entered commercial practice; yield increase in sugarcane due to 4.5 t ha^{-1} CaSiO$_3$ application in a study was about 10 t ha^{-1} of millable cane

(Fox et al., 1967). Silicate application resulted in a 3 to 13% increase in rice grain yield in Japan, Korea, and Taiwan.

3. Some of the yield increase in rice could be due to increased resistance to fungal diseases such as blast (Ishizuka and Hayakawa, 1951), to insect attack, and to Mn toxicity (Okuda and Takahashi, 1962).

4. Plants differ widely in respect of Si absorption from soil. The members of Cyperaceae (e.g. *Equisetum arvens*) and rice can accumulate >4% Si in the dry matter of their tops compared with <0.5% in dicots; dryland Graminaceae form an intermediate group (Asher, 1991).

5. Reduced pollen viability in the absence of Si has been reported in tomatoes (*Lycopersicon esculentum*, Mill), cucumbers (*Cucumis sativus* L.), soybeans [*Glycine max* (L.) Merr.], and strawberries (*Fragaria ananassa* Duchesne) (Miyake and Takahashi, 1986).

6. The addition of Si is reported to increase P availability in soils (Suehisa et al., 1963), and this can result in improved plant growth and increased crop yield.

7. The addition of Si may decrease the solubility of Al and heavy metals that may otherwise be present in toxic levels.

The above observations indicate that Si may be essential for the growth of plants of some species, but failure to complete their life cycle has not yet been demonstrated.

17.3. COBALT

Cobalt is involved in N_2-fixation and is therefore essential for legumes (Ahmed and Evans, 1960). Field responses of subterranean clover (*Trifoleum subterraneum* L.) in Australia have been reported (Ozanne et al., 1963). Narrow-leafed lupins (*Lupinus angustifolius* L.) are reported to be especially responsive to Co fertilization. Cobalt is also reported to be essential for N_2-fixation in nonlegumes, for example, in Alnus and Azolla. Essentiality of Co for plants not dependent on N_2-fixation has not been established.

Cobalt is essential for ruminants, where it is involved in the synthesis of vitamin B_{12} (Underwood, 1984). Inadequate dietary Co leads to wasting disease characterized by anemia and the loss of appetite. In the case of breeding ewes Co deficiency can lead to reduced lamb weights at birth and poor lamb survival (Norton and Hales, 1976).

17.4. NICKEL

Nickel like Al has mostly received the attention of soil scientists and agronomists from the viewpoint of its toxic effects on plant growth; the toxicity symptoms resemble Fe deficiency symptoms. A characteristic feature of Ni toxicity in cereals and grasses is a variation in the intensity of chlorosis along the length of the leaf, yielding a series of transverse bands (Anderson et al., 1973).

The discovery that the urease from jackbean (*Canavalia ensiformis* [L.] DC) is an Ni metalloenzyme (Dixon et al., 1975) elevated its status of Ni to that of a functional nutrient. (This term was coined to describe an element that plays a role in plant metabolism, whether or not that role is specific or indispensable). The role of Ni in urea metabolism of plants was reviewed by Walker et al. (1985).

Brown et al. (1987) showed that Ni is essential for grain viability in barley. At grain Ni concentrations <100 μg kg^{-1}, germination percentage decreased linearly with decreasing Ni concentrations. These findings strengthen the claim of Ni as an essential nutrient for plant growth. Mishra and Kar (1974) cited several instances where the application of Ni had improved growth of plants. The essentiality of Ni as a plant nutrient in soybeans, chickpeas (*Cicer arietinum* L.), and temperate cereals (Brown et al., 1987) has been illustrated, but its essential function in higher plants other than in urease metabolism has yet to be established.

17.5. ALUMINUM

The toxic effects of excess Al in acidic soils have been discussed in Chapter 6. Aluminum, when not present in toxic levels yet in abundant amounts, is reported to reduce the toxic level for uptake of Cu, P (Asher, 1991), and Zn.

Stimulatory effects of Al on plant growth have been observed in Al accumulators such as tea (*Camellia sinensis* [L.] Kuntze) as well as non-Al accumulators.

17.6. VANADIUM, LANTHANUM, AND CERIUM

Vanadium is reported to stimulate growth and nitrogenese synthesis in *Anabaena variabilis* cells in the absence of Mo (Yakunin et al., 1991). While excessive amounts can be toxic to microorganisms, some Va is required for their growth. In a study by Lyalikova and Yurkova (1989) 0.6 g L^{-1} of Va was found to be optimal for the growth of microorganisms.

Large-scale use of a fertilizer called Nongle ("Happy Farmer"), containing La and Ce nitrates, is reported from China (Guo, 1987). An increase in yield of the order of 5 to 15%, as well as product quality improvement, is reported in a number of annuals and perennials. Soils containing less than 5 to 10 mg kg^{-1} sodium acetate–acetic acid buffer (pH 8.0) extractable rare earths are considered responsive to Nongle application (Zhu and Liu, 1985). No evidence of essentiality of La and Ce has been produced.

REFERENCES

Ahmed, S. and H.J. Evans. 1960. Cobalt: a micronutrient element for the growth of soybean plants under symbiotic conditions. Soil Sci. 90:205–210.

Anderson, A.J., D.R. Meyer, and F.K. Meyer. 1973. Heavy metal toxicities: levels of nickel, cobalt, and chromium in the soil and plants associated with visual symptoms and variation in growth of an oat crop. Aust. J. Agric. Res. 24:557–571.

Arnon, D.I. and P.R. Stout. 1939. The essentiality of certain elements in minute quantity for plants with special reference to copper. Plant Physiol. 14:371–375.

Asher, C.J. 1991. Beneficial elements, functional elements and possible new essential elements, *Micronutrients in Agriculture*, Soil Sci. Soc. Am., Madison, WI, pp. 703–723.

Brown, P.H., R.M. Welch, and E.E. Cary. 1987. Nickel: a micronutrient essential for higher plants. Plant Physiol. 85:801–803.

Brownell, P.F. 1968. Sodium as an essential micronutrient element for some higher plants. Plant Soil 28:161–164.

Dixon, N.E., C. Gazzola, R.L. Blackley, and B. Zerner. 1975. Jack bean urease (EC 3.5.1.5). A metalloenzyme. A simple biological role for nickel. J. Am. Chem. Soc. 97:4131–4133.

El-Sheikh, A.M. and A. Ulrich. 1970. Interactions of rubidium, sodium, and potassium on the nutrition of sugarbeet plants. Plant Physiol. 46:645–649.

Flowers, T.J., P.F. Troke, and A.R. Yeo. 1977. The mechanism of salt tolerance in halophytes. Ann. Rev. Plant Physiol. 28:89–121.

Fox, R.L., J.A. Silva, O.R. Younge, D.L. Plunknett, and G.D. Sherman. 1967. Soil and plant silicon and silicate response by sugarcane. Soil Sci. Soc. Am. Proc. 31:775–779.

Guo, B. 1987. A new application of rare earths — agriculture, in *Rare Earth Horizons*, Aust. Dept. Industry and Commerce, Canberra, Australia, pp. 237–246.

Ishizuka, Y. and Y. Hayakawa. 1951. Resistance of rice plant to mthe Imodhi (rice blast) disease in relation to their silica and magnesia contents. J. Sci. Soil and Manure, Japan 21:253–260.

Joham, H.E. and L. Johanson. 1973. The effects of sodium and calcium on the translocation of ^{14}C-sucrose in excised cotton roots. Physiol. Plant. 28:121–126.

Johnson, M., C.P.L. Graf, and P.F. Brownell. 1988. The effect of sodium nutrition on the pool sizes of intermediates of the C_4 pathway. Aust. J. Plant Physiol. 15:749–760.

King, E.J., and V. Davidson. 1933. The biochemistry of silicic acid. IV. Relation of silica to the growth of phytoplankton. Biochem. J. 27:1015–1021.

Lyalikova, N.N., and N.A. Yurkova. 1989. The influence of vanadium on microorganisms and their role in the transformation of this element, in Proc. 6th Int. Trace Element Symp., Vol. I, M. Anke, Ed., Molybdenum. Vanadium, Jena, Germany, pp. 74–78.

Mishra, D. and M. Kar. 1974. Nickel in plant growth and metabolism. Bot. Rev. 40:395–449.

Miyaki, Y. and E. Takaheshi. 1986. Effect of silicon on the growth and fruit production of strawberry plants in a solution culture. Soil Sci. Plant Nutr. 32:321–326.

Norton, B.W. and J.W. Hales. 1976. A response of sheep to cobalt supplementation in southeastern Queensland. Proc. Austr. Soc. Anim. Prod. 11:393–396.

Okawa, K. 1936. Investigations on the physiological action of silicic acid for plants. I and II. J. Sci Soil and Manure, Japan 10:95–110; 216–243.

Okawa, K. 1937. Investigations on the physiological action of silicic acid for plants. III. J. Sci. Soil and Manure, Japan 11:23–36.

Okuda, A., and E. Takahashi. 1962. Effect of silicon supply on the injuries of excessive amounts of Fe, Mn, Cu, AsO3, Al, Co in barley and rice plants. J. Sci. Soil and Manure, Japan 33:1–8.

Ozanne, P.G., E.A.N. Greenwood, and T.C. Shaw. 1963. The cobalt requirement of subterranean clover in the field. Aust. J. Agric. Res. 14:39–50.

Playne, M.J. 1970. The sodium concentration in some tropical pasture species with reference to animal requirements. Aust. J. Exp. Agric. Anim. Husb. 10:32–35.

Raghavendra, A.S., I.M. Rao, and V.S.R. Das. 1976. Replacibility of potassium by sodium for stomatal opening in epidermal strips of Commelina benghalensis. Z. Pflanzenphysiol. 80:36–42.

Raleigh, G.J. 1939. Evidence for the essentiality of silicon for growth of the beet plant. Plant Physiol. 14:823–828.

Schwarz, K. 1974. Recent dietary trace element research, exemplified by tin, fluorine and silicon. Fed. Proc. 33:1748–1757.

Suehisa, R.H., O.R. Young, and D.G. Sherman. 1963. Effects of silicates on phosphorus availability to sudangrass grown on Hawaiian soils. Hawaii Agric. Expt. Stn. Tech. Bull 51.

Underwood, E.J. 1984. Cobalt, in *Nutrition Reviews Present Knowledge in Nutrition*, 5th ed, R.E. Olson, Ed., The Nutrition Foundation, Washington, D.C., pp. 528–537.

Walker, C.D., R.D. Graham, J.T. Madison, E.E. Cary, and R.M. Welch. 1985. Effects of Ni deficiency and some nitrogen metabolites in cowpeas (*Vigna unguiculata* L. Walp.). Plant Physiol. 79:474–479.

Yakumin, A.F., N. Chan Van, and I.N. Gogitov. 1991. Effect of molybdenum, vanedium and tungsten on the growth of Anabaena variabilis and its synthesis of nitrogeneses. Microbiology (New York) 60:52–56.

Zhu, Q. and Z. Liu. 1985. Soluble rare earth elements in soils, in *New Frontiers in Rare Earth Science and Application*, Xu, G. and J. Xiao, Eds., Proc. Int. Conf. Rare Earth Develop. App. Beijing, China. Science Press, Beijing, China, pp. 1511–1514.

18 NUTRIENT INTERACTIONS

Success in the control of crop production will be achieved when scientists can reliably produce the yield that the genetic potential of the plant and the solar/radiation at the site make possible. This can be achieved by gaining a full understanding of a particular crop production system—its components, the interactions that impact yields, and the properties and processes of the basic working resources (soil and inputs).

Most soils are generally deficient to some degree in more than a single essential plant nutrient. Unless all deficient elements are supplied in adequate quantities, benefits from the application of even large amounts of a single nutrient are not realized. This limiting of the expression of the effects of one plant nutrient due to limited supply of another nutrient is due to their interaction. A careful study of these interaction effects on different soils and in different crops permits balanced fertilization, which is the key for the realization of production potential of a genotype in any crop. Furthermore, the extent and severity of nutrient deficiencies increase with an increase in the intensity of agriculture in a region, and so does the practical significance of nutrient interactions. This chapter explains the interaction effect and provides some examples of the nutrient interactions in crop production.

18.1. INTERACTIONS

When the effect of one factor is influenced by the effect of another factor, the two factors are said to interact. Furthermore, when the combined effect of two factors is more than their additive effects, the interaction is said to be positive (or synergistic). When their combined effect is less than their additive effects, the interaction is said to be negative (antagonistic).

For example, in a field experiment with mustard (*Brassica juncea* L.) grain yields shown in Table 18.1 were obtained. The response of mustard to N was greater when S was applied, and the response to S was greater when N was applied. The combined effect of N and S was greater than the sum of the individual effects of N and S. Yield with both N and S was increased 1.2 Mg ha^{-1}, compared with 0.99 (0.77 plus 0.22) Mg ha^{-1} when considered separately. Standard statistical procedures are available for determining the

Table 18.1 Grain Yield of Mustard (Mg ha^{-1}) as Influenced by Nitrogen and Sulfur Fertilization

kg N ha^{-1}	kg S ha^{-1}		Response to S
	0	30	
0	0.45	0.67	0.22 (in the absence of N)
90	1.22	1.65	0.43 (in the presence of N)
Response to N	0.77 (in the absence of S)	0.98 (in the presence of N)	

From Dubey and Khan. 1993. Indian J. Agron. 38:270–276.

interaction effects and testing their significance. This is an example of a positive interaction. Recognizing the potential for positive interactions and capitalizing on them is the secret of successful crop production. Examples of negative interactions will be provided later in this chapter.

18.2. INTERACTIONS OF PRIMARY MACRONUTRIENTS

Interactions of primary nutrients have received considerable interest from the agronomists and soil scientists (Prasad et al., 1992; Dev, 1992). A number of researchers have reported data supporting positive N × P interaction. For example, data on N × P interaction in sorghum (*Sorghum bicolor*) at four locations in India having different soils are shown in Figure 18.1. In all four soils the response to N increased when P was applied; the increase was greatest on red soil. Thus in soils that are severely deficient in P, application of N alone will produce only a small increase in yield, much below the potential. Likewise, when N is provided as an ammonium or ammonium-producing fertilizer, the acidifying effect can enhance P solubility and thus provide a positive interaction. In some situations the contribution of N × P interaction can be large enough to overshadow the effects of N or P alone. As an example, the data from field experiments with sorghum and fingermillet (*Setaria italica* L.) grown under dryland agricultural conditions are shown in Table 18.2.

Tropical soils such as ultisols and oxisols are poor in soil P and K, and field experiments on such soils provide interesting data on N × K and P × K interactions. Data from Brazil (Figure 18.2) (PPI, 1988) show a positive N × K interaction in rice. A good response to K was obtained only when adequate N (90 kg ha^{-1}) was applied. Also, the response to N increased as the level of K was increased; the highest yield of rice was obtained when both N and K were applied at 90 kg ha^{-1}. Data on P × K interaction in pearl millet (*Pennisetum typhoides* L.) on similar soils are provided in Figure 18.3. Response to K was obtained when adequate P was applied. Again the highest dry matter was

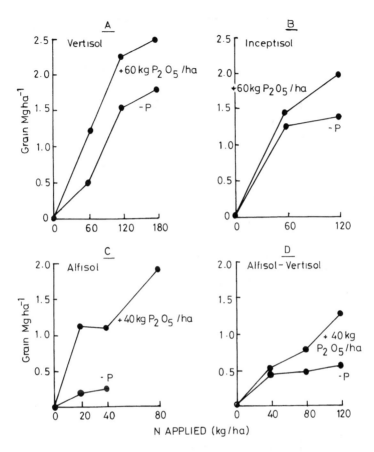

Figure 18.1. N × P interaction in sorghum at four locations in India.
A. Mishra and Singh, 1978; B. Roy and Wright, 1973; C.
Venkateswarlu and Rao, 1978; D. Nagre and Bathkal, 1979.
(From Sharma and Tandon, 1992. *Management of Nutrient
Interactions*, pp. 1–20. With permission ot Fertilizer Devel-
opment and Consultation Organization, New Delhi, India.)

Table 18.2 Response of Sorghum and Finger Millet to N and P Alone
and in Combination (N + P)

Crop	Response to			Estimated contribution of		
	N (kg ha⁻¹)	P (kg ha⁻¹)	N + P (kg ha⁻¹)	N (%)	P (%)	NP interaction (%)
Sorghum	110	490	1570	7	31	62
Finger millet	390	170	1300	30	13	57

From Tandon. 1992. *Fertilizer Management in Dryland Agriculture*, p. 59. With permission of Fertilizer Development and Consultation Organization, New Delhi, India.

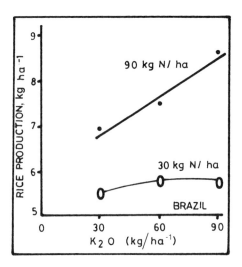

Figure 18.2. Effects of N and K fertilization on rice. (From *Better Crops International*, December 1988, p. 9. With permission from Phosphate and Potash Institute.)

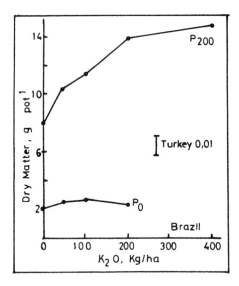

Figure 18.3. Effects of P and K on pearl millet. (From *Better Crops International*, December 1988, p. 10. With permission from Phosphate and Potash Institute.)

produced when both P and K were applied. Thompson et al. (1986) from North Dakota reported a positive response to P × K interaction in irrigated alfalfa (Table 18.3).

Table 18.3 Alfalfa Forage Yield and Return
as Influenced by P, K, and PK Application
(Averaged over 1983, 1984, and 1985)

Treatment (kg ha⁻¹)		Forage	Return
P_2O_5	K_2O	(Mg ha⁻¹)	($ ha⁻¹)
0	0	11.2	20.19
50	0	11.9	43.87
0	100	11.6	−1.50
50	100	12.7	132.37
100	250	13.3	95.75

From Thompson et al. (1986).

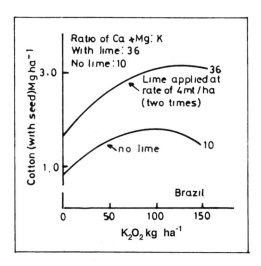

Figure 18.4. Effect of K on cotton 5 years after liming. (From *Better Crops International*, December 1988, p. 11. With permission from Phosphate and Potash Institute.)

Ca, Mg, and S may interact among themselves, but more interesting interactions are between them and NPK. N × S interaction has already been discussed (see Table 18.1). Darst et al. (1993) from Alabama reported a significant lime × P interaction in crimson clover (*Trifolium incarnatum* L.); response to lime was obtained only when adequate P was applied.

A liming and K fertilization experiment on a red latosol showed a positive interaction in cotton; the response to K was much greater when Ca + Mg:K ratio was 36 than when it was only 10 (Figure 18.4). In experiments where K application increases corn yield, generally a negative K × Ca and K × Mg interaction is observed in respect of Ca, Mg, and K concentration in leaves (Figure 18.5).

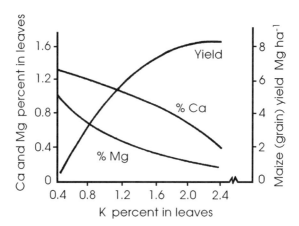

Figure 18.5. Potassium, Ca, and Mg content in corn leaves and effect on grain yield. (From *Better Crops International*, December 1988, p. 71. With permission from Phosphate and Potash Institute.)

18.3. INTERACTIONS OF MICRONUTRIENTS

Micronutrients may interact among themselves, or with secondary or primary nutrients. One of the most studied and reported interactions is P × Zn. Takkar et al. (1976) showed that increasing the level of P above 44 kg ha^{-1} significantly decreased roots, stems, leaves, shoots, and grain yield in corn — a negative interaction. This effect could be overcome only when adequate Zn was applied (Figure 18.6). Similar data for rice are shown in Figure 18.7 (Tiwari and Pathak, 1978). The reasons for phosphate-induced Zn deficiency include (1) P depressed soil-Zn availability; (2) P reduced Zn absorption by roots and subsequent retardation in Zn translocation from roots to shoots; (3) Zn dilution in plant tops arising from P induced growth response; (4) P-Zn imbalance–related metabolic disorder(s); and (5) P activated interference in Zn functioning (Olsen, 1972; Katyal et al., 1992).

Karle and Babula (1985) reported a highly positive B × S interaction in ground nut (*Arachis hypogea* L.) on a S- and B-deficient Vertisol at Parbhani, India. The contribution of B x S interaction was 22% in kernel yield and 43% in oil yield (Figure 18.8); the interaction was much more pronounced in oil yield. Sulfur increased oil yield by 73% and B by 29%, while the combined application of B and S increased oil yield by 181%.

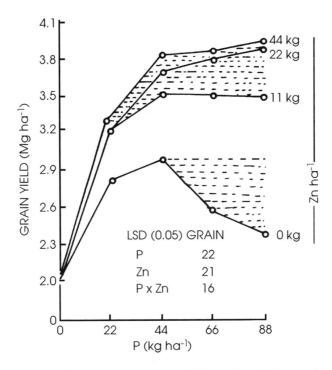

Figure 18.6. Corn grain yield (average of 3 years) as influenced by P and Zn interaction. The shaded area indicates P-induced Zn deficiency. (From Takkar et al. 1976. Agron. J. 68:942–946.)

Another example of the negative interaction is that of Cu and Mo. Application of high doses of Cu decreases Mo concentration in alfalfa, while increased Mo application results in decreased Cu concentration in rice. Because micronutrients are associated with enzymatic activities in plants, the reduced concentration of one micronutrient due to increased application of another micronutrient expresses itself in such activities. For example, excess Cu not only decreases Mo concentration in corn grown in refined sand at a low Mo level, but also inhibits its metabolic functioning by decreasing the protein-N concentrate and nitrate reductase activity and increasing peroxidase activity (Agarwal, 1989).

A negative interaction between B and Ca has been reported in wheat, pea, cotton, and alfalfa (Fox, 1968; Chauhan and Powar, 1978).

The above are only a few examples. The literature on agronomy, plant nutrition, and soil science contains numerous data on nutrient interactions. Research on intensive agriculture and farming systems will add more such examples to the literature.

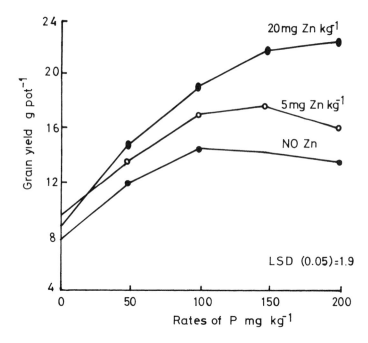

Figure 18.7. Interaction effect of Zn and P on rice grain yield. (From Tiwari and Pathak 1978. J. Indian Soc. Soil Sci. 26:385–389.)

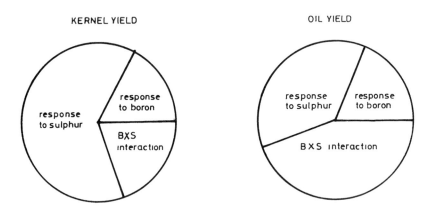

Figure 18.8. Partitioning of response of groundnut to S and B in groundnut. (From Karle and Babula, 1985; Tandon, 1992.)

REFERENCES

Agarwal, S.C. 1989. Copper-molybdenum interaction in maize. Soil Sci. Plant. Nutr. 35:435–442.

Cahuhan, R.P.S. and S.L. Powar. 1978. Tolerance of wheat and pea to boron in irrigation water. Plant Soil 50:145–149.

Darst, B.C., W.R. Thompson, Jr., and J.T. Thompson. 1993. Forage legumes respond to lime and phosphorus. Better Crop with Plant Food 77(3):24–25.

Dev, G. 1992. Interaction of phosphorus with other nutrients and crop husbandry factors. Fertil. News 37(4):59–63.

Dubey, O.P. and R.A. Khan. 1993. Effect of nitrogen and sulfur on dry matter, grain yield and nutrient content at different growth stages of mustard (*Brassica juncea* L.) under irrigated vertisol. Indian J. Agron. 38:270–276.

Fox, R.H. 1968. The effect of calcium and pH on boron uptake from high concentrations of boron by cotton and alfalfa. Soil Sci. 106:435–439.

Karle, B.G. and A.V. Babula. 1985. Effect of B and S on yield attributes and quality of groundnut. Proceedings of Tamil Nadu Agricultural University (India) — Fertilizers and Chemicals, Travancore (TNAU-FACT) Seminar on Sulphur, Coimbatore, pp. 158–168.

Katyal, J.C., S.K. Das, K.L. Sharma, and N. Saharan. 1992. Interactions of Zn in soil and their management. Fertil. News 37(4):27–33.

Mishra, A.P. and V.S. Singh. 1978. Response of sorghum to nitrogen and phosphate as fertilizer/application under Brudelkhand region. Indian J. Agron. 23:263–365.

Nagre, K.T., and B.G. Bathkal. 1979. Studies on the NPU requirements of sorghum hybrid CSH-1 under Kharif rainfed conditions. J. Maharashtsu Agric. Univ. 4:113–115.

Olsen, S.R. 1972. Micronutrient interactions, in *Micronutrients in Agriculture*, J.J. Mortvedt, P.M. Giordano, and W.L. Lindsay, Eds., Soil Sci. Soc. Am., Madison, WI, pp. 243–264.

Phosphate and Potash Institute. 1988. Effects of N and K fertilization in rice crop, p. 9; Effects of P and K in pearl millet, p. 10; Effects of K on cotton, five years after liming, p. 11; Potassium, Ca, Mg content in corn leaves and effect on grain yield, p. 71. *Better Crops International*, December 1988 issue.

Prasad, R., S.N. Sharma, and S. Singh. 1992. Fertilizer nitrogen interactions in crop production. Fertil. News 37(4):75–83.

Roy, R.N. and B.C. Wright. 1973. Sorghum growth and nutrient uptake in relation to soil fertility. I. Dry matter accumulation patterns by various plant parts. Agron. J. 66:5–10.

Sharma, P.K. and H.L.S. Tandon. 1992. Nitrogen and phosphorus in crop production. in *Management of Nutrient Interactions*, H.L.S. Tandon, Ed., Fertilizer Development and Consultation Organization, New Delhi, pp. 1–20.

Takkar, P.N., M.S. Mann, R.L. Bansal, N.S. Randhawa, and H. Singh. 1976. Yield and uptake response of corn to zinc as influenced by phosphorus fertilization. Agron. J. 68:942–946.

Tandon, H.L.S. 1992. *Fertilizer Management in Dryland Agriculture*, Fertilizer Development and Consultation Organization, New Delhi, p. 59.

Thompson, C.R., D.L. Dodds, and B.K. Hong. 1986. Potash and phosphate increase yield and profits from irrigated alfalfa. Better Crops with Plant Food 70:6–7.

Tiwai, K.N. and A.N. Pathak. 1978. Zinc-phosphate relationship in submerged rice in an alluvial soil. J. Indian Soc. Soil Sci. 26:385–389.

Venkateswarlu, J. and U.M.B. Rao. 1978. Relative response of different crops to fertilizer P in semi-arid red soils. Indian Soc. Soil Sci. Bull. 12:404–407.

19   ORGANIC MANURES

Environmental issues and the concern over the sustainability of the present national agricultural systems have stimulated interest in the integrated use of organic manures and chemical fertilizers. China for centuries has made the best use of recycled organic matter, animal manure, night soil, and composted crop residue. Even after the introduction of high-yielding varieties of cereals and the consequent use of large amounts of chemical fertilizers (China now ranks first in the consumption of nitrogen and phosphatic fertilizer), Chinese agriculture continues to use organic manures (Figure 19.1), providing sustenance to this enterprise. Long-term experiments in India (Sarkar et al., 1989) have also clearly shown that sustainable crop production is possible only when farmyard manure (FYM) is applied along with balanced NPK fertilization and lime (on acidic soils) (Figure 19.2). The U.S. Congress appropriated over $8 million to the U.S. Department of Agriculture (USDA) in recent years for research on a program known as Low-Input Sustainable Agriculture (LISA), which includes substitution of legumes in rotation with other crops to supply N, integrated livestock enterprises to supply manure as a nutrient source for crops, and the use of mechanical-biological pest control.

Organic manures include materials largely of plant or animal origin in different states of decomposition that are added to soil to supply plant nutrients and improve soil physical properties. Organic manures include animal manure, crop residues, logging and wood manufacturing residues, industrial organic wastes such as those from paper and sugar industries, sewage sludge, and residues from the food-processing industry. In the United States estimates of organic residues are at about 694 million metric tons per year (Table 19.1). Typical values for chemical composition for a number of organic wastes are presented in Table 19.2.

19.1. CROP RESIDUES

About 1000 million tons/annum crop residues are globally produced from cereals alone. In addition, there are crop residues from fiber crops such as cotton and linseed, sugar crops such as sugarcane and sugar beet, and grain

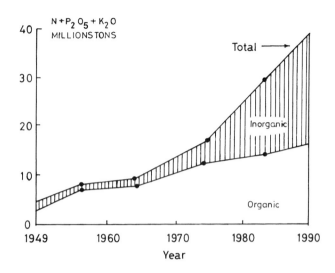

Figure 19.1. Trends in nutrients applied as organics and inorganics in
                 Chinese agriculture. (From vonUexkull and Mutert, 1993.)

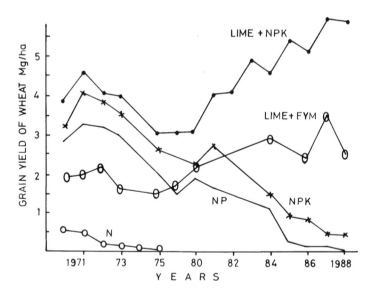

Figure 19.2. Grain yield of wheat at Ranchi (India) as affected by fertil-
                 ization with N, NP, NPK, lime + NPK, and lime + FYM.
                 (From Sarkar et al., 1990. Fert. News, 34(4):71–80.) With
                 permission.

**Table 19.1 Principal Organic Residues in the United States,
Their Annual Production, Percentage of Production Applied to Soils,
and Percentage of the Particular Residue to All Organic Residues**

Organic residue	Total production per year		
	Thousands of metric tons (dry-weight basis)	Percentage of each residue applied to soils	Percentage of all residues produced
Crop residues	448,740	68	70.4
Animal manures	158,700	90	23.0
Logging and wood manufacturing residues	32,390	5	4.7
Industrial organic residues	7,450	3	1.0
Sewage sludge and septage	3,960	23	0.5
Food-processing residues	2,900	13	0.4
Total	694,179		100

From USDA (1978).

**Table 19.2 NPK Content of Some Organic Manures/Residues
(% Dry Weight)**

Manure/residue	Nutrient content		
	N	P_2O_5	K_2O
Animal manure			
Cow dung	1.23	0.55	0.69
Buffalo dung	1.91	0.56	1.40
Pig manure	2.80	1.36	1.18
Chicken manure	3.77	1.89	1.76
Duck manure	2.15	1.13	1.15
Agricultural wastes			
Rice straw	1.70	0.37	2.92
Sugarcane	0.55	0.09	2.39
Corn cobs	0.53	0.15	2.21
Beans	2.30	0.54	2.92
Agro-industrial wastes			
Sugarcane	0.87	0.25	0.98
Saw dust	0.51	0.16	0.43
Coconut hull	0.61	0.14	2.03
Chaffs (e.g., pineapple)	1.23	4.03	1.29
Municipal waste			
Household garbage	0.98	1.04	1.06

From Schumann. 1994. Agrochemicals, News in Brief 17(2):24–31. With
permission.

legumes. With the widespread use of the combine harvesters in developed countries, crop residues remain in the field and must be managed to provide the greatest advantage possible, especially for water conservation, erosion control, and maintenance of soil organic matter. In some developed countries (United Kingdom, Canada, and Australia) wheat straw is often burned, whereas in other countries (such as West Germany) there are strict laws against burning (Staniforth, 1979).

In less affluent countries, such as those in southern Asia and Southeast Asia, grain is directly used for human consumption, and crop residues are the main source of fodder for animals. In addition, residues from crops such as pigeonpea and cotton are used as cooking fuel and thatching for dwelling huts and cattle sheds.

In many regions of the United States leaving cereal residues in situ has been utilized to control soil erosion and conserve water (Prasad and Power, 1991). Tillage and planting machinery were developed and widely used for this purpose. When adequate weed control is achieved on well drained soils, crop yields of no-tillage are usually similar to or better than those obtained with conventional tillage, especially in drier regions. In wetter and cooler regions of the world or on poorly drained soils, crop yields may be lower when crop residues are left in situ and no preplanting tillage is practiced. In general, leaving crop residues on the surface as in no-tillage farming results in an accumulation of organic carbon, total N, available P, K, and some other plant nutrients in the surface 5 cm of soil.

A number of workers have also reported improvement in many soil physical properties resulting from leaving crop residues in the field (Prasad and Power, 1991). In those countries where crop residues are routinely removed for use as fuel or bedding, soils eventually become depleted in soil organic matter and less desirable physical properties usually develop.

19.2. ANIMAL MANURES

Most countries have fairly large numbers of livestock, which excrete millions of tons of dung and urine. Global estimates indicate that there are 1707 million heads of sheep and goats, 1279 million head of cattle, 149 million head of buffalo, and 12107 million chickens (FAO, 1994). Obviously, large amounts of excreta are produced, which if properly utilized can contribute significantly toward meeting the need for plant nutrients and for building up soil fertility. In the United States, an estimated 175 million tons of manure are excreted each year by all domestic livestock and poultry (Table 19.3). Nitrogen content in animal manure (dry weight basis) varies from 3 to 4% in poultry to 1 to 2% in beef/dairy cattle, while P content varies from 1 to 2% in poultry to 0.2 to 1.0% in beef/dairy cattle (Table 19.4).

In the United States, 65% of the dairy cattle, 80% of the swine, and nearly all poultry are fed in confinement. However, because of the large number of animals involved (over 30 million head annually), confined, beef-cattle feed-

Table 19.3 Annual Total and Confined Production of N and P in Livestock and Poultry Manure in the United States and the Fertilizer Value

	Beef	Dairy	Swine	Poultry	Turkey	Total
Total production						
Number of head (million)	189.3	10.3	55.5	1044.3	73.8	
Manure produced (million ton)						
N	10.65	0.98	0.83	0.43	0.10	12.99
P	3.44	0.17	0.29	0.13	0.04	4.07
Fertilizer value (million $)[a]						
N	4696	432	368	188	44	5728
P	3487	175	294	129	37	4122
Confined production						
Number of head (million)	11.5	10.0	51.5	1021.0	73.8	
Manure produced (million ton)						
N	0.64	0.94	0.76	0.41	0.10	2.85
P	0.21	0.17	0.27	0.12	0.04	0.81
Fertilizer value (million $)[a]						
N	287	416	336	184	44	1267
P	212	166	276	120	37	811

[a] N at $0.45 kg^{-1}; P at $1.03 kg^{-1}.

From USDA (1990); White (1989).

lots are the primary source of livestock manure in the United States. Consequently, environmentally safe utilization or disposal of livestock manure is an important public concern. The safest way of utilization is incorporation in soil. Barnyard manure, because of its water content, has a fertilizer grade often less than 1-0.7-1 (Miller and Donahue, 1992). However, when applied in quantities such as 20 tons/ha/yr, manures can add substantial amounts of plant nutrients. Moreover, because they also contain substantial amounts of micronutrients, continuous use of organic manures is a good prophylactic measure against micronutrient deficiencies that may result from continuous use of high-analysis chemical fertilizers. This is well supported by the data from many long-term experiments such as those conducted in India (Nambiar, 1994).

Use of livestock manure as a nutrient source for crop production is not without problems. The low nutrient content of manure restricts greatly the

Table 19.4 Typical Composition of Selected Animal Manures (Dry-Weight Basis)

Constituent	Beef/Dairy (%)	Poultry (%)	Swine (%)	Sheep (%)
Nitrogen (N)	2–8	5–8	3–5	3–5
Phosphorus (P)	0.2–1.0	1–2	0.5–1.0	0.4–0.8
Potassium (K)	1–3	1–2	1.0–2.0	2.0–3.0
Magnesium (Mg)	1.0–1.5	2–3	0.08	0.2
Sodium (Na)	1–3	1–2	0.05	0.05
Total soluble salts	6–15	2–5	1–2	1–2

From Miller and Donahue. 1992. *Soils—An Introduction to Soils and Plant Growth*, 6th ed., pp. 196–211. With permission of Prentice Hall of India Put. Ltd., New Delhi.

distance it can be economically transported, often no more than 10 km. Chemical composition of manure is highly variable, so it is difficult to apply a specific amount of nutrients when manure is spread. Mineralization of manure N is dependent on many factors and not well controlled by the producer. Thus there is potential for nitrate leaching.

19.3. COMPOSTING

Composting is the microbiological conversion of biodegradable organic wastes to stable humus by indigenous microflora, including bacteria, fungi, and actinomycetes, which are widely distributed in nature. The major objectives in composting are to stabilize putrescible organic matter, to conserve as much of the plant nutrient and organic matter as possible, and to produce a uniform, relatively dry product suitable for use as a manure. In composting, factors such as C:N ratio of the material, water content, aeration, pH, and ambient temperature regulate the prevalence and succession of microbial population. A number of composting processes have been developed over the world (Gaur and Sadasivan, 1993; Tyler, 1970). These can be broadly classified as either aerobic or anaerobic. Anaerobic decomposition results in only a partial breakdown of organic matter and is generally associated with a disagreeable odor. Adequate aeration is essential in aerobic composting. To achieve optimum aerobic decomposition, the water content of the organic material should be between 50 and 60% (wet weight basis).

In composting there is an intrinsic relationship of temperature and pH variation with time. During the early mesophilic stage, pH decreases to about 5 and fungi are the dominant organisms. However, as the temperature of the composting mass increases (thermophilic stage), there is a corresponding increase in pH, and bacteria and actinomycetes are most active. The maximum pH rises to about 8.0, synchronized with the temperature peak during the thermophilic stage. Thereafter, pH usually levels off at values above 7.0.

Conventional methods of composting require a long period to produce good compost, often 8 weeks or more. Recent research has shown that inoculation with mesophilic cellulolytic fungi such as *Aspergillus niger* and *Penicillium* sp. can considerably hasten the process of composting (Gaur and Sadasivan, 1993). Often 25 to 50% of the carbon and 10 to 40% of the nitrogen in the original material is lost to the atmosphere in the composting process. While composts can be usefully employed in field crop production, they have found more favor with horticultural crops and commercial floriculture and in gardening.

In recent years there has been considerable interest in the use of earthworms for composting. This practice is called vermi-composting (Bhawalkar and Bhawalkar, 1993). Worms used are *Lumbricus rubellus* or some other species. The earthworms are commercially raised and multiplied in shallow wooden boxes (45 × 60 × 20 cm) provided with drainage holes. Compost pits measuring approximately 3 × 4 × 1 m deep are dug with sloping sides, and are filled with organic residues such as straw, animal manure, green wastes, or leaves. The earthworms from the wooden boxes are emptied onto the surface of the compost pits, and the worms immediately bury in the compost, helping decomposition of organic residues. When the compost is used, earthworms are removed and are either kept in wooden boxes for further breeding or are transferred to another compost pit. Worm compost is becoming quite popular in Asian countries.

Studies at Rothamstad Experimental Station in the United Kingdom have also shown that earthworms (*Eisenia foetida*) can break down organic wastes into peatlike materials rich in available nutrients and with good water-holding capacity and porosity (Edwards, 1983). These peatlike materials have considerable potential in horticulture as a plant growth medium. Methods of obtaining maximum waste turnover in 2 to 4 weeks under controlled water and temperature conditions have been developed.

19.4. ORGANIC FARMING

The management in ecofriendly sustainable agricultural systems (a broad term encompassing organic farming) includes the use of vegetative cover as an effective soil and water conservation measure. This requirement is met through the use of no-till practices, mulch farming, use of cover crops, or other such practices. Plant nutrients are provided through organic manures, compost, and legumes. Nutrient recycling mechanisms include the use of crop rotations, crop/livestock mixtures, appropriate tree/crop combinations in agroforestry, and intercropping systems involving use of legumes (Altieri, 1992). Different practices provide different pools and fluxes of carbon and nutrients in the soil and have varying effects on biological activity and biodiversity of soil organism communities. For example, Hendrix et al. (1986) showed that for no-till soils in Georgia (U.S.), biological activity was dominated by fungi and earthworms, whereas the biota of the conventional tilled soils was dominated by bacteria, nematodes, and enchytraeids.

Earthworms have received considerable attention in the context of organic farming, and farming systems (vermi-culture) have been developed (Bhawalkar and Bhawalkar, 1993). In typical agricultural systems, earthworm populations are integral to the functioning of the system, and the agronomic value of earthworms is difficult to define. Direct N contributions of earthworms as a proportion of the annual net mineral N flux ranges from 7.9 to 27.3% for arable crops in the Netherlands, 3.4 to 17.6% for a Polish pasture, and 18 to 24.5% in New Zealand grasslands (Knight et al., 1989). However, it must also be pointed out that earthworms can also contribute toward N losses by denitrification and leaching (Knight et al., 1989). One must therefore look at the overall impact for a given agroecological situation.

In the humid and subhumid/semiarid tropics, all factors involved in crop production variables often operate over a wide range of extremes. Compared with temperate regions, storm events are often more erosive, leaching due to heavy monsoons or other rains may be more intensive, dry periods limiting plant growth are often longer or more extreme, and highly weathered soils with low inherent fertility are more prevalent (Anderson, 1994). Furthermore, decomposition of organic residues is more rapid, and maintenance or enhancement of soil organic matter as a source of plant nutrients is more difficult and demands large and frequent inputs. In these regions the increased soil biological activity and community diversity that results from reduced tillage is offset by the low organic resource base maintaining the biota and soil structure (Figure 19.3). Hence, although the principles of sustainable farming practices generally apply to tropical as well as temperate regions, in practice the opportunities for many tropical farmers to optimize nutrient and organic matter management are limited by environmental as well as social and economic constraints (Anderson, 1994).

Organic farmers rely heavily on composting, manuring, crop rotation, intercropping, mulching, and hand weeding. Many Japanese organic farmers intercrop several vegetables and fruits such as potatoes, sweet potatoes, carrots, eggplants, garlic, leeks, and strawberries (Ahmed, 1994). Gardeners were probably the first to develop organic farming (Tyler, 1970). In the United Kingdom the principle guiding organic farming nutrient inputs is that mineral forms must be water soluble, and composts, manures, and slurries are also allowed (Stickland, 1990). Conventional farming by contrast uses many soluble salts of nutrients that are very easily carried by soil water to the plant.

Crop yields under organic farming are often lower than under conventional farming; in some cases they are significantly lower because inputs for the organic farmer are usually less. With good management, however, yields very close to those obtained in conventional farming can be obtained. In the United Kingdom a conventional 7.0 ton ha^{-1} farmer is expected to produce 5.5 to 6.0 ton ha^{-1} with organic farming on similar soils and with the same crop varieties (Stickland, 1990). In some Japanese studies rice grain yields were even slightly greater with organic than with conventional farming. Data from Japan on

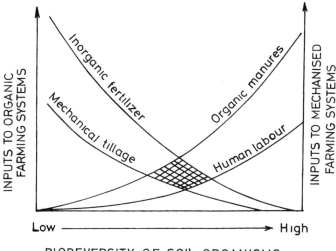

BIODEVERSITY OF SOIL ORGANISMS

Figure 19.3. Schematic representation of the emergence of soil fauna effects on soil acidity with shift from technology-based agriculture management to production based on organic sources. (From Anderson, 1994. *Soil Resilience and Sustainable Land Use***, D.J. Greenland and I. Szaboles, Eds., pp. 267–289. With permission of CAB International, Wallingford, England.)**

vegetable crops are given in Table 19.5. An important feature of organic farming is the premium on quality, especially with vegetables (Table 19.6). When carefully marketed, this can result in a higher price for produce from organic farmers, which can also compensate for lower yields. Consumers in Europe, Japan, and other countries are willing to pay higher prices for agricultural products from organic farms.

The choice of using organic vs. conventional farming depends upon a number of factors, including the availability of land; the population density and the associated demand for food, fiber, feed, and fuel; the resources for gathering and distributing animal manure and other organic residues to farm fields; and the capability of consumers to pay higher prices for farm produce. Also, many choose organic farming because they feel it is more sustainable and does less damage to the environment.

19.5. INTEGRATED NUTRIENT MANAGEMENT

The exclusive and large scale use of chemical fertilizers, occurring after the development of hybrids in the United States and high yielding hybrids/composites/varieties in other parts of the world, is at one extreme, while the concept of organic farming is at another extreme. Because of the

Table 19.5 Yields of Selected Vegetables under Organic and Conventional Farming (Mg ha⁻¹)

Vegetable	Miyoshi village (organic farming)	Chiba prefecture (conventional farming)	Miyoshi yield as % of Chiba average
Cabbage	1.3–3.8	4.8	27–79
Onion	0.6–2.0	2.3	26–87
Radish	4.7–4.8	5.0	94–95
Carrot	1.3–1.5	3.7	35–40
Tomato	1.9–3.0	4.0	47–75
Eggplant	0.8–2.8	2.9	27–96
Green pepper	0.8–1.0	3.4	23–29
Cucumber	1.8–2.7	2.6	69–104

From Ahmed. 1994. Agrochemicals News in Brief 17(2):16–23. With permission.

Table 19.6 Quality and Fresh Weight of Lettuce under Organic and Conventional Farming

Farming type	Brix (%)	Reducing sugar (%)	Total vitamin C (mg/100 g)	Nitrate N (ppm)	Fresh weight (kg)
Organic	3.4	1.91	7.5	263	300
Conventional	2.7	1.30	5.5	615	386

From Ahmed. 1994. Agrochemicals News in Brief 17(2):16–23. With permission.

increasing food demand for the world (especially in densely populated countries) and because of the perceived lack of sustainability of present intensive and highly productive farming systems, the best course to follow might be the use of integrated nutrient management systems involving judicious use of both chemical fertilizers and organic manures. This system is currently practiced in China and elsewhere. Integrated nutrient management also includes the use of biofertilizers and legumes in crop rotation. Many publications show that this conjunctive use of nitrogen and other nutrients results in the most efficient use of nutrients. Figure 19.4 shows how the nitrogen needs of agriculture in India can be partly met from sources other than chemical fertilizers.

The common assumption among some environmentalists, soil scientists, and agricultural researchers, and especially among those from privileged countries, is that the next step toward improvement of soil fertility and desired crop production is to introduce so-called low-input sustainable agriculture (LISA) technology. Some of the practices recommended include mulch farming, ley farming, alley cropping, and related practices. The point often overlooked is that low-input technologies are often knowledge-intensive and require many skills in crop management that many farmers do not possess.

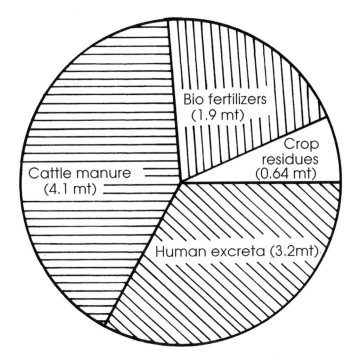

Figure 19.4. Potential of N from organic and biological souces in India. (From Gaur et al., 1984. *Organic Manures,* **p. 159. Indian Council of Agricultural Research, New Delhi.)**

The large scale use of only organic manure as the sole source of nutrients has several problems. First of all, because of low, variable, and generally unbalanced nutrient contents, it is difficult to provide the proper nutrient balance to meet crop requirements with bulky organic matter. Animal manures create unique problems in application that increase farm workload and management requirements. Bulky, organic manures can be applied only during a few weeks or months each year, usually before seeding or after harvesting a crop. Nutrient deficiencies in crops that arise during the crop growth period can seldom be corrected with bulky, organic manures. Large-scale farmers in advanced countries also find problems in scheduling the application of bulky, organic manures with seeding and other management practices in crop production.

The use of large amounts of animal manures can also lead to environmental problems, especially a buildup of nitrates and phosphates in groundwater. After careful examination of groundwater nitrate data from several agencies in the United States, Spalding and Exner (1993) concluded that the highest incidence of contamination in groundwater occurs in Iowa, Nebraska, and Kansas where nitrate-N levels exceeded 10 mg/1 in 20% or more of the samples collected. In Long Island (New York) the high density of septic tanks along with the application of fertilizers and manures on agricultural lands probably contributed to high groundwater nitrate-N. Intensive dairy operations with associated problems of manure disposal may be the primary source of nitrates in wells in southeast Pennsylvania and northern Maryland. High nitrate concentrations in Delaware and parts of North Carolina could arise from intensive livestock and poultry operations and from septic tanks (Prasad and Power, 1995). Date on leaching of nitrate-N from a bare soil at Rothamstad Experimental Station in the United Kingdom, as influenced by inorganic and organic manuring, are shown in Table 19.7. Many western European countries have observed high soluble-P concentrations in surface water and groundwater because of continued high application rates of animal manures.

These results and data from a number of other countries indicate that agricultural systems dependent mainly on animal or other bulky organic manures could be difficult to manage. Only a limited number of farmers have the management skills required to make such a system work properly. Also, the quantity of manures available within a suitable transportation distance would often be insufficient to meet the needs of all of the cultivated land. While in cooler climates the decomposition of animal manures may be very slow and may create environmental problems, in warmer, humid climates decomposition is generally too rapid to permit buildup of soil organic-matter content and the nutrient-supplying status of the soil.

The use of chemical fertilizers along with organic manures is probably the best way to keep food production level with or ahead of the increase in the population. For example, in a study at Yurimaguas, Peru (Lavell et al., 1994), organic residues with or without earthworms were studied in maize production. Crop production was sustained at acceptable levels (0.2 to 2.4 tons ha^{-1}) according to local standards, and the use of earthworms showed a definite advantage. However, there was a drastic decline in production from the third year onward (Figure 19.5). This reduction could be controlled only with the application of chemical fertilizers. After application of fertilizers with the crop residues, plots with earthworms again gave higher production. This is an example of conjunctive use wherein a mixture of organic and inorganic nitrogen sources gives better nitrogen-use efficiency than using either source alone.

We would like to conclude this chapter with two sentences from Drs. Borlaug and Christopher (1994). From their keynote lecture at the 15th World Congress of Soil Science held at Acapulco, Mexico, in 1994, we quote, "Indeed, for those concerned with trying to preserve pristine environments or protect endangered species, we would submit that human demographic changes are the greatest threat to the planet Earth in the years ahead. Indeed, if this relentless growth in human numbers goes on unabated, *Homo sapiens* will no doubt end up as an endangered species themselves."

Table 19.7 Leaching of Nitrate from Bare Soil of the Hoosfield Barley Experiment (Rothamsted) During 1986 and 1987

Treatment	Cumulative leaching loss (kg NO_3-N ha^{-1})	Range of NO_3-N concentration in soil at 110 cm during main period of leaching (mg NO_3-N L^{-1})
PK[a] + 96 kg N ha^{-1}	25	4–20
FYM[b] + 96 kg N ha^{-1}	124	40–50

[a] PK–35 kg P and 90 kg K ha^{-1} annually since 1984.

[b] FYM at 35 t ha^{-1} annually since 1843; FYM supplied 238 kg N ha^{-1} yr^{-1}.

From Powlson et al. 1989. *Nitrogen in Organic Wastes Applied to Soils*, pp. 334–345. With permission of Academic Press, Orlando, FL.

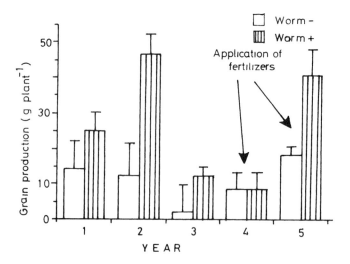

Figure 19.5. Effect of earthworm inoculation on grain production in a continuous corn crop at Yurimaguas, Peru. (Adapted from Lavell et al., 1994.)

REFERENCES

Ahmed, S. 1994. Japan's organic farming system: modern for other countries. Agrochemicals News in Brief 17(2):16–23.

Anderson, J.M. 1994. Functional attributes of biodiversity in land use systems, in *Soil Resilience and Sustainable Land Use*, D.J. Greenland and I. Szaboles, Eds., CAB International, Wallingford, U.K., pp. 267–289.

Altieri, M.A. 1992. Sustainable agricultural development in Latin America: exploring the possibilities. Agric. Ecosyst. Environ. 39:1–21.

Bhawalkar, U.S. and V.U. Bhawalkar. 1993. Vermiculture biotechnology, in *Organics in Soil Health and Crop Production*, P.K. Thampan, Ed., Peekay Tree Crops Development Foundation, Cochin, India, pp. 68–85.

Borlaug, N.E. and R.D. Christopher. 1994. Feeding a human population that increasingly crowds a fragile planet. Keynote lecture, 15th World Congress of Soil Science.

Edwards, C.A. 1983. Earthworms, organic waste and food. Span, Shell Chemical Co., U.K. 26(3):101–108.

FAO. 1994. Quarterly Bulletin of Statistics, Food and Agriculture Organization of the U.N., Rome 7(2/3/4):45–46.

Gaur, A.C., S. Neelakantan, and K.S. Dargan. 1984. *Organic Manures*, Indian Council of Agricultural Research, New Delhi, p. 159.

Gaur, A.C. and K.V. Sadasivan. 1993. Theory and practical consideration of composting organic wastes, in *Organics in Soil Health and Crop Production*, P.K. Thampan, Ed., Peekay Tree Crops Development Foundation, Cochin, India, p. 1–22.

Hendrix, P.F., R.W. Parmelee, D.A. Crossley, D.C. Coleman, E.P. Odum, and P.M. Groffman. 1986. Detritus food webs in conventional and no-tillage agro-ecosystems. Bioscience 36:374–380.

Knight, D., P.W. Elliott, and J.M. Anderson. 1989. Effects of earthworms upon transformations and movement of nitrogen from organic matter applied to agricultural soils, in *Nitrogen in Organic Wastes Applied to Soils*, J.A. Hansen and K. Hendriksen, Eds., Academic Press, London, pp. 59–80.

Lavell, P. C. Gilot, C. Fragoso, and B. Pashanasi. 1994. Soil fauna and sustainable land use in the humid tropics, in *Soil Resilience and Sustainable Land Use*, D.J. Greenland and I. Szaboles, Eds., CAB International, Wellington, U.K., pp. 291–308.

Miller, R.W. and R.L. Donahue. 1992. *Soils — An Introduction to Soils and Plant Growth*, 6th ed., Prentice Hall of India Pvt. Ltd., New Delhi, pp. 196–211.

Nambiar, K.K.M. 1994. *Soil Fertility and Crop Productivity Under Long-term Fertilizer use in India*, Indian Council of Agricultural Research, New Delhi, p. 144.

Powlson, D.S., P. R. Poulton, T.M. Addiscot, and D.S. McCann. 1989. Leaching of nitrate from soils receiving organic or inorganic fertilizers continuously for 135 years, in *Nitrogen in Organic Wastes Applied to Soils*, J.A. Hansen and K. Hendriksen, Eds., Academic Press, London, pp. 334–345.

Prasad, R. and J.F. Power. 1991. Crop residue management. Adv. Soil Sci. 15:205–251.

Prasad, R. and J.F. Power. 1995. Nitrification inhibitors for agriculture, health and environment. Adv. Agron. 54:233–281.

Sarkar, A.K., B.S. Mathur, S. Lal, and K.P. Singh. 1989. Long-term effects of manure and fertilizers on important cropping systems in sub-humid, red and laterite soils. Fert. News. 34(4):71–80.

Shumann, H.A. 1994. The production of organic and biofertilizers. Agrochemicals News in Brief 17(2):24–31.

Spalding, R.F. and M.E. Exner. 1993. Occurrence of nitrate in ground water: a review. J. Environ. Qual. 22:393–402.

Staniforth, A.R. 1979. *Cereal Straw*, Clarendon Press, Oxford.

Stickland, D. 1990. Organic farming in the U.K. — News from the Industry. Paper read before the Fertilizer Society of London, April 5, 1990, pp. 1–16.

Tyler, H. 1970. *Organic Gardening Without Poison*, Van Nostrand Reinhold Col., New York, p. 111.

USDA. 1978. Improving soils with organic manures. U.S. Dept. Agriculture. Washington, D.C.

USDA. 1990. Agricultural statistics. U.S. Govt. Printing Office, Washington, D.C.

Von Uexkull, H.R. and E. Mutert. 1993. Principles of balanced fertilization, Proc. Regional FADINAP Seminar on Fertilization and Environment. Bangkok, Thailand, pp. 17–36.

White, W. 1989. Sources of nitrogen and phosphorus. Fertilizer Research Project. U.S. Environmental Protection Agency, Washington, D.C., p. 82.

20

CROPPING SYSTEMS, SOIL FERTILITY, AND FERTILIZER USE

Cropping systems, including monocropping, sequential cropping, intercropping, mixed cropping, ley farming, and other systems, can be traced back to time immemorial. A reference to crop rotations (sequential cropping) can be found in Indian agriculture dating back to the Chalcolithic period, that is, before 1000 B.C. (Raychaudhri and Mira, 1993); in Chinese agriculture during the Han dynasty in the second century (Hsu, 1980); and, of course, in ancient Greece and Rome (White, 1970). The key to a successful crop rotation was a soil restorer crop of legumes such as beans (*Vicia faba* L.), various clovers, medicago species, lupins (*Lupinus album* L.), and vetch (*Vicia sativa* L.). These restorer crops provided sustainability to 3- to 5-year crop rotations. The famous English Norfolk rotation popular in the eighteenth century (Martin et al., 1976) consisted of turnip, barley, clover, and wheat in a 4-year sequence. Crop rotations were also practiced in colonial United States. For example, at Monticello, Thomas Jefferson followed a 5-year rotation of wheat-corn/potato-pea (*Pisum sativum* L.)-rye (*Secale cereale* L.)/wheat-clover/buckwheat (*Fagopyrum esculentum* Moench) (Karlen et al., 1994). The prevailing thought was that each of the crops in the rotation obtained their nutrients from different soil layers. Even today only 20% of the corn in the United States is grown in continuous monoculture, but most of the remaining 80% is grown in a 2-year rotation with soybean or in short (2- or 3-year rotations) with alfalfa, cotton, dry bean, or other crops (Power and Follet, 1987).

20.1. LEGUMES IN CROP ROTATIONS

As already pointed out, the soil restoring nature of legumes in crop rotation was known by ancient farmers. It was only in 1888 that Hellriegel and Wilfarth discovered *Rhizobia* in the root nodules and the N-fixation capacity of legumes. Since then there have been numerous reports from all regions of the world on the N contribution of legumes to succeeding crops. Most of these reports use the fertilizer replacement value as the method for assessing N contribution from legumes in rotation with nonlegume crops such as corn or grain sorghum. After reviewing several such studies Bullock (1992) concluded that the replacement method overestimated the N contribution from legumes grown in rotation with nonlegumes and underestimated the overall rotation effect. For example, fertilizer recommendations for corn following alfalfa in the United States credit alfalfa with N contribution of 100 to 125 kg N ha^{-1} (Fox and Pickielek, 1988) based on the fertilizer replacement methodology. The actual contribution measured with ^{15}N methodology was only 24 kg N ha^{-1} (Harris and Hesterman, 1990). Several such reports are available in the literature and suggest that benefits of growing legumes or other crops in a crop rotation include factors other than just N effects. Some of the effects could be as follows:

1. Increase in the P, K, and micronutrient concentration and uptake of the crops succeeding legumes. For example, Copeland and Crookston (1992) reported that potassium and total micronutrient content increased in corn grown in a 2-year rotation with soybean, as compared with continuous corn. One of the reasons for such effects could be the fact that legumes can feed the deeper layers of soil and part of the nutrients absorbed are left in the root mass.
2. Increase in the water use efficiency.
3. Lessening of disease and pest problems. Francis and Clegg (1990) stated that the greater the difference between crops in rotation in sequence, the better the cultural control of pests. Pests that are controlled by crop rotations have the following characteristics: (a) The pest inoculum source must be from the field itself; this includes soil and root dwelling nematodes and soilborne pathogens; (b) a narrow range of the hosts; and (c) incapability of surviving long periods without a living host.
4. Better control of weeds. For example, in Nebraska rotating corn and grain sorghum with broadleaf crops is an effective methods of controlling shattercane (*Sorghum bicolor*). Van Heemst (1985) ranked 25 crops for their ability to compete with weeds based on the mean reduction in yield resulting from controlled weed infestations. Wheat was considered most competitive and given the first rank; sorghum ranked fourth and corn seventh.

20.2. INTERCROPPING SYSTEMS

In several tropical countries the per capita availability of land is decreasing daily, creating an urgent need to produce more and more food from the same piece of land. Intercropping systems permit the growing of a bonus crop along with the main crop. Intercropping is generally recommended for crops that are widely spaced such as corn, sorghum, pigeonpea (*Cajanus cajan* L.), and pearlmillet (*Pennisetum typhoides* L.). The intercrop may be a grain legume such as mungbean (*Vigna radiata* L.), urdbean (*Vigna mungo* L.), cowpea (*Vigna unguiculata* L.), or an oilseed such as groundnut (*Arachis hypogaea* L.). The intercrops are not necessarily sown and harvested at exactly the same time, but they grow together for a large part of their growth period. The intercropping systems may be an additive series where the intercrop is grown in addition to the main crop or a replacement series where the two crops may be grown in different row ratios such as 1:1, 2:1, etc.; one crop replaces the other crop as far as the area is concerned.

Intercropping systems generally result in increased productivity from the same piece of land. Observed major advantages of intercropping systems are (1) better utilization of solar radiation; (2) better water use efficiency; (3) yield advantages under stress water conditions; and (4) better control of weeds, pests, and diseases (Chatterjee and Mandal, 1992). Because of fewer alternatives, intercropping is used less in temperate than in tropical regions.

Regarding nutrient management, legumes grown as an intercrop do not compete for nitrogen with the component crop. On the contrary, legumes may provide some nitrogen benefit to the associated crop. For example, Bandyopadhyay and De (1986) used N^{15} data to show that sorghum derives part of its nitrogen from a soil pool enriched by concurrently grown legumes. Other nutritional benefits have also been attributed to the legumes grown as an intercrop. For example, Yadava (1986) reported that the efficiency of P applied in sugarcane (*Saccharum officinarum* L.) increased from 3.5% in sole cropping to 15.8% when sugarcane was intercropped with mungbean. Experiments conducted under the All India Coordinated Project on Cropping Systems Research on fertilizer application in intercropping systems reveal that application of the recommended dose of fertilizer was adequate for cereals, but only 25% of the recommended dose was required for the legume (Chatterjee and Mandal, 1992). Thus from the soil fertility viewpoint, as well as overall productivity viewpoint, the intercropping systems hold considerable promise, especially for tropical regions of the world.

When arable crops are grown as intercrops in alleys between tree rows, the term generally used is alley cropping. For example, cowpea, sorghum, castor (*Ricinus communis* L.), and other crops can be grown in alleys of Leucaena (*Leucaena latisiliqua* [L.] Gillis; syn *L. leucocephala* [Lamb.] de Wit) (Singh et al., 1989). This system is very efficient in utilizing water resources and conserving the soil. In harsh climates the trees protect the alley crop, and cuttings from the branches can be used as a mulch.

Another term that needs to be mentioned is mixed cropping. Mixed cropping is a practice in arid regions of the world where the seeds of a number of crops such as pearlmillet, mungbean, and mothbean (*Vigna aconifolia* [Jacq.] Marechal) are mixed and sown together at the onset of rains. This practice assures the growth of at least one of these crops, regardless of the weather that follows. The farmers can depend on getting some harvest as determined by the amount and distribution of rainfall received. In regions where crop

Table 20.1 Changes in Input Response Pattern in System Requiring NPK + FYM for Maximum Yield (Example of Maize at Palampur)

Item	1973–1975	1982–1984
Total grain response to NPK + FYM (kg ha^{-1})	4825	5866
Percent contribution of N alone	51	0
Percent contribution of P (over N)	22	42
Percent contribution of K (over N + P)	0	23
Percent contribution of FYM (over NPK)	27	35

From Tandon. 1989. Fert. News 34(4):21–26. With permission.

production potential is greater, crops grown alone generally produce better than when grown in a mixture.

20.3. CROPPING SYSTEMS AND SOIL FERTILITY

Continuous cropping and removal of a harvest year after year without adequate and balanced fertilization can lead to soil fertility problems. Plant nutrients are depleted and this can show up in crop yields and response to applied fertilizer. Results from multilocation, long-term fertilizer experiments in India (Tandon, 1989) have brought out several interesting points in this respect. For example, from 1973 to 1975 at Palampur, 51% of the increase from NPK and FYM resulted from N alone. By 1983 to 1984 crop yield response to N alone dropped to 0; NPK and FYM were applied due to depletion of soil P and K (Table 20.1). The contribution of P during the same period increased from 22 to 42%, and K responses increased from 0 to 23%. Similarly, at Pantnagar (Table 20.2) in earlier years, only N produced a substantial increase in the yield of rice and wheat, but with the passage of time, P, K, and Zn also became more responsive. Such results send out important messages for continuously monitoring and revising fertilizer schedules for cropping systems as a whole.

The matter of concern is that in many cases this depletion of P and K is rather poorly reflected in soil test values over time. Further research is therefore necessary to develop more sensitive soil test methods.

20.4. FERTILIZER APPLICATION IN CROPPING SYSTEMS

While adequate fertilization of all the crops grown in a cropping sequence is the best policy, many developing countries do not have adequate indigenous fertilizer production and imported fertilizers are expensive. For this reason optimum utilization of applied fertilizer and economy in their use in a crop sequence has been the goal of several research investigations. This is especially important for nutrients like P and Zn, which are not easily lost from the soil. For example, studies in India have shown that in crop sequences such as rice-rice, rice-wheat, and maize-wheat (the first crop grown during rainy season,

Table 20.2 Changes in Input Response Pattern of Rice and Wheat in System Requiring NPK + Zn (Pautnagar)

Item	Rice		Wheat	
	1972–1974	1982–1984	1972–1974	1982–1984
Total response to NPK + Zn (kg ha^{-1})	1376	2867	2469	2132
Percent contribution of N	66	43	102	75
Percent contribution of P	–6	0	–5	7
Percent contribution of K	21	29	0	7
Percent contribution of Zn	18	28	2	11

From Tandon. 1989. Fert. News 34(4):21–26.

**Table 20.3 Some Strategies for Allocation of Fertilizer P
in Cropping Systems in India**

Cropping systems (2 or 3 crops a year)[a]	Annual input (kg P_2O_5/ha)	Allocation strategy
Wheat-mungbean-pearl millet	60	60 to wheat
	90	60 to wheat, 30 to mungbean, 30 to millet
	135	90 to wheat, 15 to mungbean, 30 to millet
Wheat-pearl millet	75	45 to wheat, 30 to millet
	120	60 to wheat, 60 to millet
Potato-wheat-rice	90	60 to potato, 30 to wheat
Cotton-wheat	90	60 to wheat, 30 to cotton
Soybean-wheat	50/100	All to soybean or 1/2 to soybean and 1/2 to wheat

[a] Wheat: November–April; mungbean: May–June; pearl millet/rice/cotton/soybean: June–July–October/November; potato: September–November/ December.

From Tandon. 1993. *Fertilizer Management in Food Crops*, p. 191. With permission of the Fertilizer Development and Consultation Organisation, New Delhi, India.

that is, July to October and the second crop during the autumn-winter season, that is, November to April), adequate P fertilization of the winter crop could be sufficient for the entire one-year crop rotation; the rainy-season crop could be grown on residual P (Kundu and De Datta, 1988; Kolar and Grewal, 1989). Some strategies for allocation of fertilizer P in different cropping systems are given in Table 20.3. Similarly, for the rice-wheat cropping system on Zn-deficient alkaline soils of Ludhiana (Punjab, India) 5 kg Zn/ha was optimum for rice, while an application of 10 kg Zn/ha to rice was required if the Zn needs of the following wheat were also to be met (Thakkar et al., 1989).

In temperate regions often responses to fertilizers are greater in rotation than in monoculture. For example, Varvel and Peterson (1990) showed that N responses by corn, as well as absolute yields, were greater when grown in rotation than when grown as a monoculture. Part of this response may be due to less weed and disease pressure in the rotation. Also, probably the soil organic N in rotations is more labile than that in monocultures.

An FAO-sponsored workshop "Expert Consultation on Fertilizer Use Under Multiple Cropping Systems" held at New Delhi in 1982 made the following fertilizer scheduling recommendations (FAO, 1983):

1. Rice-based cropping systems
   a. Irrigated rice
      (1) Rice-wheat sequential system: for alluvial soils in the Indian subcontinent, N to be applied to both crops, P to be applied to wheat, and K and Zn to be applied to rice.

(2) Rice-rice-mungbean or soybean sequential system: N to be applied to both the rice crops, while P to be applied only to one (preferably the second, dry season) rice crop together with K, S, and Zn on the basis of soil tests.

(3) Rice-jute sequential system: N to be applied to both crops; P, K, S, and Zn, if needed, to be applied to jute.

b. Rainfed rice

(1) Rice-chickpea, rice-lentil, rice-horsegram, rice-niger, rice-mustard, rice-linseed, rice-groundnut, and rice-soybean sequential systems: N, P, and other nutrients, as required, to be applied to rice crop, only 20 kg P_2O_5/ha to be applied to the sequential legume crop if moisture conditions are favorable.

(2) Rice + pigeonpea, rice + maize, rice + cassava, rice + *Leucaena leucocephala*, and rice + kenaf intercropping systems: N, P, and K to be applied to the rice crop only; Zn and Fe to be applied to rice when needed (iron as foliar spray).

2. Maize-based cropping systems

a. Humid tropics

(1) Maize-cowpea sequential system:

(2) Maize + cassava, maize + groundnut, and maize + *Phaseolus* bean intercropping systems:

(3) Maize + grain/cowpea alley cropping with *Leucaena leucocephala*.

b. Subhumid tropics

(1) Maize + pigeonpea, maize + soybean, maize + cowpea, and maize + chickpea/safflower (for deep Vertisol areas with 200 mm plant-extractable water per meter depth) intercropping systems: N to be applied to maize only; P to be applied to maize and associated legumes; K, S, and Zn to be applied to maize when needed.

3. Sorghum-based cropping systems for semiarid tropics

a. Sorghum + pigeonpea, sorghum + mungbean, sorghum + cowpea, and sorghum + groundnut intercropping systems.

b. Sorghum-yam and sorghum-chickpea/safflower sequential systems: N, P, K, S, and Zn to be applied to sorghum only.

4. Cassava-based cropping systems for humid tropics

a. Cassava + maize/beans intercropping systems: fertilizers to be applied to either crop, according to its importance in the region.

5. Inclusion of leguminous green manure or forage legume prior to an irrigated rice crop can contribute 30 to 40 kg N/ha.

6. Inclusion of grain legumes such as mungbean or cowpea in the cropping systems can contribute 20 to 25 kg N/ha.

7. Inclusion of blue-green algae/Azolla in the irrigated rice crop can contribute 20 to 25 kg N/ha.

8. Leucaena sown at 4-m spacing can contribute, from top prunings incorporated in the soil, up to 60 kg N/ha to the companion crop.
9. Fertilizer applications should be based on local experience and on the corresponding soil tests. In formulating N rates, due consideration should be given to the contributions from associated leguminous crops grown in the system.

REFERENCES

Bandyopadhyay, S. and R. De. 1986. N relationship in a legume-nonlegume association grown in an intercropping system. Fert. Res. 10:73–82.

Bullock, D.G. 1992. Crop rotation. Crit. Rev. Plant Sci. 11:309–326.

Chatterjee, B.N. and B.K. Mandal. 1992. Present trends in research on intercropping. Indian J. Agric. Sci. 62:507–518.

Copeland, P.J. and P.K. Crookston. 1992. Crop sequence affects nutrient composition of corn and soybean grown under high fertility. Agron. J. 84:503–509.

FAO. 1983. Fertilizer use under multiple croping systems — report of an expert consultation held in New Delhi, February 3–6, 1982. Food and Agriculture Organization of the United Nations, Rome, pp. 1–8.

Fox, R.H. and W.P. Pickelek. 1988. Fertilizer N equivalence of alfalfa, birdsfoot trefoil, and red clover for succeeding crops. J. Prod. Agric. 1:313–317.

Francis, C.A. and M.D. Clegg. 1990. Crop rotation in sustainable production systems, in *Sustainable Agricultural Systems*, C.A. Edwards, R. Lal, P. Madden, R.H. Miller, and G. House, Eds., Soil Water Conservation Society, Ankeny, IA, pp. 107–122.

Harris, G.H. and O.B. Hestermann. 1990. Quantifying nitrogen contribution from alfalfa to soil and two succeeding crops using nitrogen-15. Agron. J. 82:129–134.

Hsu, C. 1980. *Han Agriculture*, University of Washington Press, Seattle, p. 377.

Karlen, D.L., Varvel, G.E. Bullock, D.G., and R.M. Cruse. 1994. Crop rotations for the 21st century. Adv. Agron. 53:1–45.

Kolar, J.S. and J.S. Grewal. 1989. P. management in rice-wheat cropping system. Fert. Res. 20:27–32.

Kundu, D.K. and S.K. De Datta. 1988. Integrated nutrient management in irrigated rice. Proc. Int. Rice Res. Conf. IRRI, Los Banos, Philippines.

Martin, J.H., Leonard, W.H., and D.L. Stamp. 1976. *Principles of Field Crop Production*, 3rd ed., Macmillan, New York.

Power, J.F. and R.F. Follet. 1987. Monoculture. Soil Sci. Soc. Am. 256:78–86.

Raychaudhari, S.P. and Mira, R. 1993. Agriculture in ancient India—a report. Indian Council of Agricultural Research, New Delhi, p. 201.

Singh, R.P., C.K. Ong, and N. Saharan. 1989. Above and below ground interactions in alley-cropping in semi-arid India. Agroforestry Systems 9:259–274.

Tandon, H.L.S. 1989. Long-term fertilizer experiments in India—lessons and practical expectations. Fert. News 34(4):21–26.

Tandon, H.L.S. 1993. *Fertilizer Management in Food Crops*, Fertilizer Development and Consultation Organisation, New Delhi, India, p. 191.

Thakkar, P.N., I.M. Chibba, and S.K. Mehta. 1989. Twenty years of coordinated research on micronutrients in soils and plants. Indian. Inst. of Soil Science, Bhopal Bull No. 1, p. 314.

Van Heemst, H.D.J. 1985. The influence of weed competition on crop yield. Agric. Syst. 18:81–93.

Varvel, G.E. and T. A. Peterson 1990. Fertilizer nitrogen recovery by corn in monoculture and rotation systems. Agron. J. 82:935–938.

White, K.D. 1970. Fallowing, crop rotation and crop yields in Roman times. Agric. Hist. 44:281–290.

Yadava, R.L. 1986. Response of sugarcane to phosphorus through legume intercropping. Indian J. Sugarcane Tech. 3:24–28.

INDEX

Q

R

S